Studies in Classification, Data Analysis, and Knowledge Organization

Studies in Classification, Data Analysis, and Knowledge Organization is a book series which offers constant and up-to-date information on the most recent developments and methods in the fields of statistical data analysis, exploratory statistics, classification and clustering, handling of information and ordering of knowledge. It covers a broad scope of theoretical, methodological as well as application-oriented articles, surveys and discussions from an international authorship and includes fields like computational statistics, pattern recognition, biological taxonomy, DNA and genome analysis, marketing, finance and other areas in economics, databases and the internet. A major purpose is to show the intimate interplay between various, seemingly unrelated domains and to foster the cooperation between mathematicians, statisticians, computer scientists and practitioners by offering well-based and innovative solutions to urgent problems of practice.

Paula Brito · Francesco Denti ·
Francesca Greselin · Krzysztof Jajuga ·
Mariangela Zenga
Editors

Navigating Complexity

Statistical Methods, Data Analysis, and Machine Learning for Actionable Insights

Springer

Editors
Paula Brito
Faculty of Economics
University of Porto
Porto, Portugal

Francesco Denti
Department of Statistics
University of Padova
Padua, Italy

Francesca Greselin
Department of Statistics and Quantitative Methods
University of Milano-Bicocca
Milan, Italy

Krzysztof Jajuga
Department of Financial Investments and Risk Management
Wrocław University of Economics and Business
Wrocław, Poland

Mariangela Zenga
Department of Statistics and Quantitative Methods
University of Milano-Bicocca
Milan, Italy

ISSN 1431-8814 ISSN 2198-3321 (electronic)
Studies in Classification, Data Analysis, and Knowledge Organization
ISBN 978-3-032-32008-7 ISBN 978-3-032-32009-4 (eBook)
https://doi.org/10.1007/978-3-032-32009-4

This Springer imprint is published by the registered company Springer Nature Switzerland AG
The registered company address is: Gewerbestrasse 11, 6330 Cham, Switzerland

Preface

The present volume, *Navigating Complexity - Statistical Methods, Data Analysis, and Machine Learning for Actionable Insights*, gathers the revised, refereed versions of short papers accepted for presentation at the 2026 Conference of the International Federation of Classification Societies (IFCS). The upcoming conference, themed "*Classification, Machine Learning and Statistical Methods: from Data to Discovery*", will take place in Milan, Italy from July 14 to July 16, 2026. The event is locally organized by a joint effort of University of Milano-Bicocca and University of Milan, and two hosting scientific societies: the Classification and Data Analysis Group (CLADAG) and the Italian Statistical Society (SIS).

Founded in 1985, the International Federation of Classification Societies (IFCS) is an international scientific organization dedicated to promoting mutual communication, cooperation, and the interchange of views among all those interested in the scientific principles, numerical methods, and practice of data science, data analysis, and classification. IFCS 2026 continues this proud tradition by serving as a premier platform for dissemination and collaboration. Indeed, this year's scientific program features over 260 accepted contributions, thoughtfully organized into thematic tracks, special sessions, contributed paper sessions, and poster presentations. We are excited to welcome over 300 scholars from countries of all continents to the upcoming event. We are particularly looking forward to the six keynote lectures to be addressed by Krzysztof Jajuga (Wrocław U of Economics and Business), Sugnet Lubbe (Stellenbosch U), Brendan Murphy (U College Dublin), Mohamed Nadif (U Paris Cité), Adrian Raftery (U Washington), and Laura Sangalli (Politecnico di Milano). Additionally, the conference program will include highly anticipated tutorials: "Machine learning with functional data: from statistical methods to deep architectures" by Alberto Suárez González (U Autónoma de Madrid) and "Hidden but powerful: practical insights into latent Markov models to discover dynamics in longitudinal data" by Fulvia Pennoni (U of Milano-Bicocca). A Data Challenge has also been organized to engage young students.

The Conference Scientific Program Committee is co-chaired by Francesca Greselin (U of Milano-Bicocca) and Silvia Salini (U of Milan), and includes representatives of the IFCS member societies: Krzysztof Jajuga (IFCS President, Wrocław U of Economics and Business), Rebecca Nugent (IFCS Past President, Carnegie Mellon U), Mark de Rooij (IFCS President Elect, Leiden U), Vladimir Batagelj (U of Ljubljana), Paula Brito (U do Porto), Sonya Coleman (Ulster U), Rosa Crujeiras (U de Santiago de Compostela), Jean Diatta (U of La Réunion), Luca Frigau (U of Cagliari), Raeesa Ganey (U of the Witwatersrand), Stefano Antonio Gattone (U di Chieti "G. d'Annunzio"), Aurea Grané (U Carlos III de Madrid), Julien Jacques (U Lumière Lyon 2), Joanna Landmesser-Rusek (Warsaw U of Life Sciences), Berthold Lausen (U of Essex), Agustin Mayo Iscar (U de Valladolid), Angelos Markos (Democritus U of Thrace), Raffaella Piccarreta (Bocconi U), Riccardo Rastelli (U College Dublin), Pieter Schoonees (Erasmus U Rotterdam), Luca Scrucca (U of Bologna), Piercesare Secchi (Politecnico di Milano), Gianluca Sottile

(U degli Studi di Palermo), Yoshikazu Terada (Osaka U), Cristina Tortora (San Jose State U), Tom Wilderjans (Leiden U), Adalbert Wilhelm (Constructor U). We wish to thank them, alongside the chairs of the thematic tracks, for their invaluable cooperation in evaluating submissions and shaping an outstanding scientific program prior to the event.

The local organization is chaired by Francesca Greselin (U of Milano-Bicocca). We cordially thank all members of the Local Organizing Committee – Luca Brusa (U of Milano-Bicocca), Andrea Cappozzo (Catholic U of Milan), Alessandro Casa (Free U of Bozen-Bolzano), Francesca De Battisti (U of Milan), Francesco Denti (U of Padova), Sónia Dias (Polytechnic Institute of Viana do Castelo), Andrea Marletta (U of Milano-Bicocca), Marta Nai Ruscone (U of Genova), Giorgia Zaccaria (U of Milano-Bicocca), Mariangela Zenga (U of Milano-Bicocca) – and everyone at University of Milan and University of Milano-Bicocca who has worked actively behind the scenes to prepare for this international gathering.

Reflecting this volume's title, *Navigating Complexity*, the selected papers collected here showcase modern methodologies and real-world applications designed to extract actionable insights from intricate data. The topics span a wide range of areas within statistics, machine learning, and data science. These include model-based clustering, Bayesian methods, anomaly detection, and predictive modeling. Novel and tailored models are proposed for complex data types, such as mixed-type data, compositional data, and functional data. Furthermore, advanced statistical concepts and cutting-edge artificial intelligence techniques are explored, underscoring the interdisciplinary effort required to address complex challenges in today's data-driven landscape.

Finally, the organizers and editors would like to express their deepest gratitude to all the authors for their commitment and outstanding contributions to this volume. We are especially grateful to the colleagues who served as reviewers, whose timely and dedicated work was decisive to the scientific quality of these proceedings. Their names are reported in the next page. We also thank our generous sponsors and Springer Nature for their support in producing this book ahead of the conference.

We look forward to a stimulating, fruitful, and highly engaging meeting at IFCS 2026.

June 2026

Paula Brito
Francesco Denti
Francesca Greselin
Krzysztof Jajuga
Mariangela Zenga

Acknowledgments

The editors are deeply grateful to the reviewers, whose contributions were essential to the scientific quality of these proceedings. They are listed below in alphabetical order:

Claudio Agostinelli
Leonardo Salvatore Alaimo
Alessandro Albano
Viviana Amati
Laura Anderlucci
Eleonora Arnone
Roberto Ascari
Luigi Augugliaro
Andrzej Bąk
Lynne Billard
Luca Brusa
Andrea Cappozzo
Alessandro Casa
Sonya Coleman
Dianne Cook
Pietro Coretto
Riccardo Corradin
Laura D'Angelo
Angela Maria D'Uggento
Lucio De Capitani
Grazyna Dehnel
Agnese Maria Di Brisco
Roberto Di Mari
Marco Di Marzio
Cinzia Di Nuzzo
José G. Dias
Pedro Duarte Silva
Andrzej Dudek
Peter Filzmoser
Alice Giampino
Sabrina Giordano
Erika Grammatica
Luca Greco
Patrick Groenen
Mia Hubert
Alina Jedrzejczak
Simona Korenjak-Cerne
Joanna Landmesser-Rusek
Berthold Lausen
Eufrasio Lima Neto
Rosaria Lombardo
Pawel Lula
Paolo Mariani
Andrea Marletta
Giovanna Menardi
Sara Milito
Krzysztof Najman
M. Rosário Oliveira
Andrea Ongarato
Francesco Palumbo
Andrea Panarotto
Leo Pasquazzi
Marcin Pełka
Alfonso Piscitelli
Marialuisa Restaino
Maurizio Romano
Gilbert Saporta
Pieter Schoonees
Piercesare Secchi
Andrzej Sokolowski
Andrea Sottosanti
Gero Szepannek
Paulo Teles
Salvatore Daniele Tomarchio
Cristina Tortora
Marek Walesiak
Adalbert Wilhelm

Contents

Explainable Distance-Based Predictive Models

Marcos Álvarez[1], Eva Boj[2(✉)], and Aurea Grané[1]

[1] Statistics Department, Universidad Carlos III de Madrid, C/ Madrid 126, 28903 Getafe, Spain
marcoalv@pa.uc3m.es, aurea.grane@uc3m.es

[2] Department of Economic, Financial and Actuarial Mathematics, Universitat de Barcelona, Avinguda Diagonal 690, 08034 Barcelona, Spain
evaboj@ub.edu

Abstract. Distance-based regression models provide a flexible framework for analyzing complex data structures using inter-individual distance matrices. Recent developments include robust extensions of Gower's distance for mixed-type data, implemented in the `dbrobust` package in `R`, which improve stability in the presence of outliers and dependency relationships among variables. However, distance-based regression models behave as "black boxes" with respect to the original predictors, which motivates the use of modern interpretability techniques. In this study, we apply methods from explainable artificial intelligence to clarify the role of each predictor. In particular, we use Shapley-based explanations to identify relevant variables and understand their effects, as well as permutation feature importance methods. Our results show that these interpretability tools provide meaningful insights into the structure of distance-based models.

Keywords: dbstats · dbrobust · explainable artificial intelligence · robust predictive models

1 Introduction: Distance-Based Prediction

Statistical techniques based on distances or similarities between sample units have a long and well-established tradition. Among them, cluster analysis and multidimensional scaling (MDS) [1] are widely used. MDS is a multivariate dimensionality-reduction method that applies when the information about the data is provided through an inter-individual distance matrix. Based on the metric version of MDS, [2] introduced the Distance-Based (DB) linear model, later developed in, e.g., [3–8], and [9]. The `dbstats` package [10] in `R` [11] provides functions such as `dblm`, `dbglm`, and `dbpls` to fit DB predictive models. These models assume that an inter-individual distance matrix is available for a set of individuals, along with a continuous response variable for each of them. The main idea behind DB models is to use the principal coordinates, computed from metric

P. Brito et al. (Eds.): IFCS 2026, STUDIES CLASS, pp. 1–8, 2026.
https://doi.org/10.1007/978-3-032-32009-4_1

MDS using a given distance matrix, as explanatory variables in the regression framework.

Thus, the information in the predictor space is incorporated through the distance matrix. For mixed-type data, Gower's distance [12] is defined as the simple mean of three components: the range-normalized city-block distance (for quantitative variables), a distance associated to Jaccard's similarity coefficient (for binary variables), and Hamming distance (for multiclass categorical variables). In [9], a robust version of Gower's distance is studied for weighted mixed-type data, where a robust Mahalanobis distance is proposed for the quantitative component, and the covariance matrix is estimated by a DB-trimming procedure. This new distance is implemented in the `dbrobust` package in `R` [13] and used in the predictor space of DB models. In this study, Explainable Artificial Intelligence (XAI) techniques are employed to better understand the importance of the original predictors in DB predictive models.

2 Explainable Artificial Intelligence

XAI comprises a growing set of techniques designed to interpret or clarify the predictions of complex statistical and Machine Learning (ML) models. Since modern predictive systems often operate as "black boxes", XAI methods provide essential tools for understanding how different predictors contribute to model performance, both globally (summarising patterns across the full dataset) and locally (explaining individual predictions). This is also the case for DB models, which behave as "black boxes" with respect to the original set of predictors. We present four families of model-agnostic XAI techniques that have proven particularly useful in applied settings.

2.1 Shapley Additive Explanations

SHapley Additive exPlanations (SHAP) is one of the most influential techniques in model interpretability. Its foundations lie in cooperative game theory, where Shapley values provide a principled way to distribute the "payout" of a game among its players (see [14,15]). In the context of predictive modeling, the "game" corresponds to the model prediction, and the "players" are the predictors. The method assigns each variable a contribution value that reflects its importance for a specific prediction. A major strength of SHAP is that it satisfies desirable theoretical properties, such as consistency and symmetry, that guarantee coherent importance attributions across model variations. SHAP also provides both local and global perspectives: locally, it allows us to assess how each predictor contributes to an individual prediction relative to a baseline; globally, aggregating Shapley values across observations yields an interpretable summary of overall variable importance, interaction patterns, and effect directionality. In this work, SHAP analyses are performed using the `iml` package in `R` [16], which offers efficient and flexible implementations applicable to a wide range of models.

2.2 Local Interpretable Model-Agnostic Explanations

Local Interpretable Model-agnostic Explanations (LIME) adopts a complementary strategy for interpreting "black box" predictions [17]. Instead of decomposing predictions into contributions, as SHAP does, LIME approximates the model locally around the observation of interest. The key idea is to generate perturbed versions of that observation and fit a simple, interpretable surrogate model, typically a sparse linear model or a small decision tree, that replicates the "black box" behaviour in its immediate neighbourhood. The coefficients of this surrogate model then serve as an explanation for the original prediction. This approach provides intuitive and easily interpretable explanations, especially when the underlying model is highly complex. Because the surrogate is constructed locally, LIME is effective at identifying which variables drive a specific prediction. However, the method can exhibit some instability, as different perturbation schemes or random seeds may yield slightly different explanations. Despite this limitation, LIME remains a popular and accessible tool. In practice, LIME can be implemented in `R` using packages such as `lime` [18], although we do not apply it in the present paper.

2.3 Permutation Feature Importance

Permutation feature importance is a widely used method for assessing the global importance of predictors in a model-agnostic manner [19]. Its rationale is conceptually straightforward: if the value of a predictor is permuted at random, breaking its association with the response, a model that depends strongly on that predictor should experience a substantial loss of predictive accuracy. The magnitude of this deterioration therefore quantifies the importance of the variable. This method is appreciated for its simplicity and the fact that it applies to any fitted model without requiring retraining. It also provides a direct, performance-based measure of importance, which is often easier to interpret than coefficient-based metrics. Nevertheless, permutation importance can be computationally demanding, especially for large datasets or models with high prediction time, as multiple permutations are required to obtain stable estimates. Here, permutation feature importance is computed via the `iml` package in `R`.

2.4 Feature Ablation and Dropout Feature Importance

Feature ablation techniques extend the logic of permutation importance by removing or "turning off" a predictor altogether in order to quantify its influence on model performance. Instead of shuffling values, the method consists in re-evaluating or retraining the model after excluding the variable of interest. The resulting change in predictive accuracy indicates the degree to which the model relies on that feature. This approach is closely related to the idea of dropout loss, commonly used in model comparison and in certain neural network regularization strategies. The conceptual connection between feature ablation and classical statistical methodologies is particularly relevant for DB linear and generalized linear models. In [20] influence coefficients and bootstrap confidence intervals were proposed to quantify the effect of omitting or perturbing predictors in DB regressions. These ideas provide a natural bridge between traditional regression diagnostics and more recent XAI methods, highlighting the continuity between established statistical principles and modern interpretability frameworks.

3 Application: German Credit Data

The empirical analysis in this paper is based on the well-known German credit dataset, a benchmark resource widely used in the evaluation of credit-scoring methods and classification techniques. The data originate from a German financial institution and are available for public download through the Machine Learning Repository (MLR)[1]. Because of its heterogeneous structure and the presence of challenging features such as outliers and variable dependence, the dataset has become a standard testbed in the ML and statistical modeling communities.

The dataset contains information on 1,000 individuals applying for credit, each classified by the institution as either a *good* or *bad* credit risk. Among them, 700 individuals (70%) are labeled as good risks, while the remaining 300 (30%) are regarded as bad risks. The classification is based on a set of 20 underlying predictors (named risk factors in the context of credit risk), resulting in our study in a total of 24 coded predictors of mixed type: 10 continuous numerical variables, 6 binary indicators, and 8 nominal categorical variables. As is common in credit-scoring applications, many of these variables show non-negligible correlations or associations, and the dataset exhibits atypical observations that make predictive modeling more complex. These characteristics render the German credit data particularly suitable for studying robust methods and DB approaches. Beyond its practical relevance, the dataset has been extensively used as a benchmark for methodological comparison. According to the MLR, more than 60 published studies have relied on these data when assessing the performance of new classification algorithms. This widespread use facilitates reproducibility and enables researchers to compare results across modeling strategies under consistent conditions. For comparative purposes, we adopt the conventional classification threshold of 0.5.

We present the empirical performance of the DB logistic model using the Robust Related Multidimensional Scaling (R-RelMS) distance with a trimming level of 10% (see [9] and [13]) on the German credit dataset, together with a comparison against established ML classifiers. We illustrate the use of XAI techniques to interpret individual and global model behavior, in particular SHAP and permutation feature importance.

Baseline Predictive Performance. Figure 1 reports the accuracy and precision obtained by the DB logistic R-RelMS 10% trimmed model, and some competitors (reference ML approaches), together with confidence intervals. We

[1] http://archive.ics.uci.edu/dataset/144/statlog+german+credit+data.

observe that the DB logistic model outperforms the competitors, including classical logistic regression.

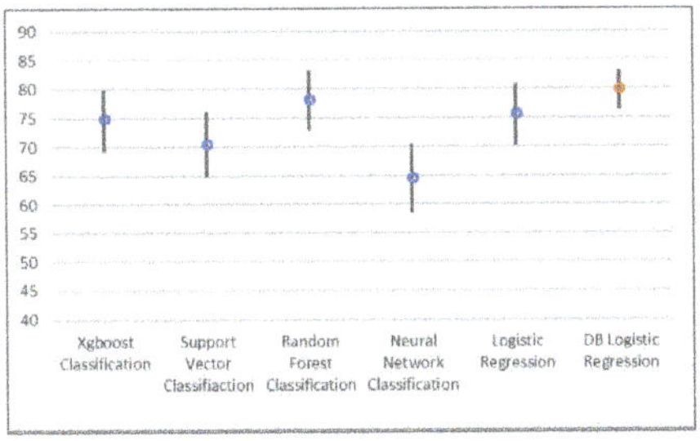

Accuracy

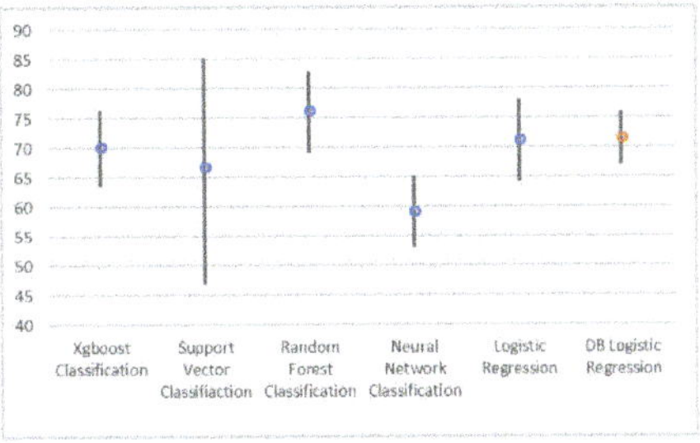

Precision

Fig. 1. Accuracy and precision for the DB logistic RRelMS 10% trimmed model and benchmark classifiers.

Local Explanations via SHAP Values. To gain insight into the behavior of the model at the individual level, we apply SHAP, which attributes to each predictor its contribution to the model's prediction for a specific observation. This enables a detailed local interpretation of the decision, helping to identify which factors contribute the most to a subject being classified as a good or bad credit risk.

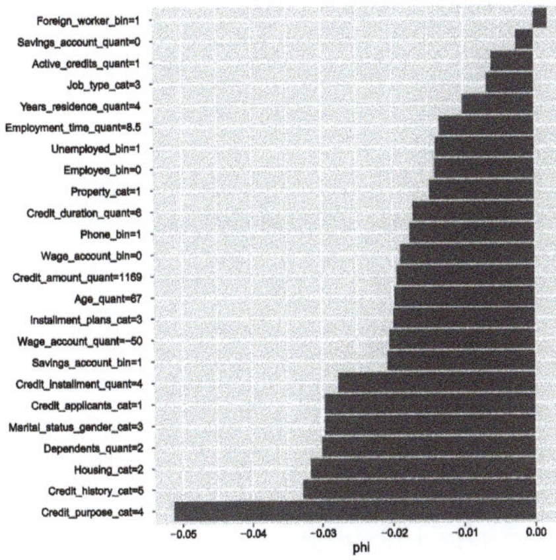

(a)

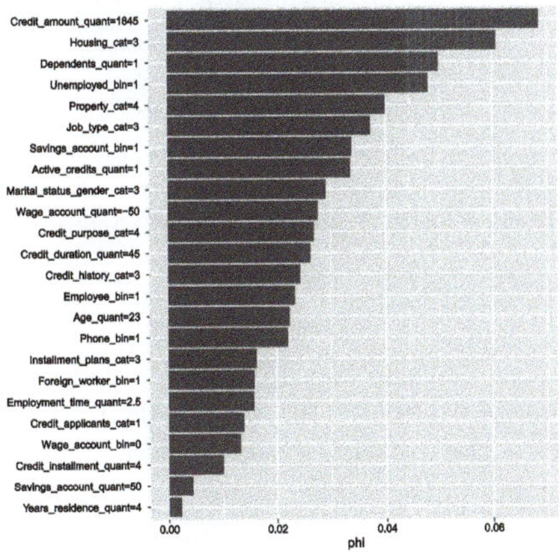

(b)

Fig. 2. SHAP explanation for (a) individual 1 (good credit risk) and (b) individual 1000 (bad credit risk)

Figure 2 shows SHAP value decompositions for two illustrative individuals. Panel (a) corresponds to a correctly classified good credit risk, whereas panel (b) corresponds to a bad credit risk. These examples demonstrate how SHAP highlights influential risk factors, quantifying their positive or negative contribution relative to the model's baseline prediction.

Global SHAP Summaries. Beyond individual explanations, SHAP values can be aggregated across all observations to provide a global assessment of variable importance and the directionality of effects. Figure 3 presents the ranked importance of predictors according to the magnitude of their SHAP values. This global view complements the local analyses and reveals consistent patterns in how certain variables influence credit risk classification across the dataset.

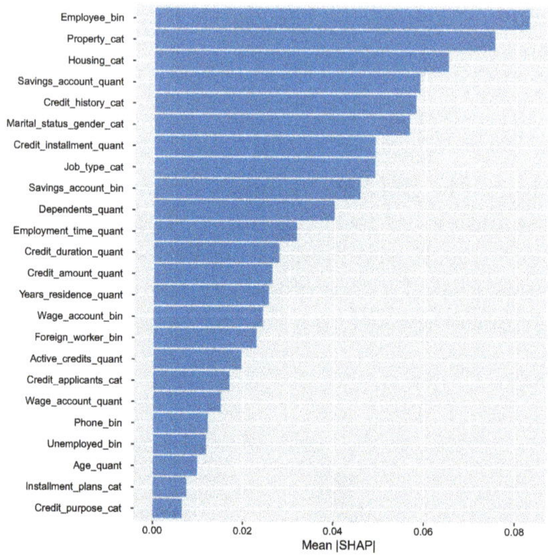

Fig. 3. Global SHAP variable importance across all observations.

Feature Importance through Model-Agnostic Methods. To further validate the relative contribution of each predictor, we compute model-agnostic feature importance using a permutation strategy. This method evaluates the increase in prediction error after destroying the information carried by each predictor, thereby quantifying its relevance. The resulting importance rankings, shown in Fig. 4, provide an additional perspective that is complementary to SHAP and consistent with the main findings obtained with the DB model.

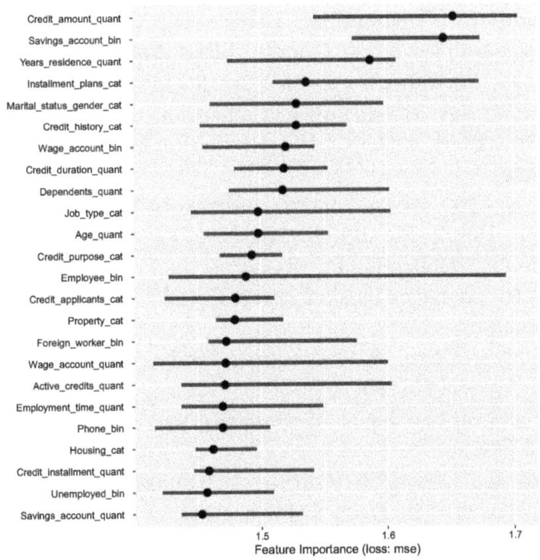

Fig. 4. Permutation based feature importance.

4 Future Research

The aim of our ongoing and future work is to develop and adapt XAI for DB predictive models (`dblm`, `dbglm`), and provide new `R` functions for practitioners that jointly use `dbstats` and `dbrobust` and scalable to the big data context.

Funding. The authors acknowledge the support of grants PID2021-123592OB-I00 funded by MICIN/AEI/10.13039/501100011033 and by "ERDF A way of making Europe".

References

1. Borg, I., Groenen, P.: Modern Multidimensional Scaling: Theory and Applications, 2nd edn. Springer, New York (2005). https://doi.org/10.1007/0-387-28981-X
2. Cuadras, C.M.: Distance analysis in discrimination and classification using both continuous and categorical variables. In: Dodge, Y. (ed.) Statistical Data Analysis and Inference, pp. 459–473. North-Holland, Amsterdam (1989)
3. Cuadras, C.M., Arenas, C.: A distance-based regression model for prediction with mixed data. Commun. Stat. A. Theory Methods **19**, 2261–2279 (1990)
4. Cuadras, C.M., Arenas, C., Fortiana, J.: Some computational aspects of a distance-based model for prediction. Commun. Stat. B. Simul. Comput. **25**, 593–609 (1996)
5. Boj, E., Claramunt, M.M., Fortiana, J.: Selection of predictors in distance-based regression. Commun. Stat. A. Theory Methods **36**, 87–98 (2007)
6. Boj, E., Delicado, P., Fortiana, J.: Local linear functional regression based on weighted distance-based regression. Comput. Stat. Data Anal. **54**, 429–437 (2010)

7. Boj, E., Delicado, P., Fortiana, J., Esteve, A., Caballé, A.: Global and local distance-based generalized linear models. TEST **25**, 170–195 (2016)
8. Boj, E., Costa, T., Fortiana, J.: Data Science. Innovative Developments in Data Analysis and Clustering. Studies in Classification, Data Analysis and Knowledge Organization. Presented at the (2017). Prediction error in distance-based generalized linear models
9. Boj, E., Grané, A.: The robustification of distance-based linear models: some proposals. Socioecon. Plann. Sci. **95**, 101992 (2024)
10. Boj, E., Caballé, A., Delicado, P., Fortiana, J.: Dbstats: Distance-Based Statistics (dbstats). R package version 2.0.3 (2025). http://CRAN.R-project.org/package=dbstats
11. R Development Core Team: R: A language and environment for statistical computing. Vienna, Austria (2026)
12. Gower, J.C.: A general coefficient of similarity and some of its properties. Biometrics **27**, 857–874 (1971)
13. Álvarez, M., Boj, E., Grané, A.: Dbrobust: Robust Distance-Based Visualization and Analysis of Mixed-Type Data. R package version 1.0.0 (2025). https://CRAN.R-project.org/package=dbrobust
14. Shapley, L.S.: A Value for n-Person Games. Contributions to the Theory of Games II. Princeton University Press, Princeton (1953)
15. Lundberg, S.M., Lee, S.-I.: A unified approach to interpreting model predictions. In: Advances in Neural Information Processing Systems (NeurIPS 30) (2017). https://proceedings.neurips.cc/paper/2017/hash/8a20a8621978632d76c43dfd28b67767-Abstract.html
16. Molnar, C., Bischl, B., Casalicchio, G.: iml: an R package for interpretable machine learning. J. Open Source Softw. **3**(26), 786 (2018). https://doi.org/10.21105/joss.00786
17. Ribeiro, M.T., Singh, S., Guestrin, C.: "Why Should I Trust You?": explaining the predictions of any classifier. In: Proceedings of the 22nd ACM SIGKDD International Conference on Knowledge Discovery and Data Mining (KDD 2016) (2016). https://dl.acm.org/doi/10.1145/2939672.2939778
18. Hvitfeldt, E., Pedersen, T.L., Benesty, M.: Lime: local interpretable model-agnostic explanations. R package version 0.5.3 (2022). https://CRAN.R-project.org/package=lime
19. Breiman, L.: Random forests. Mach. Learn. **45**(1), 5–32 (2001)
20. Boj, E., Costa, T., Fortiana, J., Esteve, A.: Assessing the importance of risk factors in distance-based generalized linear models. Methodol. Comput. Appl. Probab. **17**(4), 951–962 (2015)

Partitional Clustering of Compositional Data with Zeros

Roberto Ascari[1], Anna Maria Fiori[2(✉)], Giulia Oliosi[2], and Leo Pasquazzi[2]

[1] Department of Economics, Management and Statistics, University of Milano-Bicocca, 20126 Milano, Italy
roberto.ascari@unimib.it

[2] Department of Statistics and Quantitative Methods, University of Milano-Bicocca, 20126 Milano, Italy
{anna.fiori,leo.pasquazzi}@unimib.it, g.oliosi@campus.unimib.it

Abstract. Partitional methods for clustering compositional data (CoDa) rely on dissimilarity measures consistent with Aitchison's principles for log-ratio analysis. While these measures require strictly positive components, many observed compositions contain zero elements that prevent the application of standard CoDa techniques. Elaborating on k-prototypes clustering for mixed attributes, we introduce a novel strategy for clustering compositions in the presence of zeros. The proposed method is specifically designed to handle zeros of any type (including essential ones) concentrated in just one or a few components. We describe the method and illustrate its application on a real data set of energy mix compositions, where essential zeros naturally occur in nuclear power shares for countries that prohibit this source.

Keywords: irregular CoDa · mixed data · k-prototypes · energy mix

1 Introduction

Unlike conventional multivariate observations, compositional data (CoDa) are fundamentally about ratios: they represent the relative magnitudes of D components (parts) rather than their absolute values. Typically expressed as proportions summing to one (or percentages summing to 100%), compositions lie in a constrained sample space (i.e., the D-part simplex) where standard methods of multivariate statistics cannot be directly applied. Since the 1980s, progress in CoDa analysis has mostly followed the log-ratio methodology pioneered by Aitchison [1], where ratios of parts are analyzed on a logarithmic scale to preserve their relative information content.

Log-ratio transformations enable the application of conventional Euclidean-based clustering methods to compositional data [6]. These include k-means or Ward's method (both assuming Euclidean distances), as well as other partitional algorithms for which Euclidean distances are an option. However, the log-ratio approach requires strictly positive parts while, in practice, real compositional

P. Brito et al. (Eds.): IFCS 2026, STUDIES CLASS, pp. 9–17, 2026.
https://doi.org/10.1007/978-3-032-32009-4_2

data frequently contain zeros arising from different circumstances. Rounded zeros are associated with continuous components that fall below instrument detection limits or rounding thresholds in reported values. Count zeros may appear in discrete parts, when one or more categories remain unobserved due to limited sample sizes. Both cases can be addressed in a pre-processing phase through specific imputation strategies, usually adapted from well-established approaches to missing and censored data available in related literature [7]. Essential zeros, by contrast, represent a fundamentally different phenomenon: they indicate parts that are genuinely absent due to the nature of the data generating process, "not something recorded as zero simply because the experimental design or the measuring instrument has not been sufficiently sensitive to detect a trace of the component" [2]. Examples include absence of alcohol or tobacco expenditures in certain household budgets, plant species unable to survive in incompatible soil types, political parties with no candidates in specific regions, and many others (see, e.g., [8]). The presence of zeros prevents the direct computation of log-ratios and invalidates standard CoDa techniques. As the concept of compositional distances is not well defined for incomplete data, clustering irregular compositions (i.e., compositions containing at least one zero value) poses specific challenges. The problem is particularly relevant with essential zeros, which result from a structural process and thus are not meaningfully replaced by preliminary imputation strategies.

In this work, we propose a novel dissimilarity measure specifically designed for CoDa samples that contain zeros of any type, under a not-missing-at-random (NMAR) scenario where zeros are concentrated in one or few components. We then adapt the k-prototypes algorithm of Huang [4,5] to perform partitional clustering of irregular CoDa, handling the zero parts directly in the objective function. The methodology is described in Sect. 2 and is applied to a real-world data set in Sect. 3. Concluding remarks are included in Sect. 4.

2 Methodology

A real positive vector with D elements, $\mathbf{x} = (x_1, x_2, \ldots, x_D)^\intercal$, is compositional when its total size is irrelevant for the purpose of the analysis. The unit simplex,

$$\mathcal{S}^D = \left\{ \mathbf{x} = (x_1, x_2, \ldots, x_D)^\intercal \in \mathbb{R}_+^D : \sum_{j=1}^{D} x_j = 1 \right\}, \tag{1}$$

is therefore a suitable sample space for compositional data.

Given two compositions $\mathbf{x}, \mathbf{y} \in \mathcal{S}^D$, their Aitchison distance is defined by:

$$d_a(\mathbf{x}, \mathbf{y}) = \sqrt{\frac{1}{2D} \sum_{i=1}^{D} \sum_{j=1}^{D} \left(\ln \frac{x_i}{x_j} - \ln \frac{y_i}{y_j} \right)^2}. \tag{2}$$

This measure is scale invariant and fulfills additional properties that ensure consistency with the relative nature of compositional information [8].

It is often more convenient to express Equation (2) through log-ratio transformations. In particular, the centered log-ratio (*clr*) transformation from $\mathcal{S}^D$ to $\mathbb{R}^D$ is defined as:

$$clr(\mathbf{x}) = \left(\ln \frac{x_1}{g_m(\mathbf{x})}, \ln \frac{x_2}{g_m(\mathbf{x})}, \ldots, \ln \frac{x_D}{g_m(\mathbf{x})} \right)^\mathsf{T}, \tag{3}$$

where

$$g_m(\mathbf{x}) = \left(\prod_{j=1}^{D} x_j \right)^{1/D} = \exp\left(\frac{1}{D} \sum_{j=1}^{D} \ln x_j \right) \tag{4}$$

represents the geometric mean of the composition $\mathbf{x}$. The Aitchison distance (2) is equivalent to the Euclidean distance between *clr*-transformed compositions:

$$d_a(\mathbf{x}, \mathbf{y}) = \sqrt{\sum_{j=1}^{D} \left(\ln \frac{x_j}{g_m(\mathbf{x})} - \ln \frac{y_j}{g_m(\mathbf{y})} \right)^2} = d\left(clr(\mathbf{x}), clr(\mathbf{y})\right), \tag{5}$$

where $d(\cdot, \cdot)$ denotes the usual Euclidean distance in $\mathbb{R}^D$. This property enables a two-step approach to distance-based clustering of CoDa: first apply the *clr*-transformation to a compositional dataset, then use standard Euclidean-based clustering algorithms on the transformed data (see, e.g., [6]).

2.1 CoDa Clustering with Zeros

While the Aitchison distance is an appropriate metric for complete compositions (i.e., with no zeros), many empirical data sets contain zero elements that make it impossible to compute some ratios or their logarithms in Equation (2) or (5).

In this section, we propose a novel dissimilarity measure for irregular CoDa and adapt the k-prototypes algorithm to perform partitional CoDa clustering with zeros. The general idea of a partitional algorithm is to define:

(i) a cluster representative, or prototype, that can summarize mixed (i.e., numeric and categorical) features;
(ii) a dissimilarity measure between mixed data points and cluster prototypes;
(iii) a cost function to be minimized iteratively.

Let $\mathbf{X} = \{\mathbf{x}_1, \ldots, \mathbf{x}_n\}$ be a CoDa set where $\mathbf{x}_i$ is a composition in $\mathcal{S}^D$ describing an object or individual i, for $i = 1, \ldots, n$. We assume that zeros are concentrated in one or few irregular components, indexed by $j = p+1, \ldots, D$, while the first p parts have strictly positive observations for all i. We replace the original CoDa set $\mathbf{X}$ by a mixed data set, $\mathbf{Y} = \{Y_1, \ldots, Y_n\}$, where each object Y_i is described by two types of variables:

1. a (closed) subcomposition with $p < D$ strictly positive elements:

$$Y_i^{\text{sub}} = (y_{i,1}, \ldots, y_{i,p})^\mathsf{T} = \left(\frac{x_{i,1}}{\sum_{j=1}^{p} x_{i,j}}, \ldots, \frac{x_{i,p}}{\sum_{j=1}^{p} x_{i,j}} \right)^\mathsf{T},$$

2. a set of $D-p$ categorical attributes: $Y_i^{\text{cat}} = (y_{i,p+1}, \ldots, y_{i,D})^\intercal$, which can be either binary,

$$y_{i,j} = \begin{cases} m_0 & \text{if } x_{i,j} = 0 \\ m_1 & \text{if } x_{i,j} > 0 \end{cases},$$

or categorical with $Q+1 > 2$ levels,

$$y_{i,j} = \begin{cases} m_0 & \text{if } x_{i,j} = 0 \\ m_1 & \text{if } 0 < x_{i,j} \leq l_j^{(1)} \\ \vdots & \\ m_Q & \text{if } x_{i,j} > l_j^{(Q-1)} \end{cases}.$$

A clustering algorithm for mixed data can now be applied to $\mathbf{Y}$ to find a partition that divides the n objects Y_i into K disjoint groups. A partitional clustering criterion can be defined by the following cost function:

$$E = \sum_{i=1}^{n} \sum_{k=1}^{K} w_{i,k}\, \xi(Y_i, C_k), \tag{6}$$

where $C_k = \left(C_k^{\text{sub}}, C_k^{\text{cat}}\right) = (c_{k,1}, \ldots, c_{k,p}, c_{k,p+1}, \ldots, c_{k,D})^\intercal$ are the prototypes of each cluster ($k = 1, \ldots, K$), $\xi(\cdot,\cdot)$ is a measure of dissimilarity between Y_i and C_k, and $w_{i,k} \in \{0,1\}$ is the generic element of a $n \times K$ binary partition matrix W satisfying $\sum_{k=1}^{K} w_{i,k} = 1$, for $i = 1, \ldots, n$. We only consider hard partitions in this paper, leaving fuzzy partitions as a topic for further research.

Consistent with the types of variables in $\mathbf{Y}$, we introduce the following dissimilarity measure:

$$\xi(Y_i, C_k) = d_a^2(Y_i^{\text{sub}}, C_k^{\text{sub}}) + \lambda \sum_{j=p+1}^{D} \delta(y_{i,j}, c_{k,j}), \tag{7}$$

where the squared Aitchison distance d_a^2 is applied to compositional variables in Y_i^{sub} and the simple matching coefficient $\delta(r,s)$, which equals 0 if $r = s$ and 1 otherwise, is used for categorical variables in Y_i^{cat}. Following Huang [4], a coefficient $\lambda > 0$ is introduced to avoid favoring either type of variables. See [4] for a discussion of the influence of λ on the clustering process.

Since the Aitchison distance can be equivalently applied using the Euclidean distance between *clr*-transformed data as in (5), the dissimilarity measure in (7) admits the same formulation that Huang [4,5] adopted for k-prototypes clustering, namely a combination of squared Euclidean distance and simple matching coefficients. Consequently, our clustering problem for CoDa with zeros can be solved by Huang's k-prototypes algorithm for mixed data, with p numeric attributes corresponding to the *clr*-coefficients of the regular (strictly positive) parts and $D-p$ categorical attributes corresponding to the irregular parts. Cluster prototypes are represented by mean values for *clr*-transformed subcompositions, and modal values for categorical features (see [4] and [5] for a complete description).

3 Application to Energy Mix Compositions

We propose a preliminary application of compositional k-prototypes clustering to the energy mix of 28 European countries, aiming to identify groups with similar energy compositions and provide visual interpretations of their energy consumption patterns. Understanding where our energy comes from and how this is changing is indeed crucial to the evaluation of transition strategies toward sustainable, low-carbon economies (see [3] for a recent study of dependencies in energy mix compositions with graphical models).

We consider a 6-part compositional data set consisting of per capita primary energy consumption (measured in kilowatt-hours per person) in 2024 from six sources: oil, gas, coal, nuclear, hydropower, and renewables (wind, solar, and others). The data come from the latest release of the Energy Institute Statistical Review of World Energy (https://www.energyinst.org/statistical-review/home) and various population sources, with major processing by Our World in Data (https://ourworldindata.org/energy-mix). As a consequence of legislative prohibitions or phase-out policies against nuclear power, the data set includes 14 irregular compositions where the nuclear part is recorded as zero. All other energy sources are present in strictly positive shares.

We recode the nuclear part into a categorical variable with three levels: 0 (absent), 1 (positive and below the median), and 2 (above the median). The choice to recode the nuclear part into three levels is based on the small sample size; indeed, partitioning the 14 observations with non-zero nuclear into more than two levels may lead to non-representative modes. This categorical variable is combined with the 5-dimensional complete subcomposition of the remaining energy sources to obtain a mixed data set for k-prototypes clustering. The algorithm is implemented using the R package `clustMixType` [9] available on CRAN. To prevent over-emphasizing either numeric or categorical attributes, we use the default value of the tuning parameter λ proposed in [9], which is based on the ratio of the average variability of numeric attributes to the concentration of categorical variables (see [4,5] for a discussion). Based on preliminary analysis (i.e., scree plot and average silhouette) and cluster size considerations, we adopt a 3-group solution. The results are displayed in Table 1 and Figs. 1, 2 and 3. From the *clr*-biplot in Fig. 1, we see that the clusters appear well separated: the first principal component distinguishes between clusters C1 (negative PC.1) and C2 (positive PC.1), while all members of C3 have negative coordinates on PC.2. Moreover, large shares of hydropower consumption correspond to negative and far from zero PC.1 coordinates, large shares of coal consumption correspond to large positive coordinates on PC.1 and large negative coordinates on PC.2, while large shares of the other sources correspond to moderate positive PC.1 and PC.2 coordinates. Combining the informative content of Figs. 1 and 2 with the prototypes description in Table 1, we see that -on average and in comparison with the full sample- members of C1 depend less on coal and more on hydropower; members of C2 depend less on hydropower and more on other sources; members of C3 depend more on hydropower and coal, and less on the remaining sources.

Note also that all countries with nuclear level 2 (above median) are members of C3. The cluster membership of countries can be seen in Fig. 3.

Table 1. Cluster prototypes: centers (closed geometric means) of 5-part subcompositions and modes of nuclear levels (categorical variable). Cluster sizes in brackets

Cluster	Coal	Oil	Gas	Hydro	Renewables	Nuclear
C1 (7)	0.0072	0.4634	0.1922	0.2022	0.1350	0
C2 (11)	0.0580	0.5209	0.2246	0.0039	0.1926	0
C3 (10)	0.1281	0.4504	0.1570	0.1302	0.1344	2

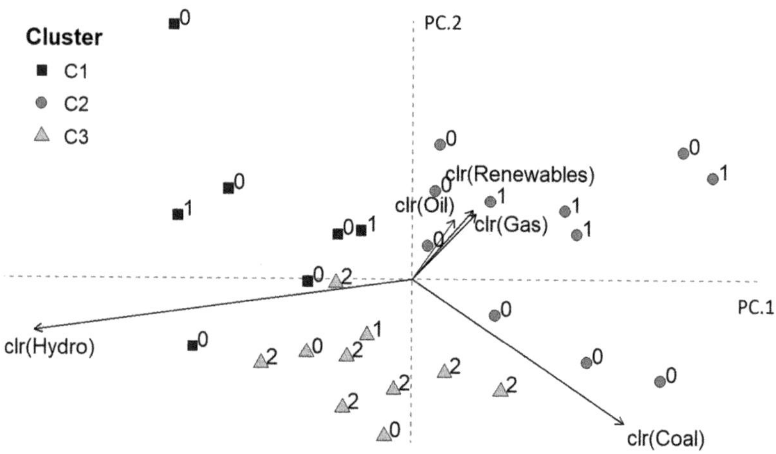

Fig. 1. *clr*-biplot, with 5-part energy mix subcompositions (oil, gas, coal, hydropower, renewables) shaped according to the 3-group solution obtained from k-prototypes clustering. The first two PC axes retain 92.6% of the total variability. The labels near points indicate categorical levels of nuclear power: 0 = absent, 1 = positive and below median, and 2 = above median

To investigate whether alternative choices of the tuning parameter λ could produce more homogeneous clusters, we computed a 3-cluster solution with different values of λ, ranging between $\lambda = 0$ (CoDa k-means based on regular parts only) and $\lambda \to \infty$ (k-modes clustering based on nuclear categories only, see [5]). To compare the different clustering profiles without favoring any of them, we adopted a "neutral" criterion that captures the overall cluster heterogeneity by averaging the within-cluster chi-squared statistics $\chi^2_k = \sum_{i=1}^{n} \sum_{j=1}^{D} w_{i,k}(x_{i,j} - \overline{x}_{j,k})^2 / \overline{x}_{j,k}$, for $k = 1, 2, 3$, where $x_{i,j}$ are the shares of the original 6-part composition (including nuclear power according to its original scale) and $\overline{x}_{j,k} = \sum_{i=1}^{n} w_{i,k} x_{i,j} / \sum_{i=1}^{n} w_{i,k}$ (the nuclear contribution is set to

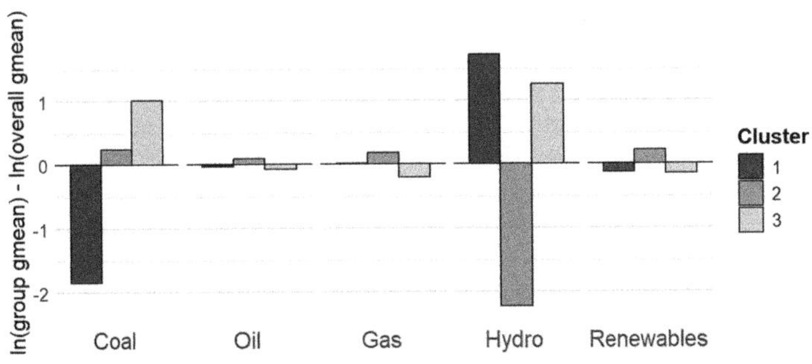

Fig. 2. Bar plot comparing the geometric means of each cluster with the overall geometric mean for the 5-part subcomposition of regular energy components

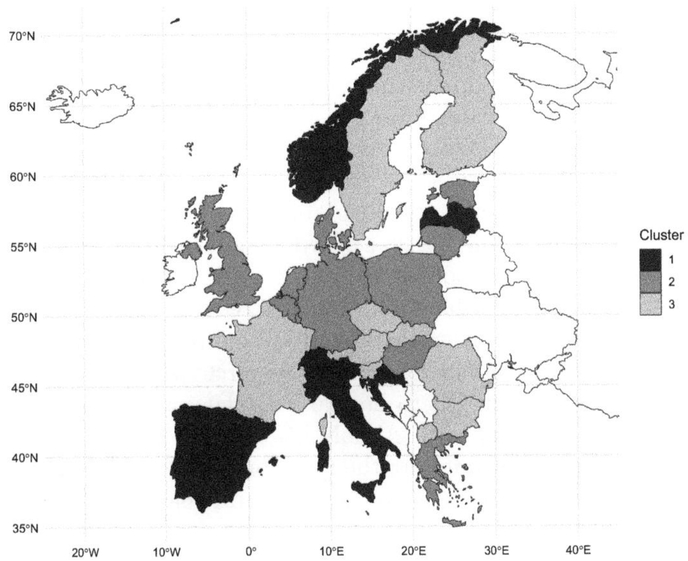

Fig. 3. Geographic distribution of the energy mix clusters obtained with k-prototypes

zero when its cluster mean is zero). As shown in Table 2, CoDa k-means (ignoring the nuclear component) delivers the worst (i.e., least homogeneous) clustering solution, followed by very small and very large values of λ which cause the nuclear part to be either overlooked or over-emphasized. Choices of λ around the value considered in Sect. 3 seem to provide a more reasonable balance between regular and irregular parts in the clustering solution for the energy mix CoDa set.

Table 2. Average within-cluster chi-squared statistic $\overline{\chi^2}$ for various choices of λ: the two "extreme" solutions correspond to CoDa k-means ($\lambda = 0$) and Huang's k-modes ($\lambda \to \infty$); $\overline{s}/h_c$ (where $\overline{s}$ is the average standard deviation of quantitative attributes and h_c is a concentration measure for the categorical variable) is the choice we made in Sect. 3 (twice and three times this value are also included for comparison); $\overline{s}, \overline{s}/3$, and $2/3\,\overline{s}$ were suggested in [4]

	k-means	$\frac{1}{3}\overline{s}$	$\frac{2}{3}\overline{s}$	$\overline{s}$	$\frac{\overline{s}}{h_c}$	$2\frac{\overline{s}}{h_c}$	$3\frac{\overline{s}}{h_c}$	k-modes
λ	0	0.366	0.732	1.098	2.197	4.394	6.590	∞
$\overline{\chi^2}$	3.8207	3.7347	3.4179	3.2878	3.2878	3.2364	3.5574	3.7892

4 Concluding Remarks

In this work, we proposed a novel strategy for partitional clustering of CoDa samples that contain zeros in just one or a few irregular parts. After recoding the irregular parts as categorical variables, we showed that the clustering problem can be effectively solved by Huang's k-prototypes algorithm for mixed attributes. Some delicate issues deserve careful consideration. First, recoding irregular components into categorical variables entails some arbitrariness in the choice of threshold values as well as loss of ordinal information. Second, the choice of λ is context-specific and requires calibration on variable scales. Finally, the proposed method is not suitable for CoDa samples that contain essential zeros in most (if not all) parts, as it would effectively collapse to a k-modes algorithm for purely categorical variables [5]. Future research will focus on elaborating the dissimilarity measure in (7) to handle irregular parts as ordinal variables without losing compatibility with Aitchison's distance for the subcomposition of regular parts. This could mitigate the need for λ calibration and improve the identification of clusters that capture the structural characteristics of the data.

References

1. Aitchison J.: The Statistical Analysis of Compositional Data. Chapman & Hall, London (1986). https://doi.org/10.1111/j.2517-6161.1982.tb01195.x
2. Aitchison, J., Kay, J.W.: Possible solution of some essential zero problems in compositional data. In: Thió-Henestrosa, S., Martín-Fernández, J.A. (eds.) Proceedings of CoDaWork'03, CD-ROM (2003)
3. Di Brisco, A.M., Ascari, R., Fiori, A.M., Nicolussi, F.: Exploring dependencies in compositional data with graphical models. In: Di Bella, E., Gioia, V., Lagazio, C., Zaccarin, S. (eds) Statistics for Innovation III. Italian Statistical Society Series on Advances in Statistics. Springer, Cham (2025). https://doi.org/10.1007/978-3-031-95995-0_22
4. Huang, Z.: Clustering large data sets with mixed numeric and categorical values. In: Proceedings of 1st Pacific Asia Knowledge Discovery Data Mining Conference, pp. 21–34. World Scientific, Singapore (1997)

5. Huang, Z.: Extensions to the k-means algorithm for clustering large data sets with categorical values. Data Mining Knowl. Discovery **2**(3), 283–304 (1998). https://doi.org/10.1023/A:1009769707641
6. Martín-Fernández, J.A., Barceló-Vidal, C., Pawlowsky-Glahn, V.: A critical approach to non-parametric classification of compositional data. In: Rizzi, A., Vichi, M., Bock, H.-H. (eds.) Advances in Data Science and Classification: Proceedings of 6th IFCS Conf, pp. 49–56. Springer-Verlag, Berlin (1998)
7. MartínFernández, J. A., PalareaAlbaladejo, J., Olea, R. A.: Dealing with zeros. In: Pawlowsky-Glahn, V., Buccianti, A. (eds) Compositional Data Analysis: Theory and Applications, pp. 43-58. John Wiley & Sons, Chichester (2011). https://doi.org/10.1002/9781119976462.ch4
8. Pawlowsky-Glahn, V., Egozcue, J.J., Tolosana-Delgado, R.: Modeling and Analysis of Compositional Data. John Wiley & Sons, Chichester (2015)
9. Szepannek, G.: clustMixType: user-friendly clustering of mixed-type data in R. R J. **10**(2), 200–208 (2018). https://doi.org/10.32614/RJ-2018-048

An Infinity Rank SAFE AI Metric for Multivariate Data

Gennaro Auricchio[1(✉)], Paolo Giudici[2], and Giuseppe Toscani[3]

[1] Department of Mathematics, University of Padua, via Trieste 36, Padova, Italy
gennaro.auricchio@unipd.it
[2] Department of Economics and Management, University of Pavia, Pavia, Italy
[3] Department of Mathematics, IMATI CNR. via Ferrata 1, University of Pavia, Pavia, Italy

Abstract. Ensuring SAFE (Sustainable, Accurate, Fair and Explainable) artificial intelligence requires the joint evaluation of accuracy, robustness, and explainability, particularly in classification problems with ordinal responses. However, the intrinsic ordering of ordinal data and the absence of a natural metric challenges standard evaluation methods. We propose a unified framework for comparing univariate and multivariate ordinal data that enables principled discrepancy assessment without arbitrary numerical assumptions. Our approach unifies the assessment of SAFE metrics and naturally connects to the Kolmogorov–Smirnov test, thus establishing a natural way to assess the significance of the measurements.

Keywords: Multivariate Analysis · Rank Graduation Metrics · Kolmogorov–Smirnov Tests for Multivariate Analysis

1 Introduction

Reliable predictive classification is a core objective of Artificial Intelligence (AI), underpinning applications such as medical diagnosis, credit scoring, and educational assessment. Many of these problems involve outputs expressed as ordered categories, making ordinal classification a central yet challenging task. Despite recent advances in machine learning, the increasing deployment of AI systems has raised concerns regarding reliability, robustness, and transparency, motivating the development of international frameworks for trustworthy AI [6,8,16,17].

Trustworthy AI is commonly assessed along three complementary dimensions: *accuracy*, *robustness*, and *explainability* [6,7,14,19,20]. Accuracy captures predictive reliability [15], robustness measures sensitivity to perturbations or adversarial interference [1], and explainability concerns the interpretability of model decisions [5]. While these dimensions are conceptually well understood, their joint assessment remains challenging, particularly for ordinal responses. Ordinal data are ubiquitous in real-world classification tasks, yet their intrinsic

P. Brito et al. (Eds.): IFCS 2026, STUDIES CLASS, pp. 18–25, 2026.
https://doi.org/10.1007/978-3-032-32009-4_3

ordering and lack of a natural metric structure often render standard discrepancy measures inadequate or potentially misleading.

To address these limitations, ranking-based evaluation criteria have received increasing attention [11]. Metrics based on concordance and ordering, such as the Area Under the Curve (AUC) [13], provide scale-invariant and distribution-free measures of predictive performance. Recently, the *Rank Graduation Accuracy* (RGA) metric has been proposed as a unifying framework that extends AUC to ordinal and continuous responses by exploiting Lorenz curves and Gini-type normalizations [12]. These Rank Graduation (RGX) metrics offer a principled way to compare predictions based on ordering information alone, making them particularly suitable for ordinal data. In this paper, we build on the Rank Graduation framework and propose a new RGX metric that extends the previous proposal while being connected to the Kolmogorov–Smirnov test.

2 Preliminaries

This section briefly introduces the notation and main concepts used throughout the paper. We consider vectors $Y = (Y_1, \ldots, Y_N) \in \mathbb{R}^N$ and denote by F_N their empirical cumulative distribution function.

The *Gini index* is a classical, dimensionless measure of dispersion based on the Lorenz curve and is widely used to quantify inequality [9,10]. For a finite sample Y, the Lorenz curve, namely L_Y, is the piecewise linear interpolation of cumulative sums of the ordered observations, that is $\left(\frac{k}{N}, \sum_{i=1}^{k} Y_{r_i}\right)$,

where r_i is the increasing reordering of Y, that is $Y_{r_1} \leq Y_{r_2} \leq Y_{r_3} \leq \cdots \leq Y_{r_N}$.

The Gini index is then defined as

$$G(Y) = 1 - \frac{2}{\sum_{i=1}^{N} Y_i} \int_0^1 L_Y(p)\, dp = \frac{1}{\sum_{i=1}^{N} Y_i} \int_0^1 |L_Y^c(t) - L_Y(t)| dt, \tag{1}$$

The Gini index ranges from 0 (perfect equality) to 1 (maximal inequality) and admits an equivalent formulation as the normalized area between the Lorenz curve and its dual.

Building on these ideas, the *Rank Graduation* metric compares the Lorenz curve of Y with a *concordance curve* obtained by ordering the components of Y according to the ranks induced by Z. Given two N-dimensional vectors $Y = (Y_1, \ldots, Y_N)$ and $Z = (Z_1, \ldots, Z_N)$, the concordance curve between Y and Z is defined as the unique piecewise linear interpolation of the points

$$\left(\frac{k}{N}, \sum_{i=1}^{k} Y_{r_i^*}\right), \tag{2}$$

where r_i^* denotes the ordering induced by Z, so that $Z_{r_1^*} \leq Z_{r_2^*} \leq \cdots \leq Z_{r_N^*}$ The Rank Graduation metric is then defined as the normalized area between the Lorenz curve of Y and the concordance curve, yielding a ranking-based measure of discrepancy that naturally extends the AUC to ordinal and continuous settings:

$$\mathrm{RGX}(Y, Z) = 1 - \frac{\int_0^1 |L_{Y,Z}(t) - L_Y(t)|\, dt}{\int_0^1 |L_Y^c(t) - L_Y(t)|\, dt}. \tag{3}$$

3 The RGX_∞ Metrics

The original RGX measure, introduced as in [4], evaluates the agreement between two vectors Y and Z by comparing the Lorenz curve of Y with the concordance curve induced by the ordering of Z, normalized by the maximal dispersion encoded by the dual Lorenz curve:

$$\mathrm{RGX}(Y, Z) = 1 - \frac{\int_0^1 |L_{Y,Z}(t) - L_Y(t)|\, dt}{\int_0^1 |L_Y^c(t) - L_Y(t)|\, dt}. \tag{4}$$

The construction behind the RGX metric can be naturally generalized to an l_p formulation

$$\mathrm{RGX}_p(Y, Z) = 1 - \frac{\int_0^1 |L_{Y,Z}(t) - L_Y(t)|^p dt}{\int_0^1 |L_Y^c(t) - L_Y(t)|^p dt}, \quad p \geq 1, \tag{5}$$

where $|\cdot|^p$ denotes the p-th power of the absolute value. Notice that RGX_p recovers the classical RGX for $p = 1$ and allows tuning the sensitivity to small versus large discrepancies between curves, as shown in [4].

One aspect that has not been studied yet is the limit for this class of metrics when $p \to \infty$. We introduce the l_∞ extension, RGX_∞, defined as

$$\mathrm{RGX}_\infty(Y, Z) = 1 - \frac{\sup_{t\in[0,1]} |L_{Y,Z}(t) - L_Y(t)|}{\sup_{t\in[0,1]} |L_Y^c(t) - L_Y(t)|}, \tag{6}$$

which measures the maximal deviation between the concordance curve and the Lorenz curve, providing a complementary perspective that emphasizes the largest mismatches rather than the average discrepancy.

Remark 1. As per the other RGX_p, the RGX_∞ metric can be expressed as the portion of sparsity explained by the ordering induced by Z over the vector Y. Indeed, as shown in [4], the denominator of RGX_∞, that is $\sup_{t\in[0,1]} |L_Y^c(t) - L_Y(t)|$, is a sparsity index closely related to the Pietra Index [18].

First, we show that this metric can be seen as the limit for $p \to \infty$ of the RGX_p metrics, thus inheriting its scale invariancy.

Theorem 1. *Let Y and Z be two N-dimensional positive vectors, then*

$$RGX_\infty(Y, Z) = 1 - \lim_{p\to\infty} \left(\frac{\int_0^1 |L_{Y,Z}(t) - L_Y(t)|^p\, dt}{\int_0^1 |L_Y^c(t) - L_Y(t)|^p\, dt} \right)^{\frac{1}{p}}. \tag{7}$$

In particular, RGX_∞ is scale invariant for positive scalars in entries.

Proof. Identity (7) follows from the fact that the l_p norm converges to the l_∞ norm as $p \to \infty$. The scale invariance property follows from the fact that for every $p \geq 1$ the ratio $\frac{\int_0^1 |L_{Y,Z}(t)-L_Y(t)|^p\,dt}{\int_0^1 |L_Y^c(t)-L_Y(t)|^p\,dt}$ is scale invariant.

Connection with the Kolmogorov–Smirnov Test

As shown in [4], the numerator of any RGX_p metric can be expressed as a Cramer-like Distance between two probability distributions μ and μ^* defined as

$$\mu = \sum_{i=1}^{N} \frac{Y_{r_i}}{N\bar{Y}} \Delta_{\frac{i}{N}} \qquad \text{and} \qquad \mu^* = \sum_{i=1}^{N} \frac{Y_{r_i^*}}{N\bar{Y}} \Delta_{\frac{i}{N}}, \tag{8}$$

where (i) $\bar{Y}$ is the average of the entries of Y, so that $N\bar{Y} = \sum_{i=1}^{N} Y_i$, (ii) r_i is the real ordering of Y, (iii) r_i^* the real ordering of Z, and (iv) $\Delta_{\frac{i}{N}}$ is the absolutely continuous probability measure whose density is defined as

$$\Delta_{\frac{i}{N}}(x) := N\mathbb{I}_{[\frac{i-1}{N},\frac{i}{N}]}(x), \tag{9}$$

where $\mathbb{I}_A$ is the indicator function of the set A.

Indeed, under these notations, we have that

$$RGX_p(Y,Z) = 1 - \frac{\int_0^1 |F_\mu(t) - F_{\mu^*}(t)|^p dt}{\int_0^1 |L_Y^c(t) - L_Y(t)|^p dt}. \tag{10}$$

A similar relation holds also for our l_∞ generalization, indeed we have

$$RGX_\infty(Y,Z) = 1 - \frac{\sup_{t\in[0,1]} |F_\mu(t) - F_{\mu^*}(t)|}{\sup_{t\in[0,1]} |L_Y^c(t) - L_Y(t)|}. \tag{11}$$

The l_∞ Rank Graduation metric thus admits a direct connection with the Kolmogorov–Smirnov (KS) test. Indeed, the numerator of RGX_∞ coincides with the KS statistic, as it is defined as the maximum vertical distance between the cumulative distribution functions of two distributions on their support, that is

$$\sup_{t\in[0,1]} |F_{\mu^*}(t) - F_\mu(t)|. \tag{12}$$

This observation allows one to assess the statistical significance of RGX_∞ by applying standard KS testing procedures to the associated curves. The normalization by $\sup_t |L_Y^c(t) - L_Y(t)|$ simply rescales the statistic to $[0,1]$ and does not affect the underlying hypothesis test. As a consequence, RGX_∞ naturally inherits a principled inferential interpretation within the KS framework.

Theorem 2. *Let Y and Z be two N-dimensional vectors, and let L_Y and $L_{Y,Z}$ denote the Lorenz curve of Y and the concordance curve of Y induced by the ordering of Z, respectively. Then the statistical significance of* $\mathrm{RGX}_\infty(Y,Z)$ *can be assessed using a Kolmogorov–Smirnov test applied to the pair of measures μ and μ^* defined as in* (8)*, so that*

$$D = \sup_{t\in[0,1]} \left|F_\mu(t) - F_{\mu^*}(t)\right|. \tag{13}$$

Proof. The Kolmogorov–Smirnov test evaluates the null hypothesis that two empirical cumulative distribution functions coincide by means of the statistic

$$D = \sup_x |F_1(x) - F_2(x)|. \tag{14}$$

Since both $L_Y = F_\mu$ and $L_{Y,Z} = F_{\mu^*}$ are empirical cumulative-type curves defined on $[0, 1]$, the numerator of $\mathrm{RGX}_\infty(Y, Z)$ coincides exactly with a Kolmogorov–Smirnov statistic computed between these two curves. The denominator

$$\sup_{t\in[0,1]} |L^c_Y(t) - L_Y(t)| \tag{15}$$

is a deterministic normalization factor depending only on Y and serves to rescale the statistic to the unit interval. Therefore, the significance of an observed value of $\mathrm{RGX}_\infty(Y, Z)$ can be directly assessed by applying the Kolmogorov–Smirnov test to F_μ and F_{μ^*}, completing the proof.

4 Extension to High–Dimensional Setting

We now extend RGX_∞ to multivariate quantities and provide a suitable statistical test tailored to the extension. To this extent, we adopt a whitening-based strategy inspired by recent multivariate generalizations of the Gini index [3]. Since no canonical multivariate Lorenz curve exists, we leverage the whitening process to reduce the dimension of the problem to one, allowing use to define the multivariate metric to be defined as a weighted combination of univariate Rank Graduation measures while preserving component-wise scale invariance.

Definition 1. *Let $Y \in \mathbb{R}^d$ be a random vector with covariance matrix Σ_Y. A whitening process is a linear transformation $Y^* = WY$, where the whitening matrix W satisfies $\mathrm{Cov}(Y^*) = W\Sigma_Y W^\top = I_d$.*

If the whitening matrix W is scale-stable, then scale-invariant properties of the underlying univariate metrics are preserved under the multivariate extension [2]. We thus propose the following multivariate extension of the RGX_∞ metric.

Definition 2. *Given two multivariate samples Y and Z, let W be a scale-stable whitening matrix associated with Y. We then define*

$$RGX_\infty(Y, Z) = \sum_{i=1}^{d} \frac{\lambda_i}{\sum_{k=1}^{d} |\lambda_k|} RGX_\infty\big((WY)_i, (WZ)_i\big), \tag{16}$$

where (i) $(WY)_i$ and $(WZ)_i$ are the vector containing the i-th components of the whitened vectors WY and WZ and (ii) λ_i is the average of the i-components of the matrix WY.

Owing to the properties of scale-stable processes, the metric defined in (16) is invariant under component-wise changes of scales [2]. Moreover, through the Bonferroni test, we are able to extend the Kolmogorov–Smirnov test to suit the multivariate setting.

Theorem 3. *Given two multivariate samples Y and Z, let*

$$RGX_\infty(Y,Z) = \sum_{i=1}^{d} \eta_i \, RGX_\infty\big((WY)_i,(WZ)_i\big), \tag{17}$$

where $\eta_i \geq 0$ and $\sum_{i=1}^{d} \eta_i = 1$. Assume that for each i, a statistical test at significance level $\eta_i \alpha$ is available for $RGX_\infty\big((WY)_i,(WZ)_i\big)$.

Then, testing each component at level $\eta_i\alpha$ ensures that the null hypothesis for the multivariate convex combination $RGX_\infty(Y,Z)$ is rejected with an overall significance level bounded by α.

Proof. By the Bonferroni inequality, the probability of falsely rejecting at least one component null hypothesis is bounded by

$$\sum_{i=1}^{d} \eta_i \alpha = \alpha. \tag{18}$$

Since $RGX_\infty(Y,Z)$ is a convex combination of the one-dimensional RGX_∞ statistics, controlling the family-wise error across the components at levels $\eta_i\alpha$ ensures that the multivariate null hypothesis is rejected with overall significance at most α.

5 Experimental Validation

To conclude, we analyse the properties of the RGX_∞ in comparison with the properties of RGX_p. In particular, we show that the RGX_∞ metric is better suited to highlight small perturbation to data. To this extent, we consider the following sythetic dataset. First, we sample 10 values from a benchmark distribution, in particular we consider the Uniform distribution over $[0,1]$, the Exponential distribution with parameter $\lambda = 2$, and the Beta distribution with parameters $\alpha = 0.5$ and $\beta = 1$. We define as the ground truth, namely Y, the sample reordered in a non-decreasing way and define Z by applying small changes to the vector Y. We consider two types of changes

- k-cycle changes: we apply a cycle transformation to the first k components of Y, so that $Z_1 = Y_k$, $Z_i = Y_{i-1}$ for every $i = 2,\dots,k$, and $Z_i = Y_i$ otherwise;
- k-simple swaps: we swap the first entry of the vector Y with the k entry of Y, so that $Z_1 = Y_k$, $Z_k = Y_1$, and $Z_i = Y_i$ otherwise.

For each couple of vectors, we compute the RGX_∞ and the standard RGX. In Table 1, we report the results of our experiments.

In accordance to with our expectations, we observe that the RGX_∞ attains smaller values on almost all the cases considered. This empirically proves that the RGX_∞ is better suited to capture small perturbations in the data. The only exception in our study is the case in which we apply the 8-simple swap. In this case, the standard RGX attains a smaller value, we believe that this is due to the fact that swapping the first and eight component induces a larger perturbation on the data, which is then noticed also by the RGX_∞.

Table 1. Values of the RGX and RGX_∞ computed of artificial data. Each value is obtained as the average of 1000 samplings.

		k-cycle				k-simple swaps			
k		2	4	6	8	2	4	6	8
Unif	RGX	0.994	0.984	0.931	0.870	0.964	0.931	0.900	0.865
	RGX_∞	0.960	0.894	0.832	0.763	0.991	0.963	0.918	0.835
Exp	RGX	0.997	0.987	0.966	0.928	0.983	0.964	0.939	0.896
	RGX_∞	0.982	0.947	0.906	0.839	0.995	0.981	0.952	0.891
Beta	RGX	0.998	0.988	0.965	0.920	0.986	0.963	0.935	0.887
	RGX_∞	0.988	0.951	0.898	0.821	0.996	0.980	0.946	0.878

6 Conclusion and Future Work

In this paper, we proposed a unified Rank Graduation framework for evaluating predictive models with ordinal responses. By extending the classical RGX metric to an l_p family, we introduced flexible and interpretable discrepancy measures that capture ordering-based agreement without relying on arbitrary numerical encodings. The framework supports the joint assessment of accuracy, robustness, and explainability, addressing central challenges in trustworthy AI for ordinal and multivariate data. A key contribution is the study of the RGX_∞ metric and its connection with Kolmogorov–Smirnov–type statistics, allowing principled hypothesis testing of ranking-based performance.

Future work includes extensions to structured outputs, theoretical analysis of asymptotic and finite-sample properties, and integration into learning and explainability pipelines, particularly in privacy-preserving and federated settings.

References

1. Aldasoro, I., Gambacorta, L., Giudici, P., Leach, T.: The drivers of cyber risk. J. Financ. Stab. **60**, 100989 (2022)
2. Auricchio, G., Brigati, G., Giudici, P., Toscani, G.: Multivariate Gini-type discrepancies. Math. Models Methods Appl. Sci. **35**(5), 1267–1296 (2025)
3. Auricchio, G., Giudici, P., Toscani, G.: Extending the Gini index to higher dimensions via whitening processes. Rendiconti Lincei **35**(3), 511–528 (2025)
4. Auricchio, G., Bernardelli, A.E., Giudici, P., Toscani, G.: On rank graduation metrics for high dimensional ordinal data. Math. Models Methods Appl. Sci. (2026). (Accepted, to appear)
5. Babaei, G., Giudici, P., Raffinetti, E.: Explainable fintech lending. J. Econ. Bus. (2022)
6. Bengio, Y.: The Singapore consensus on global AI safety. arXiv preprint: arXiv:2506.20702 (2025)
7. Bracke, P., Datta, A., Jung, C., Sen, S.: Machine learning explainability in finance: an application to default risk analysis, vol. 816. Bank of England (2019). Working Paper

8. European Commission. Artificial Intelligence Act (2022)
9. Gini, C.: Sulla misura della concentrazione e della variabilità dei caratteri. Atti del Reale Istituto Veneto di Scienze, Lettere ed Arti **73**, 1203–1248 (1914)
10. Gini, C.: Measurement of inequality of incomes. Econ. J. **31**, 124–126 (1921)
11. Giudici, P., Raffinetti, E.: SAFE artificial intelligence in finance. Financ. Res. Lett. **56**, 104088 (2023)
12. Giudici, P., Raffinetti, E.: RGA: a unified measure of predictive accuracy. In: Advances in Data Analysis and Classification (2024)
13. Hand, D., Till, R.J.: A simple generalisation of the area under the ROC curve for multiple classification problems. Mach. Learn. **45**, 171–186 (2001)
14. Lundberg, S.M., Lee, S.: A unified approach to interpreting model predictions. In: Advances in Neural Information Processing Systems, vol. 30, pp. 4768–4777 (2017)
15. Muthukumar, V., Narang, A., Subramanian, V., Belkin, M., Hsu, D., Sahai, A.: Classification vs regression in overparameterized regimes: Does the loss function matter? J. Mach. Learn. Res. **22**(222), 1–69 (2021)
16. NIST. AI risk management framework (2023)
17. Framework for the classification of AI systems, OECD (2022)
18. Pietra, G.: Delle relazioni tra gli indici di variabilità. C. Ferrari, Pavia (1915)
19. Ribeiro, M.T., Singh, S., Guestrin, C.: Why should I trust you? Explaining the predictions of any classifiers. In: Proceedings of the 22nd ACM SIGKDD International Conference on Knowledge Discovery and Data Mining, pp. 1135–1144 (2016)
20. Zuazua, E.: Machine learning and control: foundations, advances, and perspectives. arXiv preprint: arXiv:2510.03303 (2025)

A Sequence Similarity Network Approach for Fast Detection of Horizontal Gene Transfer Events

Mohammadreza Bakhtyari[1], Nadia Tahiri[2], F. Guillaume Blanchet[3], and Vladimir Makarenkov[1](✉)

[1] Computer Science, Université du Québec à Montréal, Montreal, Canada
makarenkov.vladimir@uqam.ca
[2] Computer Science, Université de Sherbrooke, Sherbrooke, Canada
[3] Biological Science, Université de Sherbrooke, Sherbrooke, Canada

Abstract. Horizontal gene transfer (HGT) is a primary driver of microbial evolution, yet existing HGT detection methods are often limited by computational scalability. Here, we present a novel framework for identifying HGT events using sequence similarity networks. Our approach constructs a bipartite network that integrates species phylogenetic tree (i.e. additive tree used to represent whole genome evolution) with clusters of homologous genes. We use a ranking-based scoring algorithm that weights sequence similarity against phylogenetic tree distance, identifying high-confidence HGT candidates via bootstrap analysis. Validated on the aminoacyl-tRNA synthetases benchmark dataset, our method successfully recovers known transfer events. The framework offers a scalable alternative to traditional gene tree-species tree reconciliation methods for investigating gene flow at the genomic and metagenomic scale. The code and data are available at: https://github.com/mrbakhtyari/simnet.

Keywords: Horizontal gene transfer · Sequence similarity network · Phylogenetic tree · Species classification

1 Introduction

The comparison of genetic sequences remains a central component of evolutionary biology, providing fundamental insights into the diversity and history of life [1]. While phylogenetic trees and networks have significantly advanced our understanding of vertical inheritance and evolutionary processes [2–4], several studies have demonstrated the profound importance of horizontal gene transfer (HGT) in shaping the evolution of life, particularly in prokaryotes [5,6].

HGT provides organisms with rapid access to genetic innovations, enabling the spread of crucial traits such as virulence factors, antibiotic resistance, and xenobiotic metabolism through microbial communities [7]. Two primary models characterize HGT: intergenic transfer, where a complete gene copy from a donor

P. Brito et al. (Eds.): IFCS 2026, STUDIES CLASS, pp. 26–34, 2026.
https://doi.org/10.1007/978-3-032-32009-4_4

organism integrates into the host genome, and intragenic transfer, where partial gene segments are incorporated into the host genome through subsequent recombination, leading to the emergence of mosaic or chimeric genes [7–9].

Despite the biological significance of HGT, current detection methods face substantial computational challenges. Traditional approaches based on gene tree - species tree reconciliation or sliding window sequence analysis [7,9,10] are computationally intensive, with prohibitive cubic time complexity, rendering them impractical for datasets comprising more than a few hundred taxa. Meanwhile, faster BLAST-based methods, though scalable, suffer from reduced reliability. This computational bottleneck has left thousands of datasets and sequenced microbial genomes unexplored [11].

In this context, sequence similarity networks (SSNs) offer a promising alternative framework for evolutionary analysis at scale [12,13]. In an SSN, each node represents a gene or a protein sequence, and edges connect nodes when their sequence similarity exceeds a predefined threshold. Unlike phylogenetic methods that require computationally expensive multiple sequence alignments, SSNs can efficiently organize millions of sequences into connected components (CCs) representing putative gene families or clusters of homologous genes (CHGs).

In this work, we present **SimNet-HGT**, a novel computational framework for detecting horizontal gene transfer events using sequence similarity networks. Our approach constructs a bipartite network linking genomes to connected components, enabling the identification of gene families shared by phylogenetically distant species. The core principle underlying our method is that unusually high sequence similarity between phylogenetically distant organisms—an anomaly under vertical inheritance—constitutes evidence of horizontal gene transfer. To quantify this, we employ a ranking-based scoring approach. Our approach naturally downweights weak similarity signals while prioritizing connections between distant taxa (the path length between taxa in a species phylogenetic tree is used to estimate the distance) with unexpectedly high similarity. High-confidence HGT candidates are then identified through bootstrap analysis of the score distribution. By leveraging SSNs, our method achieves computational efficiency that scales linearly with the number of input sequences, enabling the analysis of datasets comprising thousands of taxa (i.e. genomes) and millions of gene/protein sequences—a scale impractical for traditional tree-reconciliation approaches. The contributions of this paper are threefold:

1. **A scalable SSN-based pipeline** for HGT detection applicable to datasets comprising thousands of genomes and millions of gene/protein sequences.
2. **A ranking-based scoring method** that weights species tree distance by sequence similarity, normalized by similarity rank, with bootstrap analysis for statistically principled identification of high-confidence HGT candidates.
3. **Network reduction techniques** that exploit twin gene families and bipartite graph structure to minimize redundant computations, while preserving biologically meaningful signals.

We validate our approach on the aminoacyl-tRNA synthetases (AARSs) benchmark dataset and demonstrate that SimNet-HGT successfully identifies known HGT events, while revealing novel transfer candidates for experimental investigation. Our results indicate that SSN-based methods represent a practical and effective complement to traditional phylogenetic approaches for studying reticulate evolution at the metagenomic scale.

2 Methodology

2.1 Sequence Similarity Network Construction

The initial phase of our pipeline involves constructing an SSN, where nodes represent individual gene/protein sequences and edges account for significant sequence homology. We use MMseqs2 [14] to carry out rapid all-against-all sequence alignment. SSN edges are established between gene/protein pairs only if they meet stringent similarity criteria: a minimum sequence identity (e.g. 60%, used in our study) and minimum bidirectional alignment coverage (e.g. 80%, used in our study). Then, we identify the connected components of the SSN. Each connected component corresponds to a putative gene/protein family. Such a network-based clustering avoids the need for computationally expensive multiple sequence alignments, facilitating scalability to millions of sequences.

2.2 Bipartite Graph Representation and Reduction

To efficiently map gene/protein families into genomes, we transformed the SSN (see Fig. 1a) into a bipartite graph where one set of nodes represents genomes (species) and the other the connected components (CC) of the SSN (see Fig. 1b). Weighted edges connect a genome node to a CC if the genome contains sequences belonging to that CC, with weights corresponding to the number of gene/protein copies. A key optimization in our framework is the reduction of this bipartite graph through the identification and merging of twin gene/protein families—clusters that share identical genome membership profiles are likely to be involved in the same HGT events. These graph reduction step allows us to conduct HGT detection on unique evolutionary patterns while minimizing redundant computations.

We integrate the reduced bipartite network with a reference species phylogeny. For any two genomes A and B, we define the phylogenetic distance $L(A, B)$ as the length of the unique path connecting them in the species tree. While most of the species tree edges are assigned a weight of 1, edges separating the three domains of life (Bacteria, Archaea, and Eukarya, if any) can be assigned a higher weight (e.g. a weight of 2, as in our experimental study) to reflect a deeper evolutionary divergence. The underlying principle of our HGT detection approach is that high sequence similarity between phylogenetically distant species A and B (with $L(A, B) \gg 1$) is a strong signal of horizontal gene transfer, as such similarity is unexpected under vertical inheritance alone.

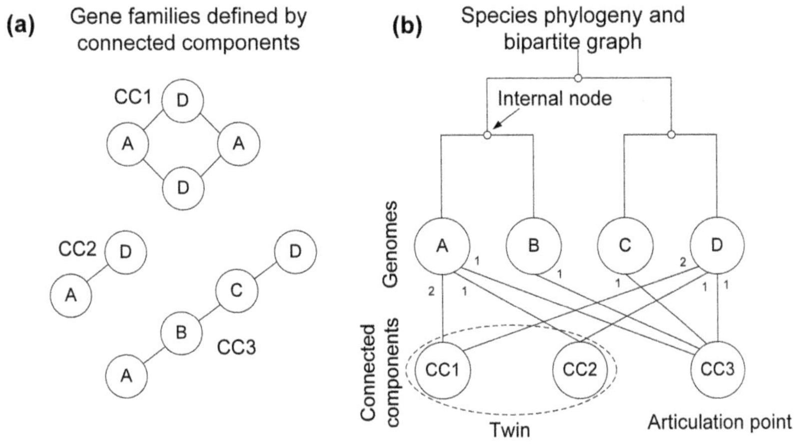

Fig. 1. (a) Sequence similarity network composed of 3 gene families defined by the related connected components CC1, CC2, and CC3 involving gene sequences belonging species A, B, C, and D; (b) Species phylogeny constructed for species A, B, C, and D (based on their genomes), and the bipartite graph including a twin, an articulation point, and the connected components CC1, CC2, and CC3.

2.3 Threshold-Based Ranking Score

To identify potential HGT events from an SSN, we employ a ranking-based scoring method. For each gene/protein a in a given CC, we rank all partner genes/proteins from the same CC, based on their sequence similarity. To accommodate minor sequence similarity variations, we implemented a robust ranking scheme in which genes/proteins with similarity differences smaller than a predefined threshold th (0.5 in our experiments) were assigned identical ranks.

Our exact ranking scheme is as follows. For a given sequence a, we first find the closest sequence to it, let's say b, from the same CC, setting the value of $Rank_{th}(a,b)$ to 1. Then, we find the second closest sequence to a, let's say c, from the same CC. If the difference in similarity values, i.e. $S(a,b)-S(a,c) \leq th$, then $Rank_{th}(a,c)$ is set to 1, otherwise, it is set to 2. For the next closest sequence to a from the same CC, let's say d, we apply the same principle: if $Rank_{th}(a,b) = Rank_{th}(a,c) = 1$, then we verify if $S(a,b) - S(a,d) \leq th$ (to see if the rank of d should be incremented compared to b and c), otherwise (i.e. if $Rank_{th}(a,c) = 2$), we verify if $S(a,c) - S(a,d) \leq th$ (to see if the rank of d should be incremented compared to c), and so on.

Obviously, this ranking measure is not symmetric, i.e. for two given sequences a and b, $Rank_{th}(a,b)$ is not necessary equal to $Rank_{th}(b,a)$.

The HGT score between species A and B related to genes/proteins a and b belonging to the same connected component CC is defined as follows:

$$HGT\_score(A,B,a,b,CC) = \frac{L(A,B) \times S(a,b,CC)}{Rank_{th}(a,b,CC)}, \quad (1)$$

where $L(A, B)$ is the distance between the species A and B in the species phylogeny, $S(a, b, CC)$ is the similarity between the sequences a and b (expressed as percent sequence identity in our study) in CC, and $\text{Rank}_{th}(a, b, CC)$ is the threshold-adjusted rank of b with respect to a in CC. This formula prioritizes high-similarity matches between distant lineages.

A primary advantage of our method is its computational efficiency. The pipeline consists of two computationally distinct steps, both designed for scalability. First, the SSN construction avoids the prohibitive $O(n^2)$ cost of all-against-all alignment by using a fast, heuristic-based local alignment tool (e.g. MMseqs2 with time complexity of $O(n)$), combined with linear-time clustering (with time complexity of $O(E \times \alpha(n))$), where n is the number of gene/protein sequences considered and E is the number of edges in the SSN. This ensures that an SSN construction scales linearly with the number of its edges (E). Second, the HGT scoring phase exploits the modularity of the bipartite network. With connected components defined, the algorithm processes each gene/protein family (i.e. each connected component) independently. The computationally intensive ranking phase depends only on the number of sequences, N_{cc}, within a given connected component CC - its time complexity is $O(N_{cc}^2 \times \log N_{cc})$. For large metagenomic datasets, we have: $N_{cc} \ll n$. Phylogenetic distances within the species tree can be precomputed once, in $O(N^2)$, where N is the total number of species, and subsequently retrieved in constant time during scoring. Thus, the overall scoring complexity scales linearly with the number of gene/protein sequences, i.e. $O(n)$. Furthermore, the independence of connected components allows for massive parallelization, distributing the workload across processors to significantly reduce runtime. Thus, our computational framework effectively overcomes the $O(nlN^3)$ time complexity bottleneck often associated with traditional tree-reconciliation methods, where l is an average sequence length.

In our real-data experiments (see the next section), we also carried out bootstrap analysis with 10,000 samples to estimate the 95th percentile of the HGT score distribution using confidence intervals. Gene/protein pairs yielding HGT scores that exceed the upper confidence bound of this threshold are flagged as statistically significant HGT candidates.

3 Experimental Results

We evaluated the proposed SimNet-HGT method using the aminoacyl-tRNA synthetases (AARSs) benchmark data originally considered by Woese et al. [15]. This extensively studied dataset [7,10] comprises protein sequences from 32 organisms spanning Bacteria, Archaea, and Eukarya (for this dataset $n = N = 32$).

Table 1. High-confidence HGT candidates identified by SimNet-HGT.

Species A	Species B	$L(A, B)$	$S(a, b)$	$Rank_{th}(a, b)$	HGT_score
T.pall	P.hori	8	60.7	1	485.6
T.pall	P.aero	8	60.5	1	484.0
B.burg	P.aero	8	60.0	1	480.0
B.burg	P.hori	8	59.5	1	476.0
B.burg	M.jann	8	59.3	1	474.4
A.fulg	H.sapi	7	66.7	1	466.9
H.sapi	A.fulg	7	66.7	1	466.9
M.ther	H.sapi	7	64.4	1	450.8
S.cere	A.fulg	7	64.2	1	449.4
C.acet	P.aeru	5	71.7	1	358.5
C.acet	R.caps	5	71.5	1	357.5

First, applying the minimum sequence identity threshold of 60%, we found two connected components for this dataset. Using them, our method identified 11 high-confidence HGT candidates exceeding the upper confidence bound of the bootstrap-derived 95th percentile threshold (see Table 1 that summarizes these candidates ranked by HGT score).

Figure 2 details the statistical properties of HGT scores. The empirical distribution approximates an exponential decay, consistent with a null model where most protein pairs reflect vertical inheritance. The bootstrap analysis establishes a 95th percentile Upper Confidence Interval (CI) of 356.0 that serves as our strict significance threshold. All 11 high-confidence candidates exceed this value.

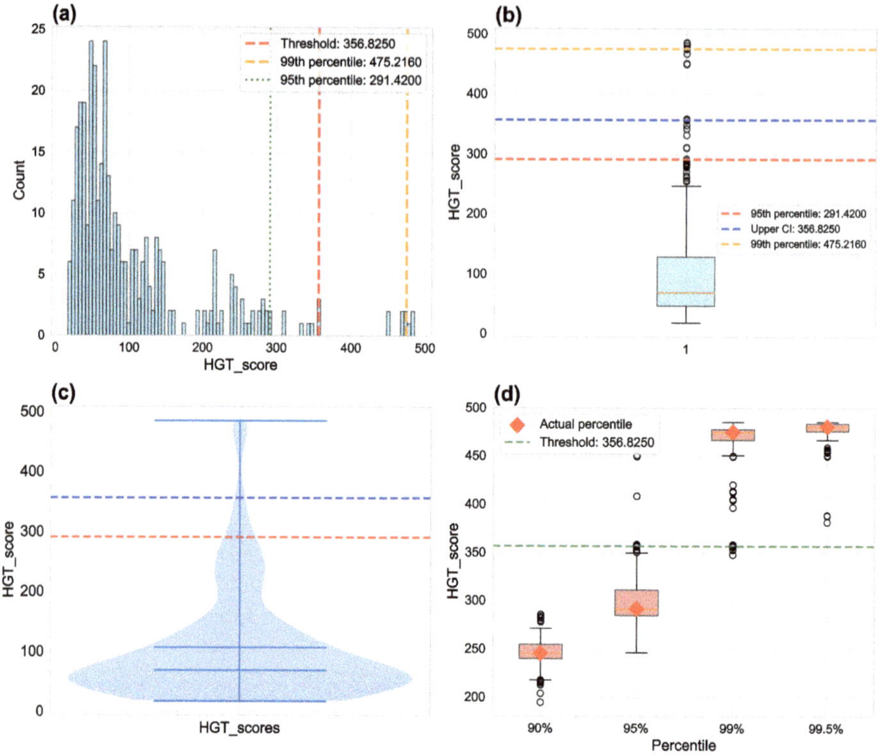

Fig. 2. Statistical analysis of HGT scores: (a) Empirical score distribution with threshold markers, (b) Box plot showing HGT scores distribution with 95th and 99th percentile thresholds, (c) Violin plot of HGT score distribution, (d) Bootstrap distribution showing the 95th percentile (291.42) and the Upper Confidence Interval (356.0). The latter was used as the cutoff for high-confidence HGT identification.

Figure 3 shows the high-confidence HGT event mapping onto the species phylogeny. The five depicted HGTs correspond to the solution suggested by biologists after their interpretation of the list of the high-confidence HGT events reported in Table 1. Interestingly, while HGTs 1, 2, and 3 between the archaeon *P. horikoshii* and the spirochetes *T. pallidum* and *B. burgdorferi*, and the archaeon *A. fulgidus* and the eukaryote cluster of *H. sapiens* and *S. cerevisiae* align with previous findings from phylogenetic reconciliation [7] and network-based [10] approaches, HGTs 4 and 5 discovered among Bacteria are novel. It is worth noting that the execution of our program on this dataset took $\sim$ 30ms on a PC equipped with an Intel(R) Core i5 11400H (2.7 GHz) processor.

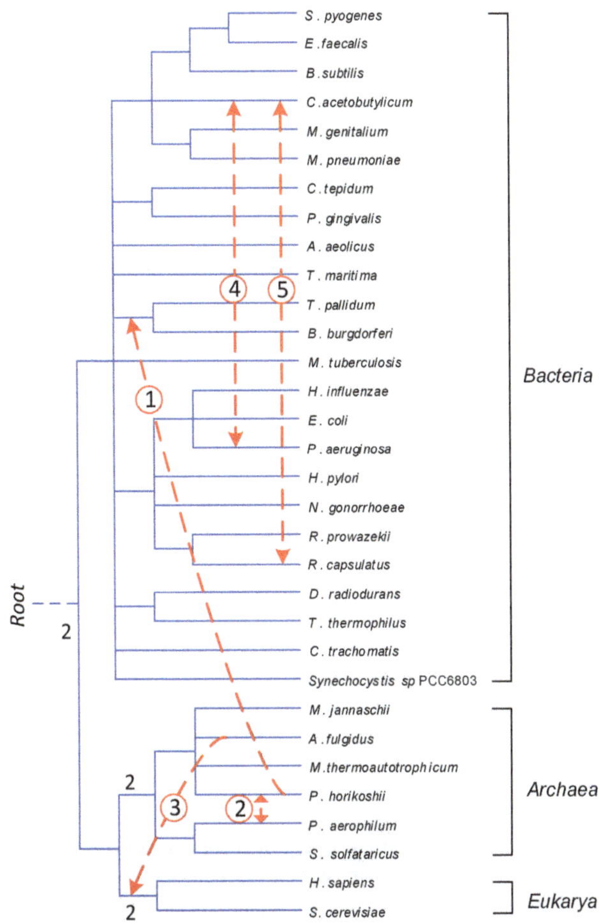

Fig. 3. Non-binary species phylogeny (undirected lines) corresponding to the NCBI taxonomic classification and 5 high-confidence HGT events (indicated by dashed arrows; when the HGT direction is uncertain the arrow is double-headed) which can be inferred from our results in Table 1.

4 Conclusion

We introduced SimNet-HGT, a scalable method for detecting horizontal gene transfer events using sequence similarity networks. By combining phylogenetic distance and sequence similarity with a ranking-based normalization, the method prioritizes high-similarity connections between evolutionarily distant lineages. The validation conducted on AARS data confirms that SimNet-HGT can accurately recover known HGTs while maintaining statistical rigor through bootstrap analysis. Despite these advantages, current implementation relies on the quality of the input sequence similarity network and weighted species phylogeny. While the method performs well on the curated benchmark, real-world metagenomic data may introduce noise from fragmentary sequences or misassemblies. Future work will extend this framework to datasets with millions of sequences to rigorously evaluate its performance at the scale of modern metagenomics.

References

1. Felsenstein, J.: Inferring Phylogenies. Sinauer Associates, Sunderland (2004)
2. Huson, D.H., Rupp, R., Scornavacca, C.: Phylogenetic Networks: Concepts, Algorithms and Applications. Cambridge University Press, Cambridge (2010)
3. Makarenkov, V., Leclerc, B.: Comparison of additive trees using circular orders. J. Comput. Biol. **7**, 731–744 (2000)
4. Makarenkov, V., Legendre, P., Desdevises, Y.: Modelling phylogenetic relationships using reticulated networks. Zoolog. Scr. **33**, 89–96 (2004)
5. Méheust, R., et al.: Hundreds of novel composite and chimeric genes with bacterial origins contributed to Haloarchaeal evolution. Genome Biol. **19**, 75 (2018)
6. Corel, E., Méheust, R., Watson, A.K., McInerney, J.O., Lopez, P., Bapteste, E.: Bipartite network analysis of gene Sharings in the microbial world. Mol. Biol. Evol. **35**, 899–913 (2018)
7. Boc, A., Philippe, H., Makarenkov, V.: Inferring and validating horizontal gene transfer events using bipartition dissimilarity. Syst. Biology **59**, 195–211 (2010)
8. Boc, A., Makarenkov, V.: Towards an accurate identification of mosaic genes and partial horizontal gene transfers. Nucleic Acids Res. **39**, e144–e144 (2011)
9. Bansal, M.S., Banay, G., Harlow, T.J., Gogarten, J.P., Shamir, R.: Systematic inference of highways of HGT in prokaryotes. Bioinformatics **29**, 571–579 (2013)
10. Sevillya, G., Adato, O., Snir, S.: Detecting horizontal gene transfer: a probabilistic approach. BMC Genomics **21**, 106 (2020)
11. Rives, A., Meier, J., Sercu, T., Goyal, S., Lin, Z., et al.: Biological structure and function emerge from scaling unsupervised learning to 250 million protein sequences. Proc. Natl. Acad. Sci. U.S.A. **118**, e2016239118 (2021)
12. Bapteste, E., van Iersel, L., Janke, A., Kelchner, S., Kelk, S., et al.: Networks: expanding evolutionary thinking. Trends Genet. **29**, 439–441 (2013)
13. Xing, H., Kembel, S.W., Makarenkov, V.: Transfer index, NetUniFrac and some useful shortest path-based distances for community analysis in sequence similarity networks. Bioinformatics **36**, 2740–2749 (2020)
14. Steinegger, M., Söding, J.: MMseqs2 enables sensitive protein sequence searching for the analysis of massive data sets. Nat. Biotechnol. **35**, 1026–1028 (2017)
15. Woese, C.R., Olsen, G.J., Ibba, M., Söll, D.: AARSs synthetases, the genetic code, and the evolutionary process. Microbiol. Mol. Biol. Rev. **64**, 202–236 (2000)

The ZED Classes of Kernels in Data Analysis

François Bavaud(✉)

University of Lausanne, Lausanne, Switzerland
fbavaud@unil.ch

Abstract. This synthetic contribution introduces the main classes of kernels encountered in Data Analysis in a weighted setting, namely dissimilarity (D), network (E) and clustering kernels (Z), collectively designated as the ZED kernels. The proposed formalism can be considered as a dual, weighted counterpart of the `RV` approach [RE76,Bav23], or as a weighted version of centered kernels of Machine Learning [MSS+98,GHS+05,JH16]. Many results (iterated dissimilarities, clustering kernels, connection with complete positivity and optimal transportation, relationship between the weighted Laplacian and the neutral kernel, reformulation and proof of the convexity properties and interpolation between the `Ncut` and the modularity objectives) are presumably original.

Keywords: complete positivity · iterated dissimilarities · optimal transportation · `Ncut` and modularity · weighted kernels

1 General Framework: Classes of Kernels

Consider n aggregated objects (such as regions, documents, species), whose importance is specified by a vector of positive relative weights $\mathbf{f} = (f_i) \in \mathbb{R}^n$ normalized to $f_\bullet = \mathbf{f}^\top \mathbf{1}_n = 1$. The usual framework of "atomic observations" naturally obtains as the uniformly weighted particular case $f_i = 1/n$.

The *diagonal weight matrix* $\mathbf{\Pi}$ and the *centering matrix* $\mathbf{H}$ are respectively defined as

$$\mathbf{\Pi} = \text{diag}(\mathbf{f}) \qquad\qquad \mathbf{H} = \mathbf{I}_n - \mathbf{1}_n\mathbf{f}^\top \ . \tag{1}$$

In this contribution, a kernel $\mathbf{K} = (K_{ij}) \in \mathbb{R}^{n\times n}$ is a symmetric matrix obeying $\mathbf{K}\sqrt{\mathbf{f}} = \mathbf{0}$.

The *neutral kernel* $\mathbf{K}_0 := \mathbf{I}_n - \sqrt{\mathbf{f}\mathbf{f}^\top} = \mathbf{\Pi}^{\frac{1}{2}}\mathbf{H}\mathbf{\Pi}^{-\frac{1}{2}}$ obeys $\mathbf{K}_0\mathbf{K} = \mathbf{K}\mathbf{K}_0 = \mathbf{K}$ for any kernel $\mathbf{K}$.

1.1 Dissimilarity Kernels

A pair dissimilarity $\mathbf{D} = (d_{ij}) \in \mathbb{R}^{n\times n}$ is a matrix obeying

$$d_{ij} \geq 0 \quad , \quad d_{ij} = d_{ji} \qquad \text{and} \qquad d_{ii} = 0 \ .$$

P. Brito et al. (Eds.): IFCS 2026, STUDIES CLASS, pp. 35–43, 2026.
https://doi.org/10.1007/978-3-032-32009-4_5

The *dissimilarity kernel* $\mathbf{K}$ associated to $\mathbf{D}$, squared Euclidean (s.e.) or not, is the quantity

$$\mathbf{K}_D = -\frac{1}{2}\mathbf{\Pi}^{\frac{1}{2}}\mathbf{H}\mathbf{D}\mathbf{H}^\top\mathbf{\Pi}^{\frac{1}{2}} \ . \tag{2}$$

Note that $\mathbf{K}_D = \mathbf{\Pi}^{\frac{1}{2}}\mathbf{B}_D\mathbf{\Pi}^{\frac{1}{2}}$ where $\mathbf{B}_D = -\frac{1}{2}\mathbf{H}\mathbf{D}\mathbf{H}^\top$ is the well-known matrix of *scalar products* entering in the Torgerson-Gower procedure for the classical, unweighted multidimensional scaling, originally based upon the fact that $\mathbf{K}_D$ and $\mathbf{B}_D$ are positive semidefinite (p.s.d.) iff $\mathbf{D}$ is s.e., with components

$$B^D_{ij} = -\frac{1}{2}\big(d_{ij} - \partial_i - \partial_j + 2\Delta\big) \text{ with } \boldsymbol{\partial} = \mathbf{D}\mathbf{f} \text{ and } \Delta = \frac{1}{2}\mathbf{f}^\top\mathbf{D}\mathbf{f} = \mathrm{Tr}(\mathbf{K}_D) \ . \tag{3}$$

Conversely, the dissimilarities can be recovered from (2) as

$$d_{ij} = B^D_{ii} + B^D_{jj} - 2B^D_{ij} = \frac{K^D_{ii}}{f_i} + \frac{K^D_{jj}}{f_j} - 2\frac{K^D_{ij}}{\sqrt{f_i f_j}} \tag{4}$$

provided $d_{ij} \geq 0$. To the neutral kernel $\mathbf{K}_0$ corresponds the *neutral dissimilarity*

$$d^0_{ij} = \begin{cases} \frac{1}{f_i} + \frac{1}{f_j} & \text{for } i \neq j, \\ 0 & \text{otherwise.} \end{cases} \tag{5}$$

1.2 Iterated Dissimilarities

In general, $\mathbf{D}$ is not s.e., and $\mathbf{K}_D$ is not p.s.d. But $\mathbf{K}^2_D$, possessing the same eigenvectors as $\mathbf{K}_D$, and whose eigenvalues are the square of those of the latter, is p.s.d by construction, making the corresponding *iterated dissimilarity*(4) $\mathbf{D}^{(2)} = (d^{(2)}_{ij})$ s.e. Interestingly, $\mathbf{D}^{(2)}$ can be analytically constructed from $\mathbf{D}$:

Theorem 1. *The iterated dissimilarity defined above reads*

$$d^{(2)}_{ij} = \frac{1}{4}\sum_{k=1}^{n} f_k(d_{ik} - d_{jk})^2 - \frac{1}{4}\Big(\sum_{k=1}^{n} f_k(d_{ik} - d_{jk})\Big)^2 \ . \tag{6}$$

Proof. Identities $\mathbf{B}^{(2)}_D = \mathbf{\Pi}^{-\frac{1}{2}}\mathbf{K}^2_D\mathbf{\Pi}^{-\frac{1}{2}}$, $\mathbf{H}^\top\mathbf{\Pi}\mathbf{H} = \mathbf{\Pi} - \mathbf{f}\mathbf{f}^\top$ and $\mathbf{D}\mathbf{f} = \boldsymbol{\partial}$ entail

$$4\mathbf{B}^{(2)}_D = \mathbf{H}\mathbf{D}\mathbf{H}^\top\mathbf{\Pi}\mathbf{H}\mathbf{D}\mathbf{H}^\top = \mathbf{H}\mathbf{D}(\mathbf{\Pi} - \mathbf{f}\mathbf{f}^\top)\mathbf{D}\mathbf{H}^\top = \mathbf{H}(\mathbf{D}\mathbf{\Pi}\mathbf{D} - \boldsymbol{\partial}\boldsymbol{\partial}^\top)\mathbf{H}^\top \ . \tag{7}$$

Terms of the form $a_i + a_j$ occurring in $\mathbf{B}^{D(2)}_{ij}$ do not contribute to $d^{(2)}_{ij}$ in view of (4): such are the contributions arising from the parts $\mathbf{1}_n\mathbf{f}^\top$ of $\mathbf{H}$ and/or $\mathbf{f}\mathbf{1}^\top_n$ of $\mathbf{H}^\top$ in (7). That is, in the computation of $d^{(2)}_{ij}$ from (7), $\mathbf{H}$ and $\mathbf{H}^\top$ can be replaced by their identity part $\mathbf{I}_n$, and the active contributions of $4\mathbf{B}^{(2)}_D$ are of the form $\mathbf{D}\mathbf{\Pi}\mathbf{D} - \boldsymbol{\partial}\boldsymbol{\partial}^\top$ with components $\sum_k f_k d_{ik} d_{jk} - \partial_i\partial_j$. Substitution in (4) achieves the proof. □

The s.e. nature of $\mathbf{D}^{(2)}$ can also be proved directly by invoking the *conditionally negative definite* condition $\mathbf{h}^\top\mathbf{D}^{(2)}\mathbf{h} \leq 0$ for any $\mathbf{h} \in \mathbb{R}^n$ with $\mathbf{1}_n^\top\mathbf{h} = 0$. Also, applying Jensen inequality on (6) directly establishes the non-negativity of $d_{ij}^{(2)}$.

Since the eigenvalues of $\mathbf{K}^2$ are the square of those of $\mathbf{K}$, iterating a s.e. configuration $(\mathbf{f}, \mathbf{D}) \rightarrow (\mathbf{f}, \mathbf{D}^{(2)})$ will *flatten* the configuration, that is inflate eigen-directions with large eigenvalues at the expense of the other directions.

1.3 Exchange Kernels: Markov and Network Kernels

A $n \times n$ non-negative row-normalized matrix constitutes the transition matrix of a Markov chain $\mathbf{W} = (w_{ij})$. We consider *weight-compatible* (w.c.) Markov chains, possessing $\mathbf{f}$ as stationary distribution and *reversible*. That is:

$$\mathbf{W} \geq 0 \ , \ \ \mathbf{W}\mathbf{1}_n = \mathbf{1}_n \ , \ \ \mathbf{W}^\top\mathbf{f} = \mathbf{f} \quad \text{and} \quad \mathbf{\Pi}\mathbf{W} = \mathbf{W}^\top\mathbf{\Pi} =: \mathbf{E} = (e_{ij}) \ . \tag{8}$$

The symmetric *exchange matrix* $\mathbf{E}$ [BS57] constitutes a joint probability of selecting the pair (ij), with margins $e_{i\bullet} = e_{\bullet i} = f_i$. Assigning weight e_{ij} to edge (ij) and weight f_i to node i defines a unoriented weighted network on the n objects.

The *exchange kernel* $\mathbf{K}$ associated to $\mathbf{W}$ or $\mathbf{E}$ is the quantity

$$\mathbf{K}_E = \mathbf{\Pi}^{\frac{1}{2}}\mathbf{W}\mathbf{\Pi}^{-\frac{1}{2}} - \sqrt{\mathbf{f}}\sqrt{\mathbf{f}}^\top = \mathbf{\Pi}^{-\frac{1}{2}}(\mathbf{E} - \mathbf{f}\mathbf{f}^\top)\mathbf{\Pi}^{-\frac{1}{2}} \quad \text{i.e.} \quad K_{ij}^E = \frac{e_{ij} - f_i f_j}{\sqrt{f_i f_j}} \tag{9}$$

made of the *Pearson residuals*of the exchange matrix, and related to the better known *normalized graph Laplacian* $\mathscr{L}\mathbf{E} = \mathbf{I}_n - \mathbf{\Pi}^{-\frac{1}{2}}\mathbf{E}\mathbf{\Pi}^{-\frac{1}{2}}$ [Chu97] as

$$\mathbf{K}_E = \mathbf{K}_0 - \mathscr{L}\mathbf{E} \ . \tag{10}$$

$\mathbf{K}_E$ defines dissimilarities from (4)

$$d_{ij}^E = \frac{e_{ii}}{f_i^2} + \frac{e_{jj}}{f_j^2} - 2\frac{e_{ij}}{f_i f_j} \tag{11}$$

provided $d_{ij}^E \geq 0$ for all pairs. This is always the case for *diffusive networks* characterized by $\mathbf{E}$ p.s.d., or equivalently by $\mathbf{W}$ p.s.d.

1.4 Clustering and Categorical Kernels

A soft clustering of n objects into m groups is specified by a $n \times m$ non-negative row-normalized *membership matrix* $\mathbf{Z} = (z_{ig})$, whose components can be interpreted as the probability that object $i = 1, \ldots, n$ belongs to group $g = 1, \ldots, m$. The weight $\dot{f}_g$ of group g is made of the object weights it contains, namely $\dot{f}_g = \sum_i f_i z_{ig}$, obeying $\dot{f}_\bullet = \dot{\mathbf{f}}^\top\mathbf{1}_m = 1$. We assume $\mathbf{Z}^\top\mathbf{1}_n > 0$ (no void group), implying $\dot{\mathbf{f}} > 0$. In matrix form

$$\mathbf{Z} \geq 0 \ , \quad \mathbf{Z}\mathbf{1}_m = \mathbf{1}_n \qquad \text{and} \qquad \mathbf{Z}^\top\mathbf{f} = \dot{\mathbf{f}} \ .$$

Define the components of the $m \times n$ dual *emission matrix*$\mathbf{A} = (a_{gi})$ as the probability that group g contains object i, that is $a_{gi} = f_i z_{ig} / \dot{f}_g$ by Bayes rule. In matrix form, with $\dot{\mathbf{\Pi}} := \text{diag}(\dot{\mathbf{f}})$,

$$\mathbf{A} \geq 0 \quad , \quad \mathbf{A} = \dot{\mathbf{\Pi}}^{-1}\mathbf{Z}^\top\mathbf{\Pi} \quad , \quad \mathbf{A}\mathbf{1}_n = \mathbf{1}_m \qquad \text{and} \qquad \mathbf{A}^\top\dot{\mathbf{f}} = \mathbf{f} \ .$$

Selecting a group from an object followed by the selection of another object from the same group defines the transition matrix of a Markov chain $\mathbf{W}_Z$ as

$$\mathbf{W}_Z = \mathbf{Z}\,\mathbf{A} = \mathbf{Z}\dot{\mathbf{\Pi}}^{-1}\mathbf{Z}^\top\mathbf{\Pi} = \mathbf{\Pi}^{-1}\mathbf{A}^\top\dot{\mathbf{\Pi}}\,\mathbf{A} \tag{12}$$

which is easily checked to be w.c. The corresponding exchange matrix reads

$$\mathbf{E}_Z = \mathbf{\Pi}\mathbf{W}_Z = \mathbf{\Pi}\mathbf{Z}\dot{\mathbf{\Pi}}^{-1}\mathbf{Z}^\top\mathbf{\Pi} = \mathbf{A}^\top\dot{\mathbf{\Pi}}\,\mathbf{A}, \text{ i.e. } e^Z_{ij} = f_i f_j \sum_{g=1}^m \frac{z_{ig}z_{jg}}{\dot{f}_g} = \sum_{g=1}^m \dot{f}_g a_{gi} a_{gj} \tag{13}$$

yielding by (9), $\mathbf{H}\mathbf{Z} = \mathbf{Z} - \mathbf{1}_n\dot{\mathbf{f}}^\top$ and $\mathbf{A}\mathbf{H} = \mathbf{A} - \mathbf{1}_m\mathbf{f}^\top$ the *clustering kernel*

$$\mathbf{K}_Z = \begin{cases} \mathbf{\Pi}^{\frac{1}{2}}\mathbf{H}\mathbf{Z}\dot{\mathbf{\Pi}}^{-1}\mathbf{Z}^\top\mathbf{H}^\top\mathbf{\Pi}^{\frac{1}{2}} \\ \mathbf{\Pi}^{-\frac{1}{2}}\mathbf{H}^\top\mathbf{A}^\top\dot{\mathbf{\Pi}}\,\mathbf{A}\mathbf{H}\mathbf{\Pi}^{-\frac{1}{2}} \end{cases} \text{ i.e. } K^Z_{ij} = \begin{cases} \sqrt{f_i f_j}\,[\sum_{g=1}^m \frac{z_{ig}z_{jg}}{\dot{f}_g} - 1] \\ \frac{1}{\sqrt{f_i f_j}} \sum_{g=1}^m \dot{f}_g a_{gi} a_{gj} - \sqrt{f_i f_j} \end{cases} \tag{14}$$

and, from (4), the dissimilarity

$$d^Z_{ij} = \sum_{g=1}^m \dot{f}_g \big(\frac{z_{ig}}{\dot{f}_g} - \frac{z_{jg}}{\dot{f}_g}\big)^2 = \sum_{g=1}^m \dot{f}_g \big(\frac{a_{gi}}{f_i} - \frac{a_{gj}}{f_j}\big)^2 \tag{15}$$

which is the *chi-square dissimilarity*between the rows of a $n \times m$ contingency table with components $f_i z_{ig} = \dot{f}_g a_{gi}$ [LS14]

Hard clusterings assign each object i to a unique group $\eth[i]$, that is $z^2_{ig} = z_{ig} = \delta_{g\,\eth[i]}$, and exactly define a *categorical variable*with m non-empty categories. The corresponding categorical kernel and dissimilarity read

$$K^Z_{ij} = \sqrt{f_i f_j}\,[\frac{\delta_{\eth[i]\,\eth[j]}}{\dot{f}_{\eth[i]}} - 1] \qquad\qquad d^Z_{ij} = \begin{cases} \frac{1}{\dot{f}_{\eth[i]}} + \frac{1}{\dot{f}_{\eth[j]}} & \text{for } \eth[i] \neq \eth[j], \\ 0 & \text{otherwise.} \end{cases} \tag{16}$$

1.5 Intersections Between Classes

A dissimilarity kernel constitutes an exchange kernel as well iff (comparing (4) and (11)) the quantity $w_{ij} = (1 + B^D_{ij})f_j$ is non-negative, that is iff $\min_{ij} B^D_{ij} \geq -1$, a condition insured by possibly shrinking the dissimilarity scale as $2\beta d_{ij}$ (see (19) below) where $\beta > 0$ is small enough.

Conversely, condition $d^E_{ij} \geq 0$ in (11), insuring an exchange kernel to be a dissimilarity kernel, always holds for $\mathbf{E}$ positive semi-definite, never holds for $\mathbf{E}$ with a null diagonal, with non-trivial characterizations lying in-between.

Clustering kernels do constitute exchange kernels (by (12)) as well as dissimilarity kernels (by (15)). By (14), the quantity $\mathbf{\Pi}^{\frac{1}{2}}\mathbf{K}_Z\mathbf{\Pi}^{\frac{1}{2}} + \mathbf{f}_n\mathbf{f}_n^\top = \mathbf{E}_Z$ is *completely positive* (c.p.), that is of the form $\mathbf{C}\mathbf{C}^\top$ where $\mathbf{C} = (c_{ig}) \in \mathbb{R}^{n \times m} \geq 0$

for some m. By (14) and $\mathbf{E}_Z = \mathbf{C}\mathbf{C}^\top$, the solution $\mathbf{C} = \mathbf{\Pi}\mathbf{Z}\dot{\mathbf{\Pi}}^{-\frac{1}{2}}$ determines entirely $\dot{f}_g = c_{\bullet g}^2$ and $z_{ig} = c_{ig}c_{\bullet g}/f_i$, obeying $z_{i\bullet} = \sum_g c_{ig}c_{\bullet g}/f_i = e_{i\bullet}/f_i = 1$, as is must.

Hence, the exchange kernel $\mathbf{K}_E$ represents a clustering kernel iff $\mathbf{E}$ is c.p., of the form $\mathbf{C}\mathbf{C}^\top$. Determining whether a given matrix is c.p. and what is the minimal rank m of $\mathbf{C}$ is NP-hard [DG14]. C.p. matrices are necessarily *doubly positive*, that is p.s.d. with non-negative components, but the converse does not hold for $n \geq 5$ [DJL94].

Finally, the dissimilarity kernel $\mathbf{K}_D$ is a clustering kernel iff $\mathbf{J}_n + \mathbf{B}_D$ is c.p.

1.6 Optimal Transportation and Gravity Model

The above derivation of an unoriented network $\mathbf{E} = \mathbf{\Pi}(\mathbf{J} + 2\beta\mathbf{B}_D)\mathbf{\Pi}$ from dissimilarities, valid if $\beta > 0$ is small enough, can be advantageously extended to any $\beta > 0$ by considering the *regularized optimal transportation problem*(ROT) [Cut13]

$$\min_{\mathbf{E}\in\mathcal{E}} [\sum_{ij} e_{ij} \ln \frac{e_{ij}}{f_i f_j} + \beta \sum_{ij} e_{ij} d_{ij}] \qquad \mathcal{E} = \{\mathbf{E} \,|\, \mathbf{E} = \mathbf{E}^\top \geq 0\,,\ \mathbf{E}\mathbf{1}_n = \mathbf{f}\} \quad (17)$$

where the first term is the Kullback-Leibler divergence of $\mathbf{E}$ from independence, with solution, computable by iterative fitting,

$$e_{ij}(\beta) = a_i(\beta)a_j(\beta)\exp(-\beta d_{ij}) \qquad a_i(\beta)\sum_j a_j(\beta)\exp(-\beta d_{ij}) = f_i \ . \quad (18)$$

Plugging the expansion $a_i(\beta) = b_i + c_i\beta + O(\beta^2)$ in (18) and comparing with (3) yields $b_i = f_i$ and $c_i = f_i(\partial_i - \Delta)$, and finally

$$e_{ij}(\beta) = f_i f_j[1 + \beta(\partial_i + \partial_j - 2\Delta - d_{ij})] + O(\beta^2) = f_i f_j[1 + 2\beta B_{ij}^D] + O(\beta^2) \quad (19)$$

which reveals the initial proposal $\mathbf{E} = \mathbf{\Pi}(\mathbf{J} + 2\beta\mathbf{B}_D)\mathbf{\Pi}$ as the small β expansion of the ROT, a.k.a. gravity model in quantitative geography [Wil69].

2 Kernel Covariances

Kernels $\mathbf{K} = (K_{ij})$ express *affinities*, *proximities* or *similarities* between pairs of objects (which permits to distinguish between $\mathbf{K}$ and $-\mathbf{K}$), and *kernel covariances* $\mathrm{Tr}(\mathbf{K}_A\mathbf{K}_B)$, amenable to testing [Bav23], express *association* between kernels. First, (2), (3) and (9) yield

$$\begin{aligned}\mathrm{Tr}(\mathbf{K}_D\mathbf{K}_E) &= \mathrm{Tr}(\mathbf{B}_D(\mathbf{E} - \mathbf{f}\mathbf{f}^\top)) = \frac{1}{2}\mathrm{Tr}(\mathbf{D}(\mathbf{f}\mathbf{f}^\top - \mathbf{E})) \\ &= \frac{1}{2}\sum_{ij} f_i f_j d_{ij} - \frac{1}{2}\sum_{ij} e_{ij} d_{ij} = \Delta - \Delta_{\text{loc}} \end{aligned} \quad (20)$$

where $\Delta_{\text{loc}} := \frac{1}{2}\sum_{ij} e_{ij}d_{ij}$, the *local inertia*, measures the average dissimilarity between pairs accordingly to their joint selection in the network (Sect. 1.3). The quantity $\delta := (\Delta - \Delta_{\text{loc}})/\Delta$ has recently been proposed [Bav24, Biv25] as a index of spatial autocorrelation, providing a direct generalization of Morann's I in the weighted, multivariate setting.

Next, replacing $\mathbf{E}$ in (20) by its clustering expression (13) yields

$$\text{Tr}(\mathbf{K}_D\mathbf{K}_Z) = \Delta - \Delta_W \quad \text{with} \quad \Delta_W := \sum_{g=1}^{m} \dot{f}_g \Delta_g \ , \ \Delta_g := \frac{1}{2}\sum_{ij} a_{gi}a_{gj}d_{ij} \quad (21)$$

where Δ_g is the average dispersion within group g, and Δ_W the within-group inertia. Both Δ and Δ_W are positive, but, unless $\mathbf{D}$ is s.e., it might happen that $\Delta_W > \Delta$, making the between-group inertia $\Delta_B := \Delta - \Delta_W$ negative: to prove the assertion, define

$$\dot{d}_{gh} := -\frac{1}{2}\sum_{ij}(a_{gi} - a_{hi})(a_{gj} - a_{hj})\, d_{ij} = \sum_{ij} a_{gi}a_{hi}d_{ij} - \Delta_g - \Delta_h \quad (22)$$

which expresses the s.e. distance between group centroids when $\mathbf{D}$ is s.e., and can, by the conditionally negative definiteness condition, be negative for dissimilarities which are not s.e. By construction,

$$\Delta_B := \frac{1}{2}\sum_{g,h=1}^{m} \dot{f}_g\dot{f}_h\dot{d}_{gh} = \frac{1}{2}\sum_{ij} f_i f_j d_{ij} \underbrace{\sum_{g,h=1}^{m} z_{ig}z_{jh}}_{=1} - \sum_{g} \dot{f}_g\Delta_g = \Delta - \Delta_W \ . \ \square$$

Finally, replacing d_{ij} in (20) by its chi-square expression (15) yields

$$\text{Tr}(\mathbf{K}_E\mathbf{K}_Z) = \sum_{g=1}^{m} \frac{\dot{e}_{gg}}{\dot{f}_g} - 1 = m - 1 - \texttt{Ncut} \quad (23)$$

where [SM00, VL07]

$$\dot{e}_{gh} := \sum_{ij} e_{ij}z_{ig}z_{jh} \quad , \quad \texttt{Ncut} := \sum_{g} \frac{\dot{e}_{g\bar{g}}}{\dot{f}_g} \quad \text{and} \quad \dot{e}_{g\bar{g}} := \sum_{h\,|\,h\neq g} \dot{e}_{gh} = \dot{f}_g - \dot{e}_{gg} \ .$$

The kernel covariance (23) is: *maximum* with value $m-1$ for $\texttt{Ncut} = 0$, attained for a network made of m disconnected components defining m sharp groups; *minimum* with value -1, attained when the non-zero edges $e_{ij} > 0$ only connect distinct, sharp groups; and *zero* for "random" memberships of the form $z_{ig} = \dot{f}_g$, making $\texttt{Ncut} = m-1$ irrespectively of the network E.

2.1 Alternative Kernels for Network Clustering: Kappa and Laplacian Kernels

Define, for $\kappa \in [0,1]$, the modified *kappa clustering* p.s.d. kernel

$$\mathbf{K}_\kappa = \mathbf{\Pi}^{\frac{1}{2}}\mathbf{H}\mathbf{Z}\dot{\mathbf{\Pi}}^{-\kappa}\mathbf{Z}^\top\mathbf{H}^\top\mathbf{\Pi}^{\frac{1}{2}} \qquad K_{ij}^{\kappa} = \sqrt{f_i f_j}\sum_{g=1}^{m} \frac{(z_{ig} - \dot{f}_g)(z_{jg} - \dot{f}_g)}{\dot{f}_g^{\kappa}} \quad (24)$$

(where $\dot{\mathbf{\Pi}}^{-\kappa} = \text{diag}(\dot{\mathbf{f}}^{-\kappa})$) whose covariance with the exchange kernel reads

$$\text{Tr}(\mathbf{K}_E\mathbf{K}_\kappa) = \sum_{g=1}^{m} \frac{\dot{e}_{gg} - \dot{f}_g^2}{\dot{f}_g^\kappa} \tag{25}$$

One recovers the standard covariance (23) for $\kappa = 1$, while $\kappa = 0$ yields an expression proportional to the so-called (weighted) *modularity* [New04].

One can also consider an alternative exchange kernel, the *Laplacian kernel* defined in view of (10) as

$$\mathbf{K}_{\mathscr{L}E} = -\mathscr{L}\mathbf{E} = \mathbf{\Pi}^{-\frac{1}{2}}\mathbf{E}\mathbf{\Pi}^{-\frac{1}{2}} - \mathbf{I}_n \qquad K_{ij}^{\mathscr{L}E} = \frac{e_{ij}}{\sqrt{f_i f_j}} - \delta_{ij} \tag{26}$$

whose covariance with the kappa clustering kernel (24) reads

$$\text{Tr}(\mathbf{K}_{\mathscr{L}E}\mathbf{K}_\kappa) = -\frac{1}{2}\sum_{g=1}^{m} \frac{1}{\dot{f}_g^\kappa} \sum_{ij} e_{ij}(z_{ig} - z_{jg})^2 \tag{27}$$

For a fixed network $\mathbf{E}$, a good clustering $\mathbf{Z}$ should *maximize* the covariances (25) and (27), and thus *minimize*the quantities

$$\mathcal{C}_\kappa[\mathbf{Z}] := -\text{Tr}(\mathbf{K}_E\mathbf{K}_\kappa) \qquad \mathcal{G}_\kappa[\mathbf{Z}] := -\text{Tr}(\mathbf{K}_{\mathscr{L}E}\mathbf{K}_\kappa) \tag{28}$$

respectively referred to as the *generalized cut* and the *generalized discontinuity* in [CB17], focused on image segmentation. It turns out that $\mathcal{C}_\kappa[\mathbf{Z}]$ is *concave* in $\mathbf{Z}$ when $\mathbf{E}$ is p.s.d., that is *diffusive*, thus attaining is minimum on the boundary of the set of admissible clusterings, that is for a *sharp clustering* $\mathbf{Z}$. On the other hand, $\mathcal{G}_\kappa[\mathbf{Z}]$ is always *convex*, generically reaching its minimum on *soft clusterings*.

Proof : define $\mathbf{C} := -\mathbf{\Pi}^{\frac{1}{2}}\mathbf{K}_E\mathbf{\Pi}^{\frac{1}{2}}$, which is negative semi-definite (n.s.d.) for $\mathbf{E}$ p.s.d., and $\mathbf{G} := -\mathbf{\Pi}^{\frac{1}{2}}\mathbf{K}_{\mathscr{L}E}\mathbf{\Pi}^{\frac{1}{2}}$, which is always p.s.d. Simple algebra yields $\mathcal{C}_\kappa[\mathbf{Z}] = \text{Tr}(\mathbf{Y}^\top\mathbf{C}\mathbf{Y})$ and $\mathcal{G}_\kappa[\mathbf{Z}] = \text{Tr}(\mathbf{Y}^\top\mathbf{G}\mathbf{Y})$ where $\mathbf{Y} := \mathbf{Z}\dot{\mathbf{\Pi}}^{-\frac{\kappa}{2}}$. Using

$$\frac{\partial y_{kg}}{\partial z_{ig}} = \alpha(\delta_{ik} + \beta f_i z_{ig}) \qquad \text{and} \qquad \frac{\partial^2 y_{kg}}{\partial z_{ig}\partial z_{jg}} = \alpha(\beta\delta_{ik}f_j + \beta\delta_{jk}f_i + \gamma f_i f_j z_{kg}) \ ,$$

where $\alpha := \dot{f}_g^{-\kappa/2}$, $\beta := -\kappa\dot{f}_g^{-1}/2$ and $\gamma := \kappa(\kappa+2)\dot{f}_g^{-2}/4$, and exploiting the symmetry of $\mathbf{C}$ yields after a bit of rearrangement

$$\frac{\partial^2\mathcal{C}_\kappa[\mathbf{Z}]}{\partial z_{ig}\partial z_{jh}} = \frac{\delta_{gh}}{\dot{f}_g^{\kappa+2}}[2\mathbf{T}_g\mathbf{C}\mathbf{T}_g^\top + \kappa(1-\kappa)\mathbf{S}_g\mathbf{C}\mathbf{S}_g^\top]_{ij} \tag{29}$$

where $\mathbf{T}_g = (t_{ik}^g)$ with $t_{ik}^g = \delta_{ik}\dot{f}_g - \kappa f_i z_{kg}$ and $\mathbf{S}_g = (s_{ik}^g)$ with $s_{ik}^g = f_i z_{kg}$. Hence, $\mathcal{C}_\kappa[\mathbf{Z}]$ is concave since its Hessian is n.s.d. Similarly,

$$\frac{\partial^2\mathcal{G}_\kappa[\mathbf{Z}]}{\partial z_{ig}\partial z_{jh}} = \frac{\delta_{gh}}{\dot{f}_g^{\kappa+2}}[2\mathbf{T}_g\mathbf{G}\mathbf{T}_g^\top + \kappa(1-\kappa)\mathbf{S}_g\mathbf{G}\mathbf{S}_g^\top]_{ij} \tag{30}$$

with parallel conclusions. □

Non-diffusive networks constitute a real challenge for clustering, and it may be advisable in that case to replace $\mathbf{E}$ by the *iterated network*with exchange matrix $\mathbf{E}^{(2)} := \boldsymbol{\Pi}\mathbf{W}^2$, which is diffusive. For instance, in a bipartite network, impossible to cluster satisfactorily by any standard, the lowest eigenvalue $\mathbf{W}$ in attains its minimum value -1, but $\mathbf{W}^2$ is p.s.d by construction, and its Fiedler eigenvalue, defined as the largest eigenvalue besides the trivial Perron eigenvalue 1, attains its maximum value 1, permitting a perfect clustering into two disconnected components:

$$\mathbf{W} = \left(\begin{array}{c|c} \mathbf{O} & \mathbf{W}_a \\ \hline \mathbf{W}_b & \mathbf{O} \end{array}\right) \qquad \mathbf{W}^2 = \left(\begin{array}{c|c} \mathbf{W}_a\mathbf{W}_b & \mathbf{O} \\ \hline \mathbf{O} & \mathbf{W}_b\mathbf{W}_a \end{array}\right) .$$

3 Conclusion

This concise contribution provides a synthesis of definitions and results relating to the main classes of kernels encountered in multivariate analysis. Kernel covariances, quantifying the interplay between classes, have been shown to express prominent classical association indexes of Data Analysis, in the continuation of program [RE76], but in the dual, objects space. The search for alternative kernels able of expressing other indices seems worth pursuing.

References

[Bav23] Bavaud, F.: Exact first moments of the RV coefficient by invariant orthogonal integration. J. Multivar. Anal. **198**, 105227 (2023)

[Bav24] Bavaud, F.: Measuring and testing multivariate spatial autocorrelation in a weighted setting: a kernel approach. Geogr. Anal. **56**(3), 573–599 (2024)

[Biv25] Bivand, R.: Implementing a weighted measure of multivariate spatial autocorrelation. Cahiers du Centre de Linguistique et des Sciences du Langage **69**, 19–35 (2025)

[BS57] Berger, J., Snell, J.L.: On the concept of equal exchange. Behav. Sci. **2**(2), 111–118 (1957)

[CB17] Ceré, R., Bavaud, F.: Soft image segmentation: on the clustering of irregular, weighted, multivariate marked networks. In: GISTAM 2017, pp. 85–109 (2017)

[Chu97] Chung, F.R.: Spectral Graph Theory, vol. 92, American Mathematical Soc (1997)

[Cut13] Cuturi, M.: Lightspeed computation of optimal transportation distances. Adv. Neural. Inf. Process. Syst. **26**(2), 2292–2300 (2013)

[DG14] Dickinson, P.J., Gijben, L.: On the computational complexity of membership problems for the completely positive cone and its dual. Comput. Optim. Appl. **57**(2), 403–415 (2014)

[DJL94] Drew, J.H., Johnson, C.R., Loewy, R.: Completely positive matrices associated with M-matrices. Linear Multilinear Algebra **37**(4), 303–310 (1994)

[GHS+05] Gretton, A., Herbrich, R., Smola, A., Bousquet, O., Schölkopf, B.: Kernel methods for measuring independence. J. Mach. Learn. Res. **6**, 2075–2129 (2005)

[JH16] Josse, J., Holmes, S.: Measuring multivariate association and beyond. Stat. Surv. **10**, 132 (2016)

[LS14] Lebart, L., Saporta, G.: Historical elements of correspondence analysis and multiple correspondence analysis. In: Visualization and Verbalization of Data, pp. 31–44 (2014)

[MSS+98] Mika, S., Schölkopf, B., Smola, A., Müller, K.R., Scholz, M., Rätsch, G.: Kernel PCA and de-noising in feature spaces. Adv. Neural. Inf. Process. Syst. **11** (1998)

[New04] Newman, M.E.: Analysis of Weighted Networks. Phys. Rev. E **70**(5) (2004)

[RE76] Robert, P., Escoufier, Y.: A unifying tool for linear multivariate statistical methods: the RV-coefficient. J. R. Stat. Soc. C **25**(3), 257–265 (1976)

[SM00] Shi, J., Malik, J.: Normalized cuts and image segmentation. IEEE Trans. Pattern Anal. Mach. Intell. **22**(8), 888–905 (2000)

[VL07] Von Luxburg, U.: A tutorial on spectral clustering. Stat. Comput. **17**(4), 395–416 (2007)

[Wil69] Wilson, A.G.: The use of entropy maximising models, in the theory of trip distribution, mode split and route split. JTEP **3**, 108–126 (1969)

GENEOmap: Binary Classification of Similar and Dissimilar Protein Binding Sites

Giovanni Bocchi[1(✉)], Alessandra Micheletti[1], Carmen Gratteri[2], and Carmine Talarico[3]

[1] University of Milan, Via Giovanni Celoria 10, 20124 Milan, Italy
{giovanni.bocchi1,alessandra.micheletti}@unimi.it
[2] LIGHT S.c.a.r.l, Via Branze 45, 25123 Brescia, Italy
carmen.gratteri.ext@exscalate.eu
[3] Dompè Farmaceutici S.p.A, Via Tommaso de Amicis 95, 80123 Naples, Italy
carmine.talarico@dompe.com

Abstract. In medicinal chemistry, assessing the similarity between protein binding sites is a fundamental task, as it enables the drug repurposing process, in which drugs that have already been approved and undergone toxicity testing are used to treat different therapeutic targets. Existing datasets usually consist of binary representations, typically lists of pairs of similar and dissimilar sites, which enable the training of binary classifiers. In this study, we introduce GENEOmap, a method based on Group Equivariant Non-Expansive Operators, which embeds sites into a d-dimensional Euclidean space by computing geometric and chemical descriptors. Then, a binary classifier is trained using a linear transformation of the descriptors, learned to maximize the AUC for prediction, to generate a similarity score. Our method was evaluated against state-of-the-art tools using the ProSPECCTs benchmark and demonstrated superior performance while accounting for uncertainty in predictions.

Keywords: pocket similarity · binary classification · AUC · GENEO

1 Introduction

Building upon recent developments in mathematical modeling, Group Equivariant Non-Expansive Operators (GENEOs) [1] are being leveraged to create novel machine learning tools for analyzing data sets. Specifically, GENEOs enable the development of transparent and trustworthy models that can be used to tackle complex problems in various fields. One notable application of GENEOs is GENEOnet [3], a model designed to detect protein pockets (for comprehensive reviews of protein pocket detection refer to [5]) with high accuracy. The use of GENEOs made GENEOnet an intrinsically explainable model and, moreover, its robustness and trustworthiness have already been investigated in [2,4]. This framework has been further improved by integrating GENEOmap, an extension

P. Brito et al. (Eds.): IFCS 2026, STUDIES CLASS, pp. 44–51, 2026.
https://doi.org/10.1007/978-3-032-32009-4_6

that facilitates the comparison of protein pocket structures. By embedding each pocket into a lower-dimensional Euclidean space using curated chemical and geometric features, GENEOmap enables the use of traditional statistical and machine learning methods for biological inference. The focus of this work is on evaluating the effectiveness of GENEOmap in classifying similar and dissimilar pockets through binary classification. We trained a classifier that leverages the sup norm distance between GENEOmap embeddings after applying a learnable linear transformation to them. Our approach yields promising results, especially when compared to state-of-the-art methods on the ProSPECCTs benchmark.

2 GENEOmap method

2.1 GENEOs

In the realm of Artificial Intelligence, Group Equivariant Non-Expansive Operators have emerged as a versatile class of mathematical tools that enable the construction of network models to maintain structural integrity under various geometric transformations of the data. This intrinsic property, known as equivariance, ensures that the model's output is coherent with respect to a group of transformations of the input data. Furthermore, GENEOs' non-expansivity property confers stability and robustness on these models, allowing them to withstand minor perturbations in the input data. The union of equivariance and non-expansivity underpins the high trustworthiness of models constructed using GENEOs. Notably, these powerful models often exhibit remarkable performance despite their relatively simple architectures. To formally define GENEOs, we start with a space Φ of functions from a set X to $\mathbb{R}$ and a group G of transformations of X. We also assume that Φ is equipped with the topology induced by the distance defined by the L_∞-norm. Next, we have another pair (Φ, H) and a group homomorphism $T\colon G \to H$. (For a more detailed understanding of GENEOs and their properties, please refer to [1]).

Definition 1. *A map $F : \Phi \to \Psi$ is called a Group Equivariant Non-Expansive Operator (GENEO) if the following conditions hold:*

1. *$F(\varphi \circ g) = F(\varphi) \circ T(g)$ for every $\varphi \in \Phi$, $g \in G$ (equivariance);*
2. *$\|F(\varphi) - F(\varphi')\|_\infty \leq \|\varphi - \varphi'\|_\infty$ for every $\varphi, \varphi' \in \Phi$ (non-expansivity).*

2.2 Data Sources

The datasets we will consider in this work are all annotated and arranged as a list of protein pairs with a binary label denoting whether the binding sites in the pair can be considered similar or not.

- **ProSPECCTs** [6]. The benchmark known as ProSPECCTs comprises eight data sets conceived to test different aspects of the comparison between protein pockets. Such datasets vary in size and in the proportion between similar and dissimilar pairs. A list of ProSPECCTs datasets is given in Table 1 together with the number of pairs in each class.

- **TOUGH-M1** [7]. The TOUGH-M1 is another popular dataset to evaluate pocket similarity from a binary perspective. From the point of view of the number of pairs, TOUGH-M1 comprises 505116 similar and 556810 dissimilar pairs. The peculiarity of TOUGH-M1 is that the sites can be identified both using the co-crystallized ligand or a pocket finding algorithm.
- **Vertex** [8]. The Vertex dataset comprises 1461668 similar and 102935 dissimilar pairs for comparisons in total. In the case of Vertex, similar pairs are made by proteins that actively bind the same ligand structure, while dissimilar pairs are made by proteins that do not share any common active ligand.

Table 1. Details about the ProSPECCTs datasets. Refer to [6] for more details.

Data set	Description	Similar pairs	Dissimilar pairs
D1	Structures having identical sequences	13430	92846
D1.2	Structures with identical sequences and similar ligands	241	1784
D2	NMR structures	7729	100512
D3	Discrimination between binding sites having different physicochemical properties	13430	13430
D4	Discrimination between binding sites having different physicochemical and geometrical properties	13430	13430
D5	Kahraman dataset	1320	8680
D6	Barelier dataset	19	43
D7	Binding site within known set of proteins	115	56284

2.3 GENEOmap

The GENEOmap method for embedding protein pockets into the latent Euclidean space $\mathbb{R}^d$ involves the following key steps. Below, we provide an overview of this procedure.

1. Read the input protein file.
2. Isolate the binding site using the crystallized ligand.
3. Represent the molecular surface as a mesh.
4. For each vertex, compute a set of local chemical features.
5. Calculate shape descriptors that remain invariant under rigid motions. Assemble all descriptors into a single vector $\eta \in \mathbb{R}^d$.
6. Standardize the vectors.
7. Apply a learnable linear map $A : \mathbb{R}^d \to \mathbb{R}^d$ with $||A||_\infty = 1$ to the vectors.

The infinity norm of a $d \times d$ matrix $A = (a_{ij})_{i,j=1,\ldots,d}$ is defined as: $||A||_\infty = \max_{i=1}^{d} \sum_{j=1}^{d} |a_{ij}|$. Notably, we emphasize that we are employing the infinity norm because the condition $||A||_\infty = 1$ means that for two sites represented by vectors $\boldsymbol{\eta}$ and $\boldsymbol{\eta}'$, the inequality

$$||A\boldsymbol{\eta} - A\boldsymbol{\eta}'||_\infty \leq ||A||_\infty ||\boldsymbol{\eta} - \boldsymbol{\eta}'||_\infty \leq ||\boldsymbol{\eta} - \boldsymbol{\eta}'||_\infty \quad (1)$$

is satisfied. Thus, since the vector η is, by construction, invariant with respect to rigid motions of the space and thanks to Inequality 1, the operator that takes a binding site to its vector $A\boldsymbol{\eta}$ is invariant with respect to rigid motions and non-expansive, and thus it is a GENEO.

3 Binary Classification

To quantify protein pocket similarity, we can leverage the output of our GENEOmap method. For two proteins, P and P', we can compute a similarity score by taking the sup norm between their respective GENEOmap embeddings. This score is calculated as the sup norm between GENEOmap outputs $A\boldsymbol{\eta}, A\boldsymbol{\eta}'$: $s(P, P') = ||A(\boldsymbol{\eta} - \boldsymbol{\eta}')||_\infty$. Our goal is to develop a binary classification framework that distinguishes proteins with similar pockets from those with dissimilar ones. The desiderata is that proteins with similar pockets have a value of $s(P, P')$ close to zero, and those with dissimilar pockets have a large value of $s(P, P')$. Thus, we derive a classifier by setting a threshold value $\tau > 0$ greater than 0. When the similarity score $s(P, P')$ exceeds this threshold, we assign a label of 1 (dissimilar pockets); otherwise, we label it as 0 (similar pockets). To evaluate the performance of our classifier, we use a benchmark dataset consisting of protein pairs with known ground truth labels. Each pair is represented by (P, P', y), where y indicates whether the proteins have similar $(y = 0)$ or dissimilar $(y = 1)$ pockets. We can then calculate standard metrics such as precision, recall, and F-score to gauge the classifier's accuracy.

Another possibility is to consider the ROC curve that plots the True Positive Rate (TPR) as a function of the False Positive Rate (FPR) for different values of the threshold τ. A more comprehensive measure of goodness is provided by the Area Under Curve (AUC) metric, which estimates the area under the ROC curve. While a random classifier would yield an AUC value of 0.5, indicating no discernible difference from chance, higher values suggest that our classifier outperforms chance and approaches perfect performance (i.e., an AUC of 1).

To train our method to maximize the AUC, we applied the approach outlined in Sect. 2.3 by determining an appropriate linear transformation A to tailor the score for the specific use case. To learn A from the data, we undertook the following steps:

1. For each of the 6 datasets (D4, D5, D6, D7, TOUGH-M1, Vertex), we split the pocket pairs into a training and a validation part, where the validation part consists of 20% of the total.

2. For each dataset, n similar and n dissimilar pairs ($2n$ overall) were sampled (with possible replacement, in the case of datasets whose training split contains less than n examples) using only the training part.
3. This set of $N = 12n$ (possibly repeated) pairs constitutes the final training set for tuning our method and learning the linear transformation A.
4. Using such a global training set, we used Adam optimizer for the loss function:

$$\mathcal{L}(A) = \frac{1}{N}\sum_{i=1}^{N}\Big(s((P,P')_i) - y_i\Big)^2 + \lambda\Big(||A||_\infty - 1\Big)^2 \qquad (2)$$

where the first term is related to the data, and the second one is a penalty term due to the constraint $||A||_\infty = 1$.

Moreover, steps from 1. to 4., were repeated 50 times while setting the learning rate of the optimizer to 1e-4, the batch size to 32, the number of epochs to 600 and $n = 500$. The value of λ was fixed to 1/2 for each simulation based on preliminary computations.

As part of the GENEOmap evaluation process, we assessed each model's performance on various datasets by calculating the AUC metric. We considered two scenarios: the entire dataset, which may include both training and testing data for that specific dataset, and a test set devised to minimize bias. To account for potential bias due to protein sequence similarity, we computed the pairs sequence identity $I(P, P')$. Sequence identity measures the proportion of identical amino acids in aligned protein sequences. High sequence identity values often indicate strong structural similarities when proteins are folded in 3D. In the field of ML for protein analysis, it's becoming increasingly common to filter out test proteins with high sequence identity to any training one to prevent data leakage. However, in the case of protein pairs, there is no universally accepted criterion for this process, as noted in [11]. To address this issue, we developed a conservative approach that discards a protein pair (P, P') if both proteins have an excessively large sequence identity with any protein Q of a training pair, i.e. if $\min\{\max_Q I(P,Q), \max_Q I(P',Q)\} > \tau_I$.

4 Results

We have evaluated the performance of our GENEOmap model on the ProSPECCTs benchmark. Our model was trained using pairs from only the last four datasets, whereas the first four were excluded from training. The results are presented as average AUC values with 95% confidence intervals for a sample of 50 models that were trained on (partially) different training sets. Our findings, summarized in Table 2, are compared to those of three other methods, Site2Vec [9], DeeplyTough [11], and SiteMine [10], as reported by their respective authors.

To compare our GENEOmap model with Site2Vec, DeeplyTough, and SiteMine, we need to consider some important factors. Firstly, let's examine Site2Vec. The authors trained their model on the PLIC dataset to generate

Table 2. AUC values for ProSPECCTs benchmark: each row represents one of the datasets and columns correspond to individual models. Note that blank cells in the second column are due to an inability to compute the AUC value for certain datasets, as all pairs were discarded due to high sequence identity. Moreover, the empty cell in the SiteMine column corresponds to no results provided for D1.2. In addition, the GENEOmap columns also report 95% confidence intervals calculated based on multiple runs (50 replicas each) of our models

Dataset	Model				
	GENEOmap ($\tau_I = 1.0$)	GENEOmap ($\tau_I = 0.6$)	Site2Vec	DeeplyTough	SiteMine
D1	0.880±0.002	0.879±0.007	1.00	0.95	1.00
D1.2	0.985±0.001		0.94	0.98	
D2	0.930±0.002	0.903±0.002	1.00	0.90	1.00
D3	0.798±0.002	0.836±0.010	0.99	0.76	0.69
D4	0.823±0.002	0.819±0.010	0.99	0.75	0.72
D5	0.929±0.002	0.936±0.005	0.86	0.67	0.78
D6	0.909±0.007	0.885±0.013	0.53	0.54	0.61
D7	0.950±0.002		0.66	0.83	0.91
Mean	0.889±0.003	0.876±0.007	0.87	0.80	0.82

200-dimensional site embeddings. However, they didn't mention discarding protein pairs due to high sequence identities with proteins in their training set. To make a fair comparison, we'll use our results with $\tau_I = 1.0$ as a benchmark (no filtering). Interestingly, GENEOmap outperforms Site2Vec when averaging AUC values across all ProSPECCTs datasets. Remarkably, as noted in [10], Site2Vec's low AUC value for D7 raises questions about its applicability in real-world drug discovery scenarios. In contrast, GENEOmap performs better than the other three methods on this dataset, highlighting its potential efficacy in real-world applications. Next, let's consider DeeplyTough. The authors trained their model using the entire TOUGH-M1 dataset after removing proteins with high sequence identity to any protein in the ProSPECCTs dataset. To fairly compare GENEOmap and DeeplyTough, we will consider our results with $\tau_I = 0.6$. GENEOmap outperforms DeeplyTough when considering the average AUC values. The reader may criticise the fact that there are two empty cells in our second column, due to the fact that the discarding of pairs completely exhausts one of the two classes of the dataset. In any case, if we look at the first column, we see that these two datasets are those for which GENEOmap achieves the highest AUCs. Therefore, it is reasonable to assume that the model's performance would not deteriorate drastically after removing similar pairs, supposing more data were available. Lastly, let's examine SiteMine, which has been optimized using a subset of the ROCS Structures dataset. Although SiteMine is not a traditional ML method, its authors didn't mention discarding protein pairs with high sequence identities. This might have a minor impact on the evaluation, but we can still compare GENEOmap and SiteMine as done for Site2Vec.

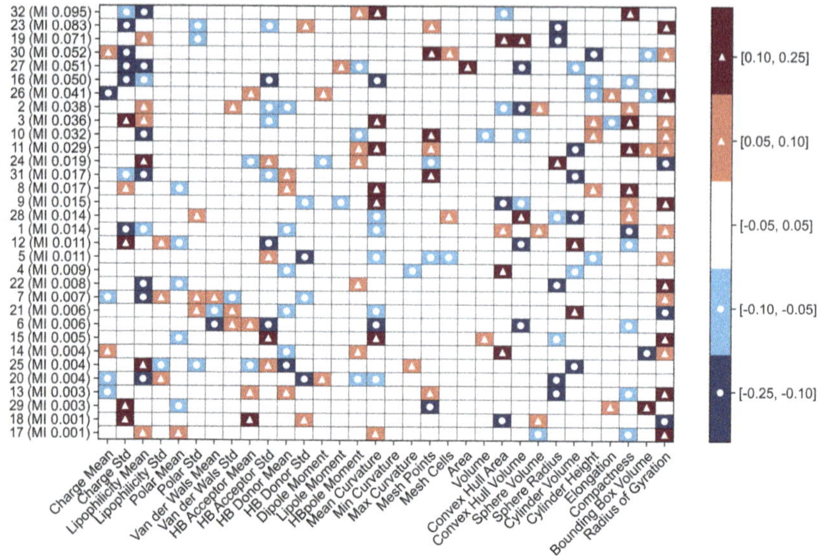

Fig. 1. A plot of matrix A is presented, where entries are colored according to their magnitude. The columns correspond to the original descriptors, whereas the rows represent their linear combinations. Notably, the rows have been ordered by permutational feature importance in the context of classifying a combined ProSPECCTs dataset.

Another relevant aspect of GENEOmap involves the inspection of the matrix A for elucidation of the classification. Figure 1 displays the entry values of A, revealing that many entries are small and contributing to a sparse structure in A. Furthermore, by ordering the mapped features (i.e., the rows of A) according to their permutational importance, it becomes possible to interpret the action of A. For instance, the most important mapped feature exhibits strong dependence on Compactness, Mean Curvature (both geometric in nature), and Lipophilicity Mean. Similarly, analogous considerations may be done for the remaining top-ranked features to provide further insight into the drivers of the GENEOmap classification outcome.

5 Conclusions

We have introduced GENEOmap, a tool that leverages the GENEO framework and allows protein pockets to be embedded into a latent space. By learning the optimal linear transformation of GENEOmap's descriptors, we have demonstrated that it outperforms many existing methods in classifying pocket pairs as similar or dissimilar. In particular, all the approaches compared rely on crystallographic data to identify pockets; future research will study extensions of GENEOmap to use predicted pockets instead, such as those from GENEOnet, thus increasing its usefulness in drug repurposing scenarios.

References

1. Bergomi, M.G., Frosini, P., Giorgi, D., et al.: Towards a topological-geometrical theory of group equivariant non-expansive operators for data analysis and machine learning. Nat. Mach. Intell. **1**(9), 423–433 (2019)
2. Bocchi, G., Frosini, P., Micheletti, A., et al.: GENEOnet: statistical analysis supporting explainability and trustworthiness. Statistics **59**(4), 1037–1062 (2025)
3. Bocchi, G., Frosini, P., Micheletti, A., et al.: GENEOnet: GENEOnet: a breakthrough in protein binding pocket detection using group equivariant non-expansive operators. Sci. Rep. **15**, 34597 (2025)
4. Bocchi, G., Frosini, P., Micheletti, A., et al.: A geometric XAI approach to protein pocket detection. In: Joint Proceedings of the xAI 2024 Late-breaking Work, Demos and Doctoral Consortium, vol. 3793, pp. 217–224 (2024)
5. Di Palma, F., Abate, C., Decherchi, S., et al.: Ligandability and druggability assessment via machine learning. Wiley Interdiscip. Rev.-Comput. Mol. Sci. **13**(5), e1676 (2023)
6. Ehrt, C., Brinkjost, T., Koch, O.: A benchmark driven guide to binding site comparison: an exhaustive evaluation using tailor-made data sets (ProSPECCTs). PLoS Comput. Biol. **14**(11), e1006483 (2018)
7. Govindaraj, R.G., Brylinski, M.: Comparative assessment of strategies to identify similar ligand-binding pockets in proteins. BMC Bioinformatics **19**, 91 (2018)
8. Chen, Y.-C., Tolbert, R., Aronov, A.M., et al.: Prediction of protein pairs sharing common active ligands using protein sequence, structure, and ligand similarity. J. Chem. Inf. Model. **56**, 1734–1745 (2016)
9. Bhadra, A., Yeturu, K.: Site2vec: a reference frame invariant algorithm for vector embedding of protein-ligand binding sites. Mach. Learn.: Sci. Technol. **2**(1), 015005 (2021)
10. Reim, T., Ehrt, C., Graef, J., et al.: SiteMine: large-scale binding site similarity searching in protein structure databases. Archiv Pharmazie **357**(5), 2300661 (2024)
11. Simonovsky, M., Meyers, J.: DeeplyTough: learning structural comparison of protein binding sites. J. Chem. Inf. Model. **60**, 2356–2366 (2020)

A Generalized $\mathcal{L}_{p_1}/\mathcal{L}_{p_2}$ Approach to Measure the Interpretability of Latent Variable Models

Mariaelena Bottazzi Schenone, Tiziano Iannaccio(✉), Ilaria Mozzetta, and Maurizio Vichi

Department of Statistical Sciences, Sapienza University, Rome, Italy
tiziano.iannaccio@uniroma1.it

Abstract. Interpretability in latent variable modeling is defined by a fundamental trade-off between structural sparseness and explained variance. However, in practice, parsimony of the loading structure is frequently sacrificed to maximize fit, due to a lack of formal evaluation metrics. To address this, we propose a synthetic measure that explicitly integrates explanatory power and structural simplicity. By extending an established metric of sparseness based on the $\mathcal{L}_1/\mathcal{L}_2$ norm ratio into a flexible $\mathcal{L}_{p_1}/\mathcal{L}_{p_2}$ formulation, we define interpretability as the geometric mean of explained variance and component-wise sparseness. Through simulations and an application to economic growth data, we show how the metric facilitates model selection within two frameworks: standard dimensionality reduction (PCA vs. Disjoint PCA) and simultaneous unit-clustering (Reduced K-Means vs. Clustering Disjoint PCA).

Keywords: interpretability measurement · $\mathcal{L}_p$-norm sparseness · dimensionality reduction

1 Introduction

Optimizing the trade-off between model accuracy and structural simplicity represents a fundamental challenge in data analysis. While sophisticated models can capture intricate patterns in high-dimensional data, their complexity often limits practical usefulness in settings where interpretability is essential. In the context of dimensionality reduction, interpretability is closely tied to the structure of the loading matrix, which governs how variables contribute to latent dimensions. Sparse loading vectors are generally easier to interpret than dense ones, hence the development of sparse decomposition methods. Hoyer [1] formalized this intuition through a sparseness measure based on the ratio of $\mathcal{L}_1$ and $\mathcal{L}_2$ norms, a formulation widely adopted in non-negative matrix factorization [2,3] and extended to tensor decompositions [4–6]. However, enforcing strong sparseness often reduces explained variance, leading to an "interpretability paradox" in which highly sparse models fail to capture meaningful data structure.

P. Brito et al. (Eds.): IFCS 2026, STUDIES CLASS, pp. 52–59, 2026.
https://doi.org/10.1007/978-3-032-32009-4_7

Several approaches have been proposed to address this issue: multi-objective frameworks explicitly balance sparseness and model fit [7], while structured regularization methods incorporate domain-specific constraints [8]. Other works compare the sensitivities of different sparseness measures [9].
Existing solutions often rely on heuristics rather than data-driven approaches, requiring external tuning parameters. This work addresses this gap with two contributions: (1) a generalized sparseness measure optimized for dimensionality reduction contexts, and (2) a composite interpretability metric. This new metric formally captures the trade-off between model fit and sparseness, providing a clearer model evaluation.

In this paper: Sect. 2 generalizes the $\mathcal{L}_{p_1}/\mathcal{L}_{p_2}$ ratio to define the interpretability index. Section 3 validates the metric through simulations comparing different structural constraints. Section 4 applies the framework to real-world data, and Sect. 5 provides concluding remarks.

2 Interpretability

To quantify the sparseness of a vector $\mathbf{v} = (v_1, \ldots, v_d)^\top \in \mathbb{R}^d$, Hoyer leveraged the fact that the $\mathcal{L}_2$ norm decreases more rapidly than the $\mathcal{L}_1$ norm as the elements of $\mathbf{v}$ become more concentrated. By scaling the ratio of these norms, Hoyer was able to map $\mathbf{v}$ onto the unit interval. This measure assigns 0 to uniform vectors and 1 to vectors containing only a single non-zero element.

$$sparseness(\mathbf{v}) = \frac{\sqrt{d} - \frac{\sum_{i=1}^{d} |v_i|}{\sqrt{\sum_{i=1}^{d} v_i^2}}}{\sqrt{d} - 1} \tag{1}$$

While Hoyer's measure captures concentration in a purely mathematical way, it was not specifically designed with dimensionality reduction in mind. A loading vector is perceived as highly interpretable if it is dominated by a single primary variable. However, the $\mathcal{L}_1/\mathcal{L}_2$ ratio is overly sensitive to "noise" (in our case, cross-loadings). The presence of several negligible but non-zero values can significantly deflate sparseness. For instance, a vector in which 95% of the mass is concentrated in a single entry –a structure a statistician would consider nearly ideal– scores at 0.7, failing to register the high clarity of the loading pattern.
To adapt Hoyer's measure to our goals, we propose a generalization based on higher-order norm ratios. For any two orders $1 \leq p_1 < p_2 < \infty$, we define the generalized $\mathcal{L}_{p_1}/\mathcal{L}_{p_2}$ sparseness ($gs(\cdot, p_1, p_2)$) of $\mathbf{v}$ as:

$$gs(\mathbf{v}, p_1, p_2) = \frac{d^{\left(\frac{1}{p_1} - \frac{1}{p_2}\right)} - \frac{\|\mathbf{v}\|_{p_1}}{\|\mathbf{v}\|_{p_2}}}{d^{\left(\frac{1}{p_1} - \frac{1}{p_2}\right)} - 1}, \tag{2}$$

where $\|\mathbf{v}\|_p = \left(\sum_{i=1}^{d} |v_i|^p\right)^{1/p}$. This formulation preserves the desirable features of Hoyer's measure while offering greater flexibility in penalizing negligible elements. Hoyer's measure is recovered by setting $p_1 = 1$ and $p_2 = 2$:

$$gs(\mathbf{v},1,2) = \frac{d^{1-\frac{1}{2}} - \frac{\|\mathbf{v}\|_1}{\|\mathbf{v}\|_2}}{d^{1-\frac{1}{2}} - 1} = \frac{\sqrt{d} - \frac{\sum_{i=1}^{d}|v_i|}{\sqrt{\sum_{i=1}^{d} v_i^2}}}{\sqrt{d} - 1}. \tag{3}$$

Moreover, $gs(\cdot, p_1, p_2) \in [0,1]$ for all $1 \leq p_1 < p_2 < \infty$, since:

$$d^{\left(\frac{1}{p_1} - \frac{1}{p_2}\right)} \|\mathbf{v}\|_{p_2} \;\geq\; \|\mathbf{v}\|_{p_1} \;\geq\; \|\mathbf{v}\|_{p_2}. \tag{4}$$

The above result follows from the relationship between $\mathcal{L}_p$ norms in $\mathbb{R}^d$. For $0 < p_1 < p_2$, the following inequality holds:

$$\|\mathbf{v}\|_{p_2} \leq \|\mathbf{v}\|_{p_1} \leq d^{\left(\frac{1}{p_1} - \frac{1}{p_2}\right)} \|\mathbf{v}\|_{p_2}. \tag{5}$$

The lower bound is a consequence of the monotonicity of $\mathcal{L}_p$ norms with respect to p. The upper bound is derived from Hölder's inequality, where the constant $d^{(1/p_1 - 1/p_2)}$ serves as the necessary normalization factor. These bounds ensure that gs is mapped strictly to the $[0,1]$ interval: $gs = 1$ for a vector with a single non-zero entry (maximum sparseness), and $gs = 0$ for a uniform vector where all entries are equal (minimum sparseness).

Given that the normalization constant $d^{(1/p_1 - 1/p_2)}$ is an increasing function of $(1/p_1 - 1/p_2)$, higher-order ratios downplay the contribution of smaller loadings, thus better mirroring what analysts perceive as "simple".

In this study, we will reduce said contributions by setting $p_1 = 2$ and $p_2 = 3$. This selection aligns with traditional results of latent variable modeling: the L_2 norm links to the variance-based framework of PCA, while the L_3 penalty effectively filters numerical noise by emphasizing primary loadings. Users may adjust these parameters to tune the metric's sensitivity to small loadings depending on the specific requirements of their analysis.

As anticipated, interpretability ultimately depends on both the loading structure and the explanatory power. Therefore, given a $(J \times Q)$ loading matrix $\mathbf{A} = [\mathbf{a}_1, ..., \mathbf{a}_Q]$, we propose to quantify the interpretability $I(\cdot, p_1, p_2)$ as follows:

$$I(\mathbf{A}, p_1, p_2) = \sqrt{\mathrm{EV}(\mathbf{A}) \cdot \overline{\mathrm{GS}}(\mathbf{A}, p_1, p_2)}\,, \tag{6}$$

where $\mathrm{EV}(\mathbf{A}) \in [0,1]$ is the proportion of variance explained by the model, and

$$\overline{\mathrm{GS}}(\mathbf{A}, p_1, p_2) = \frac{1}{Q} \sum_{q=1}^{Q} gs(\mathbf{a}_q, p_1, p_2). \tag{7}$$

The choice of the geometric mean is driven by interpretability being a"non-compensatory" construct. In an additive scenario (e.g., arithmetic mean), a model with near-zero explanatory power could still achieve a significant interpretability score through extreme sparseness (and vice-versa). The geometric mean ensures both models must perform well in both dimensions.

3 Simulation Study

This simulation compares the interpretability of the most established dimensionality reduction procedure (Principal Component Analysis; PCA) and its disjoint counterpart (Disjoint PCA; DPCA).

3.1 Data Generation Procedure

The generation of a synthetic dataset $\mathbf{X}$ of dimension $(N \times J)$ works as follows.

1. **Latent scores generation:** a matrix of true scores $\mathbf{Y}$ $(N \times Q)$ is sampled from a multivariate standard normal distribution:

$$\mathbf{Y} \sim \mathcal{N}(0, \mathbf{I}_Q).$$

   To isolate the effect of the loading structure, $\mathbf{Y}$ is orthonormalized using a QR decomposition so that $\mathbf{Y}^\top \mathbf{Y} = \mathbf{I}_Q$.
2. **Loading matrix generation:** the loading matrix $\mathbf{A}$ $(J \times Q)$ is constructed with a procedure that controls sparseness through a stochastic process that preserves orthonormality ($\mathbf{A}^\top \mathbf{A} = \mathbf{I}_Q$).
   - *Initialization*: a matrix is initialized with $\mathcal{N}(0, 1)$ draws, constrained to a maximally sparse structure. Then, columns are normalized to obtain random starting disjoint loadings.
   - *Sparseness control*: a sequence of Givens rotations [10] is applied to random pairs of rows. The number of rotations K and the maximum rotation angle are defined as functions of a parameter $S \in [0, 1]$ through the following relations:

$$K = (1 - S)f(J) \tag{8}$$

$$\theta_{max} = (1 - S)\frac{\pi}{4} \tag{9}$$

     where $f(J)$ is a scaling function of the number of variables J.
   - *Continuum:* for $S = 1$, the matrix remains perfectly disjoint. As S decreases toward 0, the number of rotations and the maximum angle increase (up to $\pi/4$), progressively distributing the weights across components and decreasing the sparseness of the matrix while maintaining orthonormality.
3. **Signal-to-Noise projection:** the observed data matrix $\mathbf{X}$ is generated by adding noise to the projected scores in the original space:

$$\mathbf{X} = \mathbf{Y}\mathbf{A}^\top + \mathbf{E}, \tag{10}$$

   where $\mathbf{E}$ has been scaled to satisfy a pre-defined Signal-to-Noise Ratio (SNR). Given a white noise matrix $\mathbf{E}_{raw}$ and a target explained variance R^2:

$$\mathbf{E} = \mathbf{E}_{raw} \cdot \sqrt{\frac{\mathrm{Var}(\mathbf{Y}\mathbf{A}^\top)}{\mathrm{Var}(\mathbf{E}_{raw})} \cdot \frac{1 - R^2}{R^2}}. \tag{11}$$

3.2 Simulation Results

While PCA is the baseline method for linear dimensionality reduction, DPCA restricts each original variable to load on a single component. A well-established procedure to enhance the interpretability of PCA is to apply an orthogonal rotation to the loading matrix. In particular, the Varimax rotation forces components to have fewer high-loading variables. For this reason, we included the Varimax-rotated PCA (PCAVar) in the analysis. The primary goal was to assess how each method handles interpretability under varying degrees of sparseness in the true loading matrix $\mathbf{A}$.

The models were fitted to the data matrix $\mathbf{X}$, and the resulting performance metrics were computed on the estimated loading matrices $\hat{\mathbf{A}}_{\text{PCA}}, \hat{\mathbf{A}}_{\text{PCAVar}}$ and $\hat{\mathbf{A}}_{\text{DPCA}}$. Parameters were fixed at $N = 1000$ observations, $J = 10$ variables, $Q = 4$ components, and a noise level of $1 - R^2 = 0.05$. For each parameters' combination, 100 datasets were generated, and the results were averaged. While sensitivity analyses were also performed by varying Q and R^2, the results followed predictable patterns: smaller values of Q reduced sparseness, since variables were forced into a more condensed loading structure. On the other hand, lower R^2 values naturally decreased the explained variance, as the models are intrinsically unable to account for increased residual noise.

Table 1. Comparison of Explained Variance (EV), Average Sparseness (AvgSpar) and Interpretability (Interp) for PCA, Varimax-rotated PCA, and DPCA across three levels of simulated sparseness.

True Sparseness	Method	EV	AvgSpar	Interp
Low ([0.50, 0.60))	PCA	0.941	0.456	0.652
	PCAVar	0.941	0.641	0.777
	DPCA	0.840	0.766	0.802
Moderate ([0.60, 0.70))	PCA	0.913	0.520	0.687
	PCAVar	0.913	0.699	0.799
	DPCA	0.862	0.771	0.815
High ([0.70, 0.80])	PCA	0.888	0.602	0.727
	PCAVar	0.888	0.747	0.814
	DPCA	0.877	0.771	0.822

Table 1 highlights the critical trade-off between variance maximization and structural recovery. As expected, standard PCA consistently yields the highest explained variance but the lowest interpretability. However, as the sparseness of the true loading matrix $\mathbf{A}$ increases, DPCA effectively closes the explanatory gap. For example, in the 'High' Sparseness condition, DPCA achieves an explained variance of 0.877, which is remarkably close to the 0.888 achieved by PCA, consequently providing a much higher interpretability (0.822 vs 0.727).

Furthermore, DPCA demonstrates superior consistency: its average sparseness remains stable regardless of the underlying data structure, whereas PCA and PCAVar are highly sensitive to the true loading density. Overall, DPCA provides the most interpretable recovery of the true **A**, outperforming PCAVar across all simulated conditions.

4 Application to Digitalization and Economic Growth Dataset

To validate the practical utility of the proposed measure, we evaluate the interpretability of four established multivariate techniques using a real-world dataset. The data explores the interplay of digital technology adoption and macroeconomic performance through key indicators like broadband subscriptions and gross fixed capital formation. To extend our analysis to procedures that simultaneously perform dimensionality reduction of units and variables, Reduced K-Means (RKM, [11]) and Clustering Disjoint PCA (CDPCA, [12]) are also considered. The former combines clustering with standard PCA, while the latter represents a more constrained setting in which a disjoint loading structure is enforced. According to Kaiser's rule and the Pseudo-F statistic, we considered $Q = 4$ variable components and $K = 9$ clusters.

Table 2. Comparison of Explained Variance (EV), Average Sparseness (AvgSpar) and Interpretability (Interp) of PCA, Varimax-rotated PCA, DPCA, RKM, Varimax-rotated RKM, and CDPCA for Digitalization and Economic Growth data.

Model	EV	AvgSpar	Interp
PCA	0.694	0.495	0.586
PCAVar	0.694	0.611	0.651
DPCA	0.665	0.678	0.672
RKM	0.556	0.447	0.498
RKMVar	0.556	0.614	0.584
CDPCA	0.527	0.737	0.623

Results in Table 2 confirm what was already highlighted in the simulation. Standard PCA naturally achieves the highest explained variance (0.694). However, the poor sparseness (0.495) translates to a lower interpretability score (0.586). This result underlines a typical limitation of PCA: loadings are often dense, and the explanatory power is "wasted" when components have to be labeled. Adding a Varimax rotation preserves the same explained variance (0.694), but substantially increases sparseness (from 0.495 to 0.611). This directly improves interpretability (from 0.586 to 0.651) and confirms the well-known advantages of orthogonal rotations.

DPCA yields the most constrained solution and, as expected, exhibits a modest reduction in explained variance (0.665) relative to PCA. In return, it achieves a higher level of sparseness (0.678), and therefore the highest interpretability score (0.672). Interestingly, the small difference in both explained variance and sparseness between PCAVar and DPCA places this real-data application in the 'High' Sparseness scenario of the previous simulation. This is also true when clustering is incorporated in the analysis: RKMVar offers little advantages in terms of explanatory power when compared to CDPCA, which gains about 12% sparseness and achieves a higher overall interpretability.

The alignment of the Digitalization and Economic Growth data with the 'High' Sparseness simulation scenario suggests a broader meta-analytical opportunity. It would be highly insightful to systematically evaluate the frequency with which empirical datasets move toward high-sparseness structures. While these two instances are not statistically significant, a large-scale analysis could provide profound insights into the intrinsic architecture of latent components in social and economic phenomena. Specifically, if real-world data consistently mirrors high-sparseness simulations, it would imply that latent structures are naturally more modular and "disjoint" than the ones that standard PCA assumes.

5 Final Remarks

This paper has introduced a unified measure of interpretability that rigorously captures the fundamental compromise between empirical fit and structural simplicity. Our results confirm that maximizing statistical fit does not necessarily yield the highest interpretability of the solution. Once interpretability is understood as actionable information, models can be compared using the same logic traditionally applied to explained variance.

Future work will move toward a more dynamic formulation. Specifically, we are developing a weighted index that scales each component's sparseness by its relative share of explained variance, ensuring that the most dominant dimensions are held to the highest standards of clarity. Furthermore, we aim to integrate this measure directly into the model's objective function. By maximizing interpretability during the estimation process, we aim to relax the disjoint assumption and allow for a more flexible, data-driven approach to sparse dimensionality reduction. Finally, the asymptotic behavior of the interpretability measure will be studied in order to build a selection tool for p_1 and p_2.

Data Availability. The data that support the findings of this study are openly available at the following link: https://data.mendeley.com/datasets/ctm7vvpp7n/1.

References

1. Hoyer, P.O.: Non-negative matrix factorization with sparseness constraints. J. Mach. Learn. Res. **5**, 1457–1469 (2004)

2. Kim, Y., Sun, W.W., Li, L.: Provable sparse tensor decomposition. J. Roy. Stat. Soc.: Ser. B (Stat. Methodol.) **79**(3), 899–916 (2017)
3. Mairal, J., Bach, F., Ponce, J., Sapiro, G.: Online learning for matrix factorization and sparse coding. J. Mach. Learn. Res. **11**, 19–60 (2010)
4. Acar, E., Kolda, T.G., Dunlavy, D.M.: All-at-once optimization for coupled matrix and tensor factorizations. Statist. Anal. Data Min. **4**(2), 149–170 (2011)
5. Cichocki, A., Zdunek, R., Phan, A.H., Amari, S.I.: Nonnegative Matrix and Tensor Factorizations: Applications to Exploratory Multi-way Data Analysis and Blind Source Separation. Wiley, Chichester (2009)
6. Bottazzi Schenone, M., Iannaccio, T., Mozzetta, I., Vichi, M.: Three-way data analysis with explainable tucker3 clustering (XT3Clus). Appl. Stoch. Model. Bus. Ind. **42**(1), e70069 (2026)
7. Sun, W.W., Li, L.: STORE: sparse tensor response regression and neuroimaging analysis. J. Mach. Learn. Res. **18**(1), 4908–4944 (2017)
8. Lock, E.F., Li, G.: Supervised multiway factorization. Electron. J. Stat. **12**(1), 1150–1180 (2018)
9. Hurley, N., Rickard, S.: Comparing measures of sparseness. IEEE Trans. Inf. Theory **55**(10), 4723–4741 (2009)
10. Givens, W.: Computation of plain unitary rotations transforming a general matrix to triangular form. J. Soc. Ind. Appl. Math. **6**(1), 26–50 (1958)
11. De Soete, G., Carroll, J.D.: K-means clustering in a low-dimensional Euclidean space. In: Diday, E., et al. (eds.) New approaches in classification and data analysis, pp. 212–219. Springer, Heidelberg (1994)
12. Vichi, M., Saporta, G.: Clustering and disjoint principal component analysis. Comput. Stat. Data Anal. **53**(8), 3194–3208 (2009)

Machine Learning and Territorial Clustering for Detecting Mismatches Between Population Ageing and Healthcare Resources

Simona Cafieri(✉)

Italian National Institute of Statistics, Naples, Italy
cafieri@istat.it

Abstract. Population ageing poses sustainability challenges for healthcare systems, requiring frameworks capable of capturing heterogeneous dynamics. This study develops an AI-enhanced, territorially statistical framework to assess healthcare sustainability under demographic ageing in Italy. The analysis relies on a dataset from Istat survey data, covering demographic structure, healthcare supply, service utilization, socio-economic conditions, and household characteristics for Italian provinces. Methodologically, the study compares regression approaches (OLS, Ridge, LASSO) with machine learning models (Random Forest and XGBoost) to identify determinants of healthcare demand and nonlinear relationships. Model performance is evaluated. To address territorial disparities, clustering detects mismatches between ageing pressure and healthcare capacity across provinces. Clusters are interpreted in terms of demographic, social, and healthcare profiles, highlighting the role of social and household factors, such as living arrangements and isolation, in healthcare utilization. The findings provide policy implications, suggesting differentiated interventions aimed at reducing social isolation and strengthening healthcare infrastructure in fragile territories.

Keywords: Sustainability · Healthcare · Machine learning · Statistical measurement · Territorial analysis

1 Introduction

Measuring sustainability is a core challenge, particularly in healthcare, where population ageing interacts with social structures, resources, and service utilization (Stiglitz et al., 2009).

The contribution of this study lies in integrating official statistics, econometric models, and machine learning techniques within a territorially informed analytical framework designed to assess healthcare sustainability and geographical disparities (Athey, 2018).

By focusing on Italian provinces, the paper examines whether healthcare resources align with demographic and social ageing patterns and how these relationships vary across territorial units.

The work is structured as follows: Sect. 2 reviews literature on sustainability, ageing, and AI/territorial methods; Sect. 3 describes data and methods; Sect. 4 presents results; Sect. 5 discusses implications; Sect. 6 concludes the paper.

P. Brito et al. (Eds.): IFCS 2026, STUDIES CLASS, pp. 60–67, 2026.
https://doi.org/10.1007/978-3-032-32009-4_8

2 Literature Review

Sustainability measurement has traditionally relied on indicator-based approaches for monitoring trends and benchmarking performance, but these often overlook structural imbalances, interaction effects, and latent vulnerabilities (Stiglitz et al., 2009). In healthcare, outcomes depend on the interplay between demographics, institutions, and behavior. While population ageing is linked to higher healthcare utilization and expenditure, this relationship is nonlinear and heterogeneous across territories (Cutler et al., 2006). Aggregate indicators like ageing or dependency ratios capture only part of healthcare pressure, as social and household characteristics significantly mediate utilization patterns. Social vulnerability and deprivation have also been widely discussed in the welfare literature (Spicker et al., 2008), highlighting how social structures can influence health outcomes and access to services. Differences in infrastructure, accessibility, and service organization can produce uneven outcomes even among areas with similar demographics, highlighting the need for small-area and territorially disaggregated analysis (Anselin, 1995; Fotheringham et al., 2002).

Recent advances in machine learning provide flexible tools for modeling nonlinear relationships. These complement traditional econometric approaches that emphasize causal interpretation under strong assumptions (Angrist and Pischke, 2009). When embedded within informed analytical frameworks, these methods allow detection of territorial mismatches between healthcare needs and resource allocation, enhancing policy relevance and supporting place-based planning in decentralized systems (Anselin, 1995; Fotheringham et al., 2002; Athey, 2018).

3 Data and Methods

3.1 Data Sources, Scope, and Conceptual Structure

The analysis is conducted at the provincial level (NUTS-3), covering 101 Italian provinces with consistent data from official sources (Istat, Ministry of Health, 2023). This scale captures territorial heterogeneity in a decentralized healthcare system (Eurostat, 2023). Variables are organized into three conceptual blocks:

1) demographic and social indicators: ageing index, household size, and the share of single-person households (capturing demographic pressure and social isolation/informal care networks);
2) socioeconomic conditions: educational attainment and labor market participation, influencing health status and service access;
3) healthcare supply and utilization: primary care (GPs), specialist care, and hospital-based utilization (beds, hospitalization rates, emergency visits).

Modeling these targets allows the analysis to evaluate how demographic and social characteristics translate into healthcare demand. In this framework, a mismatch is defined as a provincial configuration where high ageing pressure coexists with inadequate healthcare capacity or service availability relative to population needs.

3.2 Statistical Preprocessing and Exploratory Analysis

All variables are standardized using z-scores to ensure comparability. Correlation analysis is performed to assess linear associations and potential multicollinearity. Exploratory territorial visualization examines the distribution of healthcare dimensions across provinces, providing an empirical baseline for assessing whether healthcare sustainability pressures are primarily linear or influenced by more complex socio-economic interactions. The provincial-level structure of the dataset allows the analysis to explicitly consider territorial patterns in demographic, social, and healthcare indicators, guiding the subsequent clustering and modeling steps.

3.3 Econometric Modeling and AI-Enhanced Clustering

The framework combines interpretable econometric models (LASSO and Ridge) with ensemble methods (Random Forest and XGBoost, rooted in statistical learning theory (Vapnik, 1998) to capture both linear and nonlinear effects (Breiman, 2001; Tibshirani, 1996). Regularized regressions address multicollinearity and variable selection, while machine learning (ML) methods identify complex, high-dimensional dynamics without restrictive assumptions. Given the sample size ($n = 101$), model performance is evaluated using 5-fold cross-validation to ensure robust out-of-sample estimates and mitigate overfitting risks (Aittokallio, 2010).

The subsequent geographical clustering leverages the predicted values from these ML models, integrated with standardized demographic and social indicators. By using ML outputs as input features for the K-means algorithm, the framework ensures that provinces are grouped not only by raw data but also by the latent, nonlinear healthcare demand patterns previously identified. This hybrid approach allows for a more robust identification of territorial profiles and structural mismatches that linear models would overlook. The selected target variables represent complementary dimensions of the healthcare system, including expenditure, infrastructure, and service utilization, ensuring a comprehensive representation of both demand- and supply-side dynamics.

Predicted values derived from machine learning models are used as synthetic indicators of latent healthcare demand, reducing noise and capturing nonlinear interactions before clustering.

3.4 Model Interpretation and Spatial Diagnostics

The optimal number of clusters was determined using the elbow method and silhouette score, both indicating a stable structure. Based on these criteria, $K = 3$ clusters were selected, balancing within-cluster homogeneity, between-cluster separation, and interpretability of territorial patterns. These unsupervised clusters serve as diagnostic tools for place-based healthcare planning, highlighting provinces where high demographic pressure and social fragility coincide with limited healthcare capacity, signaling elevated sustainability risk. To reinforce the spatial interpretation, Moran's I was computed for each standardized indicator, with values ranging from 0.21 to 0.34 ($p < 0.05$), confirming significant spatial autocorrelation. This demonstrates that clusters reflect geographically structured disparities rather than random groupings, supporting a territorially explicit interpretation of mismatches and guiding localized policy interventions in decentralized systems. The clustering is territorial but not contiguity-constrained, with Moran's I serving as a spatial diagnostic to validate the geographic structure.

4 Results

4.1 Descriptive Evidence and Correlation Patterns

Descriptive statistics reveal substantial heterogeneity across Italian provinces in demographic, social, and healthcare indicators. Correlation analysis and territorial patterns highlight pronounced territorial disparities. Traditional ageing indicators, such as the ageing index, show weak and unstable correlations with healthcare utilization (hospitalization rates, emergency visits, specialist consultations), whereas household and social indicators—particularly average household size and share of single-person households—display stronger, systematic associations.

Provinces with smaller households and higher social fragmentation experience higher utilization rates, reflecting reduced informal support and increased reliance on formal healthcare services.

4.2 Model Performance and Predictive Accuracy

The target variables represent different dimensions of healthcare system capacity and utilization. They include indicators of healthcare expenditure (per capita health spending), infrastructure availability (share of private hospital beds, geriatric bed availability), and service utilization (hospitalization rates, emergency visits, and specialist consultations). Modeling multiple targets allows assessment of how demographic and social variables influence different functional layers of the healthcare system.

Table 1 compares out-of-sample performance (RMSE) of regularized linear models (LASSO, Ridge) and machine learning methods (Random Forest, XGBoost) across multiple healthcare-related targets.

Machine learning models capture complex nonlinear patterns: for instance, per capita healthcare expenditure RMSE decreases from ~ 0.70 (LASSO) to 0.30 (XGBoost). Similar improvements occur for emergency visits and geriatric bed availability, indicating strong nonlinearities and interactions that linear models cannot capture. These results imply that healthcare sustainability dynamics are shaped by complex interdependencies between demographic, social, and healthcare variables. AI-based models enhance statistical measurement by uncovering latent structures crucial for policy-relevant assessment.

4.3 Key Drivers of Healthcare Utilization

Variable importance analysis (XGBoost) reveals a clear geographical divide in the drivers of healthcare utilization. In Northern provinces, the share of single-person households over 60 is the dominant predictor of formal healthcare demand, indicating that social isolation increases reliance on hospitals and emergency services. Conversely, in Southern provinces, larger average household sizes act as an informal "buffer," but limited

Table 1. Model performance comparison of six healthcare- targets

Target Variable[a]	LASSO RMSE	Ridge RMSE	Random Forest RMSE	XGBoost RMSE
percapita_health_spend_z	0.7012	0.6897	0.3091	0.2983
private_beds_pct_z	0.7113	0.6751	0.3017	0.2873
specialist_visits_rate_z	0.6984	0.6710	0.3125	0.2987
emergency_rate_z	0.6889	0.6682	0.3104	0.2958
geriatric_beds_rate_z	0.7071	0.6765	0.2835	0.2716
hosp_per_1000_z	0.7045	0.6798	0.2786	0.2781

[a] Target variables (standardized) and their definitions: percapita_health_spend_z = per capita health spending; private_beds_pct_z = share of private hospital beds; specialist_visits_rate_z = specialist consultations rate; emergency_rate_z = emergency visits rate; geriatric_beds_rate_z = geriatric beds availability; hosp_per_1000_z = hospital beds per 1,000 inhabitants."

infrastructure—such as low GP density and hospital bed availability—forces residents toward emergency services, creating a structural mismatch.

These patterns in social and infrastructure drivers anticipate the three distinct provincial clusters identified in the following paragraph.

4.4 Territorial Clustering and Mismatch Results

The clustering analysis identified three distinct territorial profiles at the provincial level, capturing structural mismatches between demographic pressure and healthcare capacity in Italy. The clusters are described as follows:

Cluster 1 – Northern provinces with high social fragility:

This cluster includes provinces such as Trieste and Genoa, characterized by high ageing indices and a high share of single-person households over 60 (exceeding 56% in dense urban areas). Despite relatively adequate healthcare infrastructure, these provinces show elevated hospitalization rates, highlighting the impact of weaker informal support on formal healthcare demand.

Cluster 2 – Southern provinces with infrastructure deficits:

Provinces in this cluster experience systemic fragility due to low GP density and limited hospital bed availability. Although average household sizes are larger and elderly isolation is lower, the lack of primary care infrastructure drives higher utilization of emergency services, indicating supply-side limitations rather than social fragility.

Cluster 3 – Balanced or intermediate provinces:

This cluster includes provinces where demographic pressure and healthcare capacity are relatively aligned, providing benchmarks for sustainable healthcare provision.

The PCA loading structure (Fig. 1) confirms these distinctions: PC1 (42%) captures social fragility, emphasizing Cluster 1; PC2 (20%) reflects healthcare capacity, highlighting Cluster 2; Cluster 3 occupies intermediate positions along both axes. K-means clustering combined with PCA visually illustrates these structural mismatches, offering a diagnostic tool for targeted, place-based healthcare planning.

To better illustrate these territorial mismatches, Fig. 1 visualizes the results of the K-means clustering combined with PCA, highlighting the structural mismatches between demographic pressure and healthcare capacity across Italian provinces.

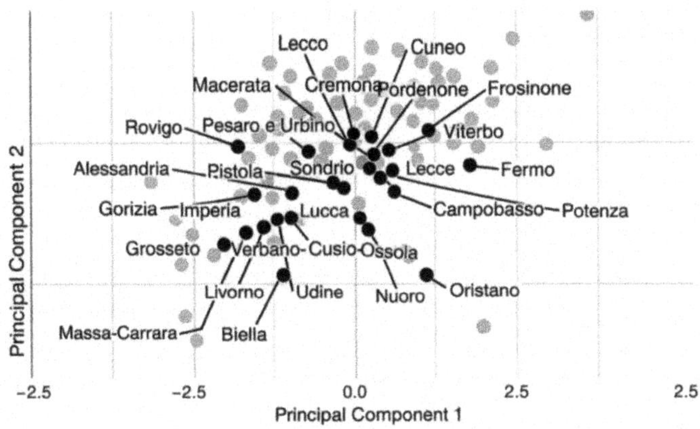

Fig. 1. PCA plot of provincial clusters.

Each point represents an Italian province. The horizontal axis (PC1) captures a composite dimension of social fragility and demographic pressure, closely associated with healthcare demand, while the vertical axis (PC2) reflects differences in healthcare capacity and supply-side constraints. Provinces located in the upper-right and lower-right quadrants exhibit higher levels of vulnerability, whereas those on the left side of the plot show more favourable socio-demographic conditions. The distribution of clusters highlights distinct territorial mismatches between healthcare demand and supply, with Cluster 1 primarily associated with high-demand contexts and Cluster 2 more clearly differentiated along the supply-side dimension, while Cluster 3 occupies intermediate positions reflecting more balanced configurations.

This pattern is consistent with the clustering results. Northern ageing-fragile provinces like Trieste or Genoa show elevated hospitalization rates, reflecting weaker informal support systems. Conversely, some southern provinces with comparable ageing but larger household sizes exhibit lower overall utilization but disproportionate reliance on emergency services—suggesting cohesive family structures buffer formal health-care pressure. Quantitatively, Cluster 1 shows hospitalization rates 15–20% above the national average, while Cluster 2 shows GP density around 20% below the mean. Additional cases illustrate further diversity. Bergamo and Aosta share high demand (PC1 +) but diverge on supply: Bergamo has moderate capacity, though pandemic pressures exposed structural weaknesses; Aosta, high on both axes, reflects inner alpine fragility with concentrated supply and geographical discontinuity. Salerno (high demand, intermediate supply) shows above-average emergency use despite cohesive family networks. Matera (intermediate demand, low supply) has contained hospitalizations but fragile proximity care due to low GP density. Overall, the joint PCA-clustering evidence confirms distinct demand- and supply-driven mismatches across Italian provinces, shaped by local interactions among demographic pressure, family networks, service organization, and seasonal dynamics.

5 Discussion

This study contributes to the literature on healthcare sustainability by showing that demographic ageing alone is not sufficient to explain healthcare demand. Consistent with previous research (Cutler et al., 2006), social and household characteristics—particularly living arrangements and isolation—emerge as key mediating factors, but their effects vary significantly across territories.

The results identify a dual-pathway mechanism. In Northern provinces, higher elderly isolation is associated with hospitalization rates approximately 15–20% above the national average, despite adequate healthcare infrastructure, indicating demand driven by weak informal care networks. In Southern provinces, larger households partially buffer demand, but structural deficits in both healthcare infrastructure and workforce availability generate supply-side constraints, mainly affecting emergency service use.

Methodologically, the integration of machine learning predictions into clustering represents a key contribution. Using predicted values as synthetic indicators of latent demand allows capturing nonlinear interactions while improving the interpretability of territorial clusters. This approach complements existing spatial methods by identifying territorial regimes rather than local parameter variation.

Policy implications are inherently place-based: interventions in the North should address social isolation, while in the South they should prioritize strengthening primary care infrastructure. Balanced provinces provide useful benchmarks for aligning demographic pressure and healthcare capacity.

Some limitations remain. Clustering is not spatially constrained, although significant Moran's I values support geographical coherence. The sample size (101 provinces) limits validation complexity, though 5-fold cross-validation mitigates overfitting risks. Future research could extend the framework using longitudinal data.

6 Conclusions

This paper presents an AI-based, territorially informed statistical framework for measuring healthcare sustainability under population ageing. By integrating indicators at the provincial level, the framework moves beyond conventional demographic approaches, offering a nuanced assessment of sustainability pressures across territories.

Social and household characteristics drive healthcare utilization, while ageing indicators alone provide only a partial view of risk.

Territorial clustering highlights provinces at elevated sustainability risk and reveals the need for place-based planning in decentralized healthcare systems.

Future research could extend the framework by incorporating temporal dynamics and applying it to other sustainability domains. The approach is particularly suited to official statistical systems supporting evidence-based policy making.

Disclosure of Interests. The author declares no competing interests relevant to the content of this article.

References

Aittokallio, T.: Dealing with missing values in large-scale studies: microarray data imputation and beyond. Brief. Bioinform. **11**(2), 253–264 (2010). https://doi.org/10.1093/bib/bbp059

Angrist, J.D., Pischke, J.-S.: Mostly Harmless Econometrics: An Empiricist's Companion. Princeton University Press, Princeton (2009). ISBN: 978-0691120355

Anselin, L.: Local indicators of spatial association. LISA Geographical Analysis 93–115 (1995). https://doi.org/10.1111/j.1538-4632.1995.tb00338.x

Athey, S.: The impact of machine learning on economics. In: The Economics of Artificial Intelligence: An Agenda, pp. 507–547. University of Chicago Press, Chicago (2018). https://doi.org/10.7208/chicago/9780226613475.003.0009

Breiman, L.: Random forests. Mach. Learn. **45**, 5–32 (2001). https://doi.org/10.1023/A:1010933404324

Cutler, D.M., Deaton, A.S., Lleras-Muney, A.: The determinants of mortality. J. Econ. Perspect. **20**(3), 97–120 (2006). https://doi.org/10.1257/jep.20.3.97

Eurostat: Sustainable Development in the European Union Monitoring Report on Progress towards the SDGs, Luxembourg (2023). https://doi.org/10.2785/313289

Fotheringham, A.S., Brunsdon, C., Charlton, M.: Geographically Weighted Regression: The Analysis of Spatially Varying Relationships. Wiley, Chichester (2002). https://doi.org/10.1002/9780470855256

Istat: Rapporto BES – Il Benessere Equo e Sostenibile in Italia. Istat, Rome (2023)

Spicker, P., Leguizamón, S.A., Gordon, D.: Poverty: An International Glossary. 2nd edn. Zed Books, London (2008). ISBN: 978-1842778234

Stiglitz, J.E., Sen, A., Fitoussi, J.P.: Report by the Commission on the Measurement of Economic Performance and Social Progress. OECD, Paris (2009)

Tibshirani, R.: Regression shrinkage and selection via the LASSO. J. R. Stat. Soc. (1996). https://doi.org/10.1111/j.2517-6161.1996.tb02080.x

Vapnik, V.: Statistical Learning Theory. Wiley, New York (1998). https://doi.org/10.1002/9780470141847

Inference for Interval Regression

YaoTong Cai[1] and Lynne Billard[2](✉)

[1] Walmart Global Tech, Sunnyvale, CA 94089, USA
[2] University of Georgia, Athens, GA 30602, USA
lynne@stat.uga.edu

Abstract. Modern large-scale data sets often necessitate aggregation into symbolic formats, such as intervals, to overcome computational limitations and preserve privacy. While various methods exist for fitting regression models to interval-valued data, formal statistical inference remains less explored. This paper utilizes the maximum likelihood principle to derive estimators for regression parameters in a simple linear interval regression framework. We can show that these estimators are unbiased and establish their distributional properties; also, we can establish the asymptotic normality of these estimators involving the Fisher information index, which allows for confidence intervals and Wald-like testing to be considered. A comprehensive simulation study, utilizing large-scale internal point-data aggregation, verifies that the proposed estimators are normally distributed with minimal mean square errors across varying noise conditions. These results provide a rigorous foundation for inference-based decision-making in large-data environments where interval aggregation is standard.

1 Introduction

Today's computer capacity brings massively large data sets, sets that are too large to analyse by conventional methods. Consequently, aggregation typically occurs (where the aggregation format is driven by the underlying scientific goals) with the aggregated data now represented as interval observations (or, histograms, lists, and the like). These aggregated formats are examples of symbolic data, first introduced in [1]. Expanded discussion of these data can be found in, e.g., [2–6], with non-technical introductions in [7–10], among others. Our focus deals with interval observations (simply, intervals).

Beyond computational efficiency, interval observations are increasingly vital in contexts where point-level data cannot be shared due to privacy constraints or where local volatility must be preserved. For instance, in high-frequency retail or financial transaction streams, aggregating data into daily or hourly intervals retains the essential spread and range of information that point-based averages discard. Establishing the inferential properties of these models is therefore critical for ensuring that high-level business and scientific decisions are based on statistically sound estimates rather than mere approximations.

P. Brito et al. (Eds.): IFCS 2026, STUDIES CLASS, pp. 68–76, 2026.
https://doi.org/10.1007/978-3-032-32009-4_9

Our particular concern is on inferential properties of the parameters for regression models when the observations are intervals. To date, several approaches have been developed for fitting linear regression models to interval data. The initial method, the so-called center method, was introduced by [11] in which interval centers were used to fit the regression and thence predictions on the response variable were found by applying the model to both lower and upper bounds of the respective predictor/regressor variables. Then, [12–14] developed the center and range method whereby they fitted regression models to both center points and ranges of intervals separately. This was modified by [5] considering these centers and ranges simultaneously as a bivariate center and range method. Later, [15] and [16] proposed a constrained method which modified the center-range method to constrain the range regression parameters to be non-negative. Subsequently, [17] proposed the symbolic method; this used the covariance functions of [18] to estimate the regression parameters. Later, [19] offered a different approach using quantile functions. These different developments were all concerned only with methodology designed to estimate the parameters and to fit designated models to interval data.

The current work looks at statistical inferential questions surrounding the fit of regression models to interval observations. In particular, it utilizes the maximum likelihood (ML) principle for estimating the model parameters. While the popular center-based method [11] provides a computationally simple heuristic by fitting a regression model, it fundamentally ignores the internal variation and boundary constraints of the interval, leading to a loss of structural information. In large-scale data environments where observations are aggregated into $[L, U]$ to preserve privacy or reduce dimensionality, the center method fails to provide a formal basis for uncertainty quantification. By contrast, our ML-based approach utilizes the full interval information to derive the maximum likelihood estimators of the model parameters ($\hat{\beta}_1$ and $\hat{\beta}_0$ when $p = 1$). Crucially, by establishing the exact variance of these estimators, we enable rigorous hypothesis testing and confidence interval construction—capabilities that are essential for reliable decision-making in automated systems but are absent in heuristic-based aggregation models. Thus, in Sect. 2, after first setting out the model and assumptions in Sect. 2.1, maximum likelihood estimators are given in Sect. 2.2 along with some distributional properties of those estimators, and the asymptotic normality of these estimators involving the Fisher information index allows for confidence intervals and Wald-like testing to be considered are in Sect. 2.3. In Sect. 3, a simulation study is presented illustrating the effectiveness of the proposed results for different parameter settings.

2 Inference Results

2.1 Model

Assume we have n observations with response variable Y and p explanatory variables $X_1, ..., X_p$. Denote X_{ij} as the ith observation of the jth variable, and Y_i as the ith response observation. The basic linear regression model is $\mathbf{Y} = \mathbf{X}\beta + \epsilon$

where the parameter vector $\beta = (\beta_0, \beta_1, ..., \beta_p)'$ with β_j being the parameter associated with the jth regression variable X_j, $\mathbf{X} = (\mathbf{X}_0, \mathbf{X}_1, \ldots, \mathbf{X}_p)$ with $\mathbf{X}_0$ being a column of ones and $\mathbf{X}_j' = (X_{1j}, \ldots, X_{np})$, $j = 1, \ldots, p$, $i = 1, \ldots, n$, $\mathbf{Y}' = (Y_1, \ldots, Y_n)$ and $\epsilon' = (\epsilon_1, \ldots, \epsilon_n)$ is the error term. For interval observations, realizations for these entities are, respectively, $X_{i0} = [1, 1]$, $X_{ij} = [X_{Lij}, X_{Uij}]$, $Y_i = [Y_{Li}, Y_{Ui}]$ and $\epsilon_i = [\epsilon_{Li}, \epsilon_{Ui}]$, with $X_{Lij} \leq X_{Uij}$, $Y_{Li} \leq Y_{Ui}$, and $\epsilon_{Li} \leq \epsilon_{Ui}$.

For notational and illustrative clarity, without loss of generality, we consider the simple linear regression model ($p = 1$); extension to $p > 1$ is routine. Thus, let us write our model as

$$[Y_{Li}, Y_{Ui}] = \beta_0 + \beta_1[X_{Li}, X_{Ui}] + [\epsilon_{Li}, \epsilon_{Ui}], \quad i = 1, ..., n. \tag{1}$$

Let us assume the error limits, ϵ_{Li} and ϵ_{Ui}, $i = 1, \ldots, n$, are independent and normally distributed $N(0, \sigma^2)$; this assumption is appropriate for data obtained by aggregation of point observations. For these error terms, both of the residuals by the lower and the upper bounds of the response are caused by the random component that fails to be explained by the linear combination of explanatory variables. Therefore, the residuals are random, and can be considered as independent both within and across observations.

Since we have the universal conditions: $Y_{Li} \leq Y_{Ui}$ and $X_{Li} \leq X_{Ui}$, the forms of Y_{Li} and Y_{Ui} by the linear regression model vary depending on the sign of the slope parameter β_1, so that, for $i = 1, \ldots, n$,

$$Y_{Li} = \beta_0 + \beta_1 X_{Li} + \epsilon_{Li}, \; Y_{Ui} = \beta_0 + \beta_1 X_{Ui} + \epsilon_{Ui}, \; \beta_1 \geq 0, \tag{2}$$
$$Y_{Li} = \beta_0 + \beta_1 X_{Ui} + \epsilon_{Li}, \; Y_{Ui} = \beta_0 + \beta_1 X_{Li} + \epsilon_{Ui}, \; \beta_1 < 0. \tag{3}$$

Therefore, when $\hat{\beta}_1 > 0$, for $i = 1, ..., n$,

$$\hat{Y}_{Li} = \hat{\beta}_0 + \hat{\beta}_1 X_{Li}, \; r_{Li} = Y_{Li} - \hat{Y}_{Li} = Y_{Li} - \hat{\beta}_0 - \hat{\beta}_1 X_{Li}, \tag{4}$$
$$\hat{Y}_{Ui} = \hat{\beta}_0 + \hat{\beta}_1 X_{Ui}, \; r_{Ui} = Y_{Ui} - \hat{Y}_{Ui} = Y_{Ui} - \hat{\beta}_0 - \hat{\beta}_1 X_{Ui} \tag{5}$$

where $[r_{Li}, r_{Ui}]$ are the residuals; and when $\hat{\beta}_1 < 0$, we have

$$r_{Li} = Y_{Li} - \hat{Y}_{Ui} = Y_{Li} - \hat{\beta}_0 - \hat{\beta}_1 X_{Ui}, \tag{6}$$
$$r_{Ui} = Y_{Ui} - \hat{Y}_{Li} = Y_{Ui} - \hat{\beta}_0 - \hat{\beta}_1 X_{Li}. \tag{7}$$

Thence, we can obtain the likelihood function of the random errors as

$$L(\epsilon_{L1}, \epsilon_{U1}, ..., \epsilon_{Ln}, \epsilon_{Un}) = (2\pi\sigma^2)^{-n} exp(-\frac{1}{2\sigma^2}\sum_{i=1}^{n}(\epsilon_{Li}^2 + \epsilon_{Ui}^2)). \tag{8}$$

2.2 Parameter Estimation

Let us consider obtaining the maximum likelihood estimator (MLE) of the parameters β_0 and β_1. By Eqs. (2) and (3), for $i = 1, ..., n$.

$$\epsilon_{Li} = Y_{Li} - \beta_0 - \beta_1 X_{Li}, \; \epsilon_{Ui} = Y_{Ui} - \beta_0 - \beta_1 X_{Ui}, \text{ when } \beta_1 \geq 0, \tag{9}$$
$$\epsilon_{Li} = Y_{Li} - \beta_0 - \beta_1 X_{Ui}, \; \epsilon_{Ui} = Y_{Ui} - \beta_0 - \beta_1 X_{Li}, \text{ when } \beta_1 < 0, \tag{10}$$

Replacing ϵ_{Li} and ϵ_{Ui} in Eq. (8) by Eq. (9) or Eq. (10) as appropriate (i.e., depending on whether β_1 is positive or negative), we can express the log-likelihood function (log$L$) of the random error as a function of the intercept and the slope parameters as

$$\log L = \log L(\epsilon_{L1}, \epsilon_{U1}, \ldots, \epsilon_{Ln}, \epsilon_{Un}) = -n \log \pi - n \log \sigma^2$$
$$- \frac{1}{2\sigma^2} \sum_{i=1}^{n} [(Y_{Li} - X_{Li}\beta_1 - \beta_0)^2 + (Y_{Ui} - X_{Ui}\beta_1 - \beta_0)^2]. \quad (11)$$

We consider each β_1 case in turn. When $\beta_1 > 0$, we can show that the maximum likelihood estimators of β_0 and β_1, respectively, are

$$\hat{\beta}_1 = \frac{\sum_{i=1}^{n}(X_{Li} + X_{Ui}) \sum_{i=1}^{n}(Y_{Li} + Y_{Ui}) - 2n \sum_{i=1}^{n}(X_{Li}Y_{Li} + X_{Ui}Y_{Ui})}{[\sum_{i=1}^{n}(X_{Li} + X_{Ui})]^2 - 2n \sum_{i=1}^{n}(X_{Li}^2 + X_{Ui}^2)}, \quad (12)$$

$$\hat{\beta}_0 = \bar{Y} - \hat{\beta}_1 \bar{X}, \quad (13)$$

$$\bar{Y} = \sum_{i=1}^{n}(Y_{Li} + Y_{Ui}), \quad \bar{X} = \sum_{i=1}^{n}(X_{Li} + X_{Ui}). \quad (14)$$

We can further show that $E(\hat{\beta}_0) = \beta_0$ and $E(\hat{\beta}_1) = \beta_1$. Also, we can derive the variances of these estimators. Hence, we can show that

$$Var(\hat{\beta}_1) = \frac{2n\sigma^2}{2n \sum_{i=1}^{n}(X_{Li}^2 + X_{Ui}^2) - [\sum_{i=1}^{n}(X_{Li} + X_{Ui})]^2} \triangleq V_{\hat{\beta}_1}, \quad (15)$$

$$Var(\hat{\beta}_0) = \frac{\sum_{i=1}^{n}(X_{Li}^2 + X_{Ui}^2)\sigma^2}{2n \sum_{i=1}^{n}(X_{Li}^2 + X_{Ui}^2) - [\sum_{i=1}^{n}(X_{Li} + X_{Ui})]^2} \triangleq V_{\hat{\beta}_0}. \quad (16)$$

Based on the derivations of $Var(\hat{\beta}_1)$ and $Var(\hat{\beta}_0)$ in Eq. (15) and Eq. (16), respectively, and the derivations of $E(\hat{\beta}_1)$ and $E(\hat{\beta}_0)$, we conclude

$$\hat{\beta}_0 \sim N(\beta_0, V_{\hat{\beta}_0}), \; \hat{\beta}_1 \sim N(\beta_1, V_{\hat{\beta}_1}). \quad (17)$$

Likewise, when $\beta_1 < 0$, we can show that the estimators of β_0 and β_1, respectively, are now

$$\hat{\beta}_1 = \frac{\sum_{i=1}^{n}(X_{Li} + X_{Ui}) \sum_{i=1}^{n}(Y_{Li} + Y_{Ui}) - 2n \sum_{i=1}^{n}(X_{Li}Y_{Ui} + X_{Ui}Y_{Li})}{[\sum_{i=1}^{n}(X_{Li} + X_{Ui})]^2 - 2n \sum_{i=1}^{n}(X_{Li}^2 + X_{Ui}^2)}, \quad (18)$$

$$\hat{\beta}_0 = \bar{Y} - \hat{\beta}_1 \bar{X} \quad (19)$$

where $\bar{X}$ and $\bar{Y}$ are as given in Eq. (14).

To ensure interval coherence, for a given $[X_{Li}, X_{Ui}]$ the predicted interval $\hat{Y}_i = [\hat{Y}_{Li}, \hat{Y}_{Ui}] = [\min\{\eta_{Li}, \eta_{Ui}\}, \max\{\eta_{Li}, \eta_{Ui}\}]$ where $\eta_{ki} = \hat{\beta}_0 + \hat{\beta}_1 X_{ki}$ ($k = L, U$). This naturally handles negative slope scenarios without loss of generality.

2.3 Some Asymptotic Properties of the Estimators

In addition to the unbiasedness of the estimators $\hat{\beta}_0$ and $\hat{\beta}_1$, we can further show these maximum likelihood estimators to be distributed as (see [21] for details)

$$\hat{\beta}_0 \sim N(\beta_0, V_{\hat{\beta}_0}),\ \hat{\beta}_1 \sim N(\beta_1, V_{\hat{\beta}_1}). \tag{20}$$

Furthermore, under standard regularity conditions for maximum likelihood estimation, the estimator $\hat{\theta} = (\hat{\beta}_0, \hat{\beta}_1, \hat{\sigma})$ is consistent and asymptotically normally distributed, with mean zero and variance equal to the inverse of the Fisher information index ($I(\theta)$). To prove this, we first recognize that the log-likelihood function (in Eq. 11) is twice differentiable with respect to θ. Then the central limit theorem is applied to the score function, to give the variance-covariance function as $I^{-1}(\theta) = -E[\nabla^2 \log L]$. Details of the derivations of the Hessian entries and formal verification of the regularity conditions are in [21].

Hence, leveraging these asymptotic results, we can construct formal inferential procedures. For example, we can derive a $(1-\alpha)100\%$ confidence interval for any parameter β_j as $\hat{\beta}_j \pm z_{\alpha/2}\widehat{se}(\hat{\beta}_j)$ where $z_{\alpha/2}$ is the upper $\alpha/2$ quantile of the standard normal distribution, and $se(\hat{\beta}_j)$ is the standard error of the estimator $\hat{\beta}_j$. Also, the Wald-statistic for hypothesis testing can be obtained. This framework allows for rigorous testing of the significance of interval predictors. The simulations of Sect. 3 demonstrate that the empirical coverage of these intervals closely matches the nominal values.

3 Simulation Study

The simulation method adopted for interval data originates from aggregating classical data. We first generate interval centers $X^{(c)}$ for the explanatory variable X by randomly sampling from a normal distribution, $N(\mu, \sigma^2)$, and ranges $X^{(r)}$ from a truncated normal distribution, $TN(a, b, \mu_0, \sigma_0^2)$ where a (resp. b) is the lower (resp. upper) bound of support, μ_0 is the mean, and σ_0 is the standard deviation. Then, the interval for X is $X = [x_L, x_U] = [x^{(c)} - 0.5x^{(r)}, x^{(c)} + 0.5x^{(r)}]$.

By the basic assumption that the distribution within each interval is uniform, for the ith observation, m values are randomly drawn from the uniform distribution $U(x_{Li}, x_{Ui})$, for $i = 1, ..., n$, where m is pre-specified, or a value drawn from a distribution with integer support for more general cases; each of the m values is denoted by x_{il}, $l = 1, \ldots, m$. We generate the random errors for the lower and upper bounds of the response, i.e., ϵ_{Li} and ϵ_{Ui}, respectively, by randomly sampling from a normal distribution $N(0, \sigma_e^2)$. Then, the interval realization for the response variable of the ith observation $Y_i = [y_{Li}, y_{Ui}]$ can be determined by

$$y_{Li} = \min_{l \in \{1,\ldots,m\}} \{\beta_0 + x_{il}\beta_1\} + \epsilon_{Li}, \quad y_{Ui} = \max_{l \in \{1,\ldots,m\}} \{\beta_0 + x_{il}\beta_1\} + \epsilon_{Ui} \tag{21}$$

where $\epsilon_{Li}, \epsilon_{Ui} \overset{iid}{\sim} N(0, \sigma_e^2)$, for $i = 1, ..., n$, and $l = 1, ..., m$.

While this simulation method is closer to how interval data arise in practice, a possible drawback is that the obtained intervals for the response variable Y_i, $i = 1, \ldots, n$, cannot be guaranteed to follow a uniform distribution internally. However, when $m \geq 3000$, [17] showed that this uniformity condition holds; see also [20] who provide a test for this internal distribution. Therefore, we take $m = 3000$ internal values.

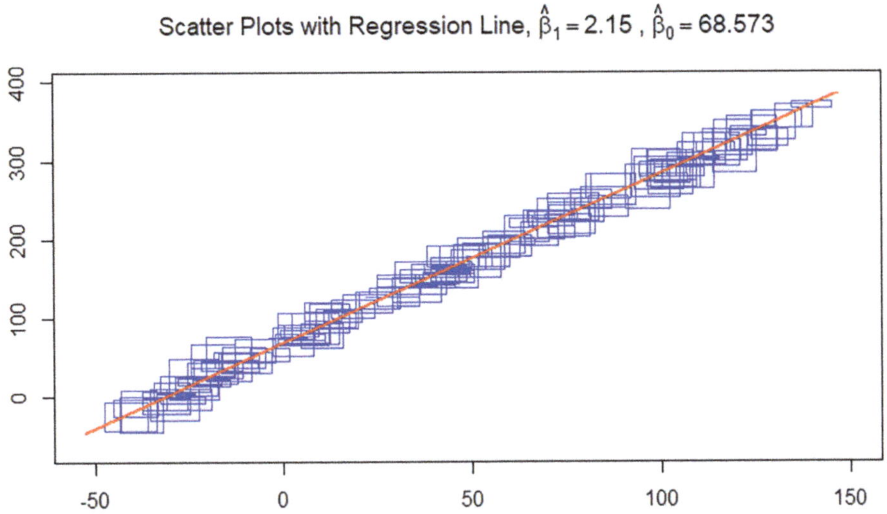

Fig. 1. Scatter plot - $\sigma_e = 10$, $\sigma_0 = 5.25$, $\beta_1 = 2.15$, $\beta_0 = 68.57$

The simulations were run for each of $B = 2000, 5000, 10000$ samples for a range of σ_e and σ_0, each with a range of negative and positive β_1 and β_0 values. A scatter plot (with the fitted regression line, in red) is shown in Fig. 1 for the particular case that $\sigma_e = 10$, $\sigma_0 = 5.25$, $\beta_1 = 2.15$, and $\beta_0 = 68.57$. This shows the interval-valued observations (i.e., rectangles in the 2-dimensional Y/X plane). This scatter plot represents one of the B samples used.

For each such sample, we can estimate the parameters β_1 and β_0. To illustrate, Table 1 displays these estimates based on the sample average of the $B = 10000$ estimates, for the cases that $\beta_1 = 0.64$ and $\beta_1 = -3.21$ with $\sigma_e = 3$, 7, 10, $\sigma_0 = 5.25, 8.07$, and $\beta_0 = 68.57$. The histogram plots for the estimates of $\hat{\beta}_0$ and $\hat{\beta}_1$, corresponding to the simulations identified in the first row of Table 1, are shown in Fig. 2 (right and left hand plots, respectively). The vertical red lines on Fig. 2 display the positions of the true parameter values, i.e., $\beta_1 = 0.64$ in (a), and $\beta_1 = -3.21$ in (b). These plots show visually that both $\hat{\beta}_0$ and $\hat{\beta}_1$ are normally distributed thus verifying the normality property of Eq. (17). From Table 1, we observe that the estimates based on the sample average of the $B = 10000$ estimates of both $\hat{\beta}_1$ and $\hat{\beta}_0$ parameters are very close to the actual parameter values. Also shown are the mean square errors (MSE) for each of these estimators $\hat{\beta}_1$ and $\hat{\beta}_0$ along with the corresponding 95% confidence intervals (CIs) of the respective estimators.

Table 1. $\hat{\beta}_1$, $\hat{\beta}_0$, $MSE(\hat{\beta}_1)$, $MSE(\hat{\beta}_0)$, and CI's for $(\hat{\beta}_1, \hat{\beta}_0)$ for $(\beta_1, \beta_0) = (0.64, 68.57)$ and $(\beta_1, \beta_0) =$ (-3.210, 68.57)

σ_e	σ_0	$\hat{\beta}_1$	$\hat{\beta}_0$	$MSE(\hat{\beta}_1)$	$MSE(\hat{\beta}_0)$	95% CI of $\hat{\beta}_1$	95% CI of $\hat{\beta}_0$
3	5.25	0.640	68.573	2×10^{-5}	0.070	[0.632, 0.647]	[68.047, 69.089]
	8.07	0.640	68.566	2×10^{-5}	0.069	[0.632, 0.648]	[68.051, 69.075]
7	5.25	0.640	68.564	8×10^{-5}	0.378	[0.622, 0.658]	[67.355, 69.799]
	8.07	0.640	68.574	9×10^{-5}	0.373	[0.622, 0.658]	[67.378, 69.767]
10	5.25	0.640	68.578	1.7×10^{-4}	0.760	[0.614, 0.665]	[68.857, 70.273]
	8.07	0.640	68.569	1.7×10^{-4}	0.766	[0.614, 0.665]	[66.810, 70.282]
3	5.25	−3.210	68.567	2×10^{-5}	0.068	[−3.218, −3.202]	[68.055, 69.076]
	8.07	−3.210	68.569	2×10^{-5}	0.069	[−3.218, −3.202]	[68.054, 69.085]
7	5.25	−3.210	68.564	8×10^{-5}	0.382	[−3.228, −3.192]	[67.355, 69.776]
	8.07	−3.210	68.583	9×10^{-5}	0.374	[−3.228, −3.192]	[67.395, 69.791]
10	5.25	−3.210	68.572	1.7×10^{-4}	0.748	[−3.236, −3.185]	[66.861, 70.289]
	8.07	−3.210	68.564	1.7×10^{-4}	0.756	[−3.236, −3.184]	[66.859, 70.223]

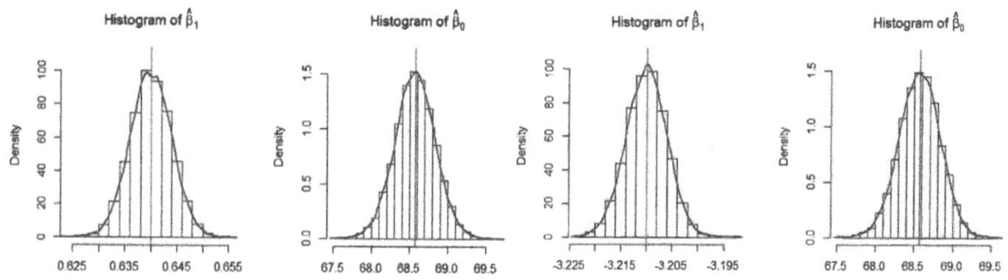

(a) $\beta_1 = 0.64$ (b) $\beta_1 = -3.21$

Fig. 2. Histograms of $\hat{\beta}_1$ and $\hat{\beta}_0$, $\sigma_e = 3$, $\sigma_0 = 5.25$, $\beta_0 = 68.57$

4 Conclusion

Inferences regarding the parameters of a linear regression model have been studied for the case that the data are in interval form. In particular, MLE estimators are shown to be unbiased and some distributional properties are derived, including confidence intervals of the parameter estimators. A simulation study verified that the proposed estimators are good with very small mean square errors and that the normality distribution and unbiasedness properties hold. The formal establishment of normality and unbiasedness for these estimators allows for the future development of autonomous decision-making agents that can reason about uncertainty ranges in commerce and finance. Without loss of generality, these results carry over to several independent variables. More complete details of the derivation of these inferential results, including an expanded simulation study covering an extensive range of situations are in [21] and will be presented elsewhere.

References

1. Diday, E. Introduction à l'approche symbolique en analyse des données. Premier Jouneles Symbolique-Numerique, CEREMADE, Université de Paris, Dauphine, pp. 21–56 (1987)
2. Bock, H.-H., Diday, E. (eds.): Analysis of Symbolic Data: Exploratory Methods for Extracting Statistical Information from Complex Data. Springer-Verlag, Berlin (2000)
3. Billard, L., Diday, E.: From the statistics of data to the statistics of knowledge: symbolic data analysis. J. Am. Stat. Assoc. **98**, 470–487 (2003)
4. Diday, E., Noirhomme-Fraiture, M. (eds.): Symbolic Data Analysis and the SODAS Software. John Wiley and Sons, Chichester (2008)
5. Billard, L., Diday, E.: Symbolic Data Analysis: Conceptual Statistics and Data Mining. John Wiley, Chichester (2006)
6. Billard, L., Diday, E.: Clustering Methodology for Symbolic Data. John Wiley, Chichester (2020)
7. Diday, E.: Thinking by classes in data science: the symbolic data analysis paradigm. WIREs Comput. Stat. **8**, 172–205 (2016)
8. Billard, L.: Brief overview of symbolic data and analytic issues. Stat. Anal. Data Min. **4**, 149–156 (2011)
9. Billard, L.: What will the present's future bring? In: Lin, X., Genest, C., Banks, D.L., Molenberghs, G., Scott, D.W., Wang, J.-L. (eds.) Past, Present, and Future of Statistical Science, pp. 323–334. Chapman and Hall, New York (2014). The past's present is now
10. Noirhomme-Fraiture, M., Brito, M.P.: Far beyond the classical data models: symbolic data analysis. Stat. Anal. Data Min. **4**, 157–170 (2011)
11. Billard, L., Diday, E.: Regression analysis for interval-valued data. data analysis, classification, and related methods, pp. 369–374. Springer (2000)
12. Lima Neto, E.A., de Carvalho, F.A.T., Tenorio, C.P.: Univariate and multivariate linear regression methods to predict interval-valued features. In: Australasian Joint Conference on Artificial Intelligence, pp. 526–537. Springer (2004)
13. de Carvalho, F.A.T., Lima Neto, E.A., Tenorio, C.P.: A new method to fit a linear regression model for interval-valued data. In: Annual Conference on Artificial Intelligence, pp. 295–306. Springer (2004)
14. Lima Neto, E.A., de Carvalho, F.A.T.: Centre and Range method for fitting a linear regression model to symbolic interval data. Comput. Stat. Data Anal. **52**, 1500–1515 (2008)

15. Lima Neto, E.A., de Carvalho, F.A.T., Freire, E.S.: Applying constrained linear regression models to predict interval-valued data. In: Annual Conference on Artificial Intelligence, pp. 92–106. Springer (2005)
16. Lima Neto, E.A., de Carvalho, F.A.T.: Constrained linear regression models for symbolic interval-valued variables. Comput. Stat. Data Anal. **54**, 333–347 (2010)
17. Xu, W.: Symbolic data analysis: interval-valued data regression. Doctoral Dissertation, University of Georgia (2010)
18. Billard, L.: Sample covariance functions for complex quantitative data. In: Mizuta, M., Nakano, J. (eds.) World Congress International Association Statistical Computing, pp. 157–163. Japanese Society of Computational Statistics, Yokohama (2008)
19. Dias, S., Brito, M.P.: Off the beaten track: a new linear model for interval data. Eur. J. Oper. Res. **258**, 1118–1130 (2017)
20. Cariou, V., Billard, L.: Generalization method when manipulating relational databases. Revue des Nouvelles Technologies de l'Information **27**, 59–86 (2015)
21. Cai, Y.: Symbolic data analysis statistical inference on interval-valued data regression. Doctoral Dissertation, University of Georgia (2018)

A Compositional Framework for Interpreting Clustering of Time Series from Neural Cultures

Cristina Campi[1,2], Francesco Porro[1(✉)], Eva Riccomagno[1], and Sara Sommariva[1]

[1] Dipartimento di Matematica, Università degli Studi di Genova, Genoa, Italy
francesco.porro@unige.it
[2] IRCCS Ospedale Policlinico San Martino, Genoa, Italy

Abstract. This work addresses the problem of identifying patterns in the electrical activity of neural cultures recorded using high-density MicroElectrode Arrays (MEAs). Pattern identification is typically performed through a two-step procedure: first, a large set of time windows is extracted from the recorded time series around each detected spike; second, clustering methods are applied to group similar time windows across all MEA sensors. We propose a compositional framework for interpreting the outcomes of this procedure. In particular, we suggest quantifying similarities between clusters and between sensors by computing distances between the row and column profiles, respectively, of an appropriate contingency table. Several distance measures are considered, including those originally developed for compositional data analysis, as well as measures commonly used in information theory and machine learning. Our results highlight the importance of carefully selecting the distance metric, especially when the aim is to assess similarities across sensors.

Keywords: Compositional data · Cluster analysis · Distance matrix · Dissimilarity measure

1 Introduction

Neurophysiological activity can be investigated both in vivo, in humans and animal models, and in vitro through experiments on neural cultures. Compared with in vivo setups, neural cultures provide a simpler and more reproducible experimental environment. At the same time they provide valuable insights into nervous system function and a reliable platform for testing therapies and treatments. In this setting, data consist of time series recorded using MicroElectrode Arrays (MEAs) [3]. Modern MEAs are equipped with a very large number of sensors -typically several thousands- and extremely high sampling rates, resulting in large-scale, high-dimensional datasets. These characteristics introduce substantial redundancy and complexity, making direct data analysis both computationally intensive and difficult to interpret.

P. Brito et al. (Eds.): IFCS 2026, STUDIES CLASS, pp. 77–84, 2026.
https://doi.org/10.1007/978-3-032-32009-4_10

An initial dimensionality reduction step is performed by discarding inactive sensors that did not record neural activity during the experiment. In the following a sensor is considered active if its spike rate exceeds 1 spike/sec, where a spike is a sharp electrical signal generated by a neuron and appears as a peak in the MEA recording. To group activity patterns with similar characteristics (such as similar waveforms), clustering techniques are applied to time windows extracted from the original time series using spike detection or spike sorting procedures. We propose a compositional framework for interpreting the outcomes of these clustering steps by analysing the row and column profiles of a contingency matrix, whose entries count, for each cluster, the number of time windows associated with each sensor.

Quantifying the similarity between two clusters or two sensors is formulated as computing an appropriate distance between two compositions. In our case study, both row and column profiles may include zero components, so it is necessary to consider dissimilarity measures capable of handling compositions that lie on the boundary of the simplex. To this end, we identify five distinct distance measures and compare their performance on a real data set.

2 The Mathematical Model

Recordings from active MEA sensors may be represented as a set of S time-series denoted as $\mathbf{m}_s = (m_s(t_1), \ldots, m_s(t_P))^\top \in \mathbb{R}^P$, being $m_s(t_p)$ the signal recorded by the s-th sensor at time t_p for all $s = 1, \ldots, S$ and $p = 1, \ldots, P$.

By applying spike detection, for each sensor we identify the set of time-points

$$\{\tau_j^s\}_{j=1,\ldots,r_s} \subset \{t_1, \ldots, t_P\}$$

where a spike occurred. For each spike, we extract a time-window

$$\mathbf{y}_j^s = \left(y_j^s(z_1), \ldots, y_j^s(z_{\tilde{P}})\right)^\top \in \mathbb{R}^{\tilde{P}}$$

centered at the time-point τ_j^s of the event. Hence, for Δt the sampling frequency of the original recordings and $\tilde{P}$ an odd integer denoting the number of time-points within each time-window, we have $y_j^s(z_i) = m_s\left(\tau_j^s + \Delta t\left(i - \frac{\tilde{P}+1}{2}\right)\right)$.

The complete collection of time-windows across sensors $\{\mathbf{y}_j^s\}_{\substack{j=1,\ldots,r_s\\ s=1,\ldots,S}}$ is provided as input to a clustering procedure that partitions them into K clusters, $\mathcal{C}_1, \ldots, \mathcal{C}_K$. The development of a clustering procedure tailored to this type of data is the subject of ongoing work; for ease of exposition, K-means clustering is adopted here. To characterise the distribution of time-windows across clusters and sensors, we summarise the clustering output as a $K \times S$ contingency table N, where each entry n_{ks} denotes the number of time-windows from sensor s assigned to cluster k. Our aim is to exploit the structure of N to identify similarities among clusters and/or sensors. In practice, N tends to be sparse, as each cluster aggregates time-windows only from a subset of active sensors.

3 Compositional Framework

From the contingency table N, the K row profiles are obtained by rescaling each row by its row sum $n_{k\cdot} = \sum_{s=1}^{S} n_{ks}$, which corresponds to the cardinality of the k-th cluster. The row profile associated with cluster k is thus defined as

$$f_{s|k}^{row} = \frac{n_{ks}}{n_{k\cdot}} \qquad s = 1, 2, ..., S.$$

The k-th row profile describes the relative frequencies of sensors associated with the detection of waveforms belonging to cluster k. Similarly, the column profiles are obtained by dividing each column of N by its column sum $n_{\cdot s} = \sum_{k=1}^{K} n_{ks}$, which corresponds to the number of time-windows r_s recorded by sensor s:

$$f_{k|s}^{col} = \frac{n_{ks}}{n_{\cdot s}} \qquad k = 1, 2, ..., K.$$

Similarities between clusters and between sensors can be quantified by computing a distance between the corresponding row and column profiles, respectively. Each row profile and each column profile is a composition. Specifically, each row profile can be seen as an S-part composition, while each column profile is a K-part composition.

In classical Compositional Data Analysis (CoDA), a D-part composition is an equivalence class of vectors in $\mathbb{R}^D$ with strictly positive components, that sum to a fixed value, usually equal to one [7]. The requirement that all the components are non null is essential for applying the classical Aitchison's geometry based on log-ratio transformations [7]. However, this condition is not met in our context. Motivated by this consideration, in this work we identify and compare five dissimilarity measures drawn from different areas of Compositional Data Analysis, Statistics, and Machine Learning, which can be applied to compositions with null parts. The selected dissimilarity measures are defined below for two generic D-part compositions $\mathbf{x} = (x_1, \ldots, x_D)^\top$ and $\mathbf{y} = (y_1, \ldots, y_D)^\top$.

Euclidean Distance. It is defined as

$$d_E(\mathbf{x}, \mathbf{y}) = \sqrt{\sum_{j=1}^{D} (x_j - y_j)^2} \,.$$

We included the Euclidean distance in our analysis since in many contexts, it is used as reference measure. However in CoDA it must be used with caution because the parts of a composition lie in the simplex as they obey a constant-sum constraint. Conversely, the Euclidean geometry assumes an unconstrained and free space, and may therefore emphasize spurious relationships arising from the negative dependence between parts [7].

Zero-adjusted Aitchison's Distance. To overcome the limitations of the Euclidean geometry for CoDA, the Aitchison's distance has been introduced. Being defined as the Euclidean distance between the Centered Log Ratio (CLR) transformed data, it better captures relative differences and results to be subcompositionally coherent [7], meaning that its value cannot increase if any of the parts of the composition is dropped. However, it cannot be applied to compositions that lie on the boundary of the simplex, as the CLR transformation involves computing logarithms. The zero-adjusted Aitchison's distance is an extension of the Aitchison's distance designed to overcome such a limitation. It is defined as

$$d_A(\mathbf{x}, \mathbf{y}) = \sqrt{\frac{1}{2D} \sum_{i=1}^{D} \sum_{j=1}^{D} \left[\ln\left(\frac{\widetilde{x}_i}{\widetilde{x}_j}\right) - \ln\left(\frac{\widetilde{y}_i}{\widetilde{y}_j}\right) \right]^2},$$

where $\widetilde{\mathbf{x}} = (\widetilde{x}_1, \ldots, \widetilde{x}_D)$ and $\widetilde{\mathbf{y}} = (\widetilde{y}_1, \ldots, \widetilde{y}_D)$ have been obtained from $\mathbf{x}$ and $\mathbf{y}$ by replacing the null parts with a small positive value ε. In this work we set ε equal to 65% of the smallest component across all observations [5,6].

Bhattacharyya's Distance (Cosine Form). It is defined as

$$d_B(\mathbf{x}, \mathbf{y}) = \arccos\left(\sum_{i=1}^{D} \sqrt{x_i y_i}\right)$$

and it is bounded in the interval $[0, \frac{\pi}{2}]$. In the literature it has been proposed as measure of the divergence (dissimilarity) between probability distributions. In CoDA it can be used to quantify the amount of overlap between the two compositions: smaller values indicate greater overlap (i.e. more similarity), while larger values suggest greater separation [1]. However, it does not satisfy the Aitchison's requirement of subcompositional coherence.

Jensen-Shannon's Divergence. It is defined as

$$d_{JS}(\mathbf{x}, \mathbf{y}) = \frac{1}{2} \sum_{i=1}^{D} \left[x_i \ln\left(\frac{x_i}{m_i}\right) + y_i \ln\left(\frac{y_i}{m_i}\right) \right],$$

where $m_i = \frac{x_i + y_i}{2}$. It is related to the well-known Kullback–Leibler divergence, but it is symmetric and always finite. The Jensen-Shannon's divergence is not defined for compositions with at least one null part; nevertheless, it can be prolonged by continuity to such compositions by setting $x \log(x) = 0$ when $x = 0$ [4]. Similarly to the Bhattacharyya's distance, the Jensen-Shannon's divergence is bounded, as ranges in $[0, \log(2)]$, and it is not subcompositional coherent.

ChiPower Distance. It is the Euclidean distance between the corresponding chiPower-transformed compositions [2], which are calculated as follows.

1. Denoted with X the data matrix, select a value $\lambda \in]0, 1]$ and compute the matrix X^λ whose entries are x_{ij}^λ.
2. Close the rows of the matrix X^λ, dividing each one by its row sum. This results in a new matrix $Y = \mathscr{C}(X^\lambda)$.

3. Apply the following Box-Cox style transformation

$$z_{ij} = \frac{1}{\lambda}\left[\sqrt{D}\frac{y_{ij}}{\sqrt{\bar{y}_j}} - 1\right],$$

being $\bar{y}_j$ the mean of the jth-column of Y. Store the z_{ij} in a new matrix Z.
4. Translate each column of Z by its mean.

The matrix Z is the chiPower-transformed data matrix of power λ. Therefore, if the compositions $\mathbf{x}$ and $\mathbf{y}$ are the h-th and the k-th rows of matrix X respectively, their chiPower distance is

$$d_{chiP}(\mathbf{x}, \mathbf{y}) = \sqrt{\sum_{j=1}^{D} (z_{hj} - z_{kj})^2}\,.$$

A key step in the computation of the chiPower transformation is the choice of the parameter λ, which introduces an additional source of variability, as results may substantially change depending on λ. In particular, it has been shown that if the original compositional dataset only contains non-null parts, the chiPower distance between two compositions converges to their Aitchison's distance when λ approaches zero. Therefore, inspired by [2], in our analysis we set λ equal to the value providing the distance matrix with the highest Pearson correlation coefficient with the zero-adjusted Aitchison's distance matrix among 20 tested values uniformly distributed on the interval [0.05, 1].

4 Results

We consider a 3-minute recording from a 64×64 MEA grid of the activity of a 2D culture. Of the 4096 sensors in the grid, $S = 183$ are active (spike rate > 1 spike/sec). From these active sensors, we extract 4873 time windows $\mathbf{y}_j^s$ of length $\tilde{P} = 41$. K-means clustering is applied with a fixed number of clusters $K = 4$, and the results are summarised using the contingency table N introduced in Sect. 2. For this dataset, approximately 26.2% of the entries of N are zero.

Pairwise dissimilarities between the row profiles and between the column profiles are then quantified using the five measures defined in Sect. 3, yielding five dissimilarity matrices of size 4×4 for the row profiles and five matrices of size 183×183 for the column profiles. Since the measures operate on different scales, their agreement is assessed by computing the Pearson and Kendall correlation coefficients between the corresponding dissimilarity matrices; results are reported in Table 1.

Several observations emerge from Table 1. All five dissimilarity measures exhibit a very similar behaviour in terms of Pearson correlation coefficients: all coefficients exceed 0.9202, with the sole exception of the pair formed by the Euclidean and the zero-adjusted Aitchison distances, for which the coefficient drops to 0.8681. Regarding rank agreement, the Kendall coefficient equals one

Table 1. Pearson (Kendall) correlation coefficient between the dissimilarity matrices obtained by applying various distances/divergences to the row profiles (upper triangular part) and to the column profiles (lower triangular part).

	Euclidean	Adj. Aitch.	Bhatt.	Jens.-Shan.	ChiPower
Euclidean		0.9752	0.9793	0.9798	0.9782
		(1)	(1)	(1)	(1)
Adj. Aitch.	0.8681		0.9991	0.9987	0.9961
	(0.6747)		(1)	(1)	(1)
Bhatt.	0.9518	0.9724		0.9997	0.9979
	(0.8082)	(0.8560)		(1)	(1)
Jens.-Shan.	0.9554	0.9442	0.9884		0.9991
	(0.8491)	(0.8173)	(0.9516)		(1)
ChiPower	0.9448	0.9202	0.9749	0.9731	
	(0.8075)	(0.7745)	(0.8885)	(0.8844)	

for the row profiles, whereas for the column profiles it ranges between 0.6747 and 0.9516, indicating that certain measures may induce different orderings of pairwise distances among column profiles.

Across both profiles, the weakest agreement is observed between the Euclidean distance and the zero-adjusted Aitchison's distance, confirming their differing sensitivity as noted in Sect. 3. While the strongest agreement is found between Bhattacharyya's distance and Jensen-Shannon's divergence, which attain the highest Pearson and Kendall correlation coefficients.

Comparing row and column profiles, the dissimilarity measures exhibit markedly greater mutual coherence when computed on row profiles, a pattern visible in both the Pearson and Kendall indices. This may be partly attributable to the difference in the size of the underlying dissimilarity matrices (183×183 for row profiles and 4×4 for column profiles).

A second analysis is summarized by the violin plot shown in Fig. 1, which compares the values and empirical distributions of the five dissimilarity measures computed on all pairwise column-profile distances. These distances capture the cluster-wise frequencies of waveforms detected by the sensors. To ensure comparability across measures with different ranges, all values are normalised using a min-max scaling

$$d' = \frac{d - \min(d_i)}{\max(d_i) - \min(d_i)},$$

where $\min(d_i)$ and $\max(d_i)$ denote the minimum and maximum values attained by measure i on the dataset. The central 50% of the values differs substantially across the five measures, with the lowest interquartile range for the Aitchison and chiPower distances, and the largest for the Jensen-Shannon's divergence.

In terms of location, the five dissimilarity measures can be grouped into two main clusters: the first one, consisting of the Euclidean, Aitchison, and Bhattacharyya distances, is characterised by a higher median value; the second one, containing the chiPower distance and the Jensen-Shannon's divergence, displays smaller median values.

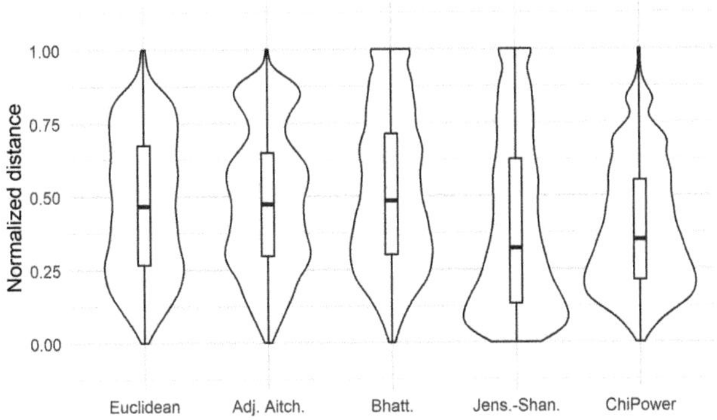

Fig. 1. Distribution of the min-max normalised values of the pairwise distances between the column profiles

5 Discussion

In this work, we compared five measures for quantifying dissimilarities between the row and column profiles of a cluster–sensor contingency table summarising the results of spike waveform clustering from MEA time-series recordings. Specifically, we considered the classical Euclidean distance, the zero-adjusted Aitchison's distance, and the chiPower distance from Compositional Data Analysis (CoDA), as well as the Bhattacharyya's distance and the Jensen–Shannon's divergence, widely adopted in machine learning.

Although limited to five dissimilarity measures, the approach presented here represents a promising tool for interpreting the outcomes of pattern identification procedures in MEA data. In particular, the analysis of column profiles has the potential to reveal similarities across sensor recordings and to characterise how such relationships change under different conditions, e.g. different drug dosages.

While all five measures provided similar results, as demonstrated by the high Pearson correlation coefficient between the associated distance matrices, the empirical distributions of the corresponding pairwise distances present some differences, especially the row profiles. In detail, the Jensen-Shannon's divergence shows the highest variability, while Euclidean, zero-adjusted Aitchison's and Bhattacharyya's distances show the highest median values. This highlights the importance of a careful selection of such a measure, especially when dealing with large datasets involving compositions with only a few parts (in our analysis we considered 183 column profiles, each one being a 4-part composition). Future work will focus on assessing the practical impact of these differences on downstream analyses, such as the identification of relationships among sensors, and on exploring the applicability of the proposed framework to other fields.

Finally, we observe that among the considered measures, the zero-adjusted Aitchison's distance and the chiPower distance were specifically devised for handling compositions with zero parts. However, both require a parameter to be set, i.e., the value ϵ replacing the zeros and the λ within the Box-Cox transformation. In this study these parameters have been fixed by following guidelines from the existing literature. However, a further analysis, not shown here for the sake of conciseness, demonstrated that the results provided by these metrics may be highly sensitive to the choice of these parameters. Therefore, developing a robust, data-driven strategy for selecting these parameters will be the subject of future work.

Acknowledgements. The authors acknowledge the financial support from the MIUR Excellence Department Project 2023–2027 awarded to the Dipartimento di Matematica of the University of Genova, CUP D33C23001110001.

References

1. Bhattacharyya, A.: On a measure of divergence between two statistical populations defined by their probability distributions. Bull. Calcutta Math. Soc. **35**, 99–109 (1943).
2. Greenacre, M.: The chiPower transformation: a valid alternative to logratio transformations in compositional data analysis. Adv. Data Anal. Classif. **18**, 769–796 (2024). https://doi.org/10.1007/s11634-024-00600-x
3. Gross, G.W., Rhoades, B.K., Azzazy, H.M., Wu, M.C.: The use of neuronal networks on multielectrode arrays as biosensors. Biosens. Bioelectron. **10**(6-7), 553–567 (1995). https://doi.org/10.1016/0956-5663(95)96931-N
4. Lin, J.: Divergences measures based on the Shannon entropy. IEEE Trans. Inf. Theory **37**(1), 145-151 (1984). https://doi.org/10.1109/18.61115
5. Lubbe, S., Filzmoser, P., Templ, M.: Comparison of zero replacement strategies for compositional data with large numbers of zeros. Chemom. Intell. Lab. Syst. **210**, 104248 (2021). https://doi.org/10.1016/j.chemolab.2021.104248
6. Martín-Fernández, J. A., Barceló-Vidal, C., Pawlowsky-Glahn, V.: Dealing with zeros and missing values in compositional data sets using nonparametric imputation. Math. Geol. **35**(3), 253–278 (2003). https://doi.org/10.1023/A:1023866030544
7. Pawlowsky-Glahn, V., Egozcue, J.J., Tolosana-Delgado, R.: Modeling and Analysis of Compositional Data. John Wiley & Sons, Hoboken (2015). https://doi.org/10.1002/9781119003144

Deliberation, Representation, and Civic Learning: A Multivariate Study of Democratic Reasoning Among High School Students

Theodore Chadjipadelis(✉) and Georgia Panagiotidou

Aristotle University of Thessaloniki, Thessaloniki, Greece
{chadji,gvpanag}@polsci.auth.gr

Abstract. This study examines how a civic education intervention shapes high school students' democratic reasoning, focusing on deliberation and representative choice. Using data from students in Thessaloniki who participated in student-led civic education workshops during the 2024–2025 academic year, the analysis applies a multistage multivariate framework. Principal Component Analysis reveals the multidimensional structure of students' understandings of deliberation, while Hierarchical Cluster Analysis identifies distinct deliberation profiles. A conjoint experiment is subsequently used to model representative choice in student council elections, highlighting the relative importance of communication skills, representational priorities, and leadership style. The findings demonstrate that civic education fosters differentiated democratic perspectives and illustrate the value of combining semantic mapping and conjoint analysis for studying democratic attitudes in educational settings.

Keywords: Multivariate Data Analysis · Principal Component Analysis · Hierarchical Clustering · Conjoint Analysis · Deliberation · Civic Engagement

1 Introduction and Analytical Framewrok

Civic education is widely regarded as a key mechanism through which democratic systems reproduce themselves, fostering political knowledge, participatory norms, and civic responsibility among future citizens [1,8]. Research on civic education consistently highlights its role in shaping political socialization, knowledge, and participation across different contexts (e.g., Torney-Purta et al.; Campbell; Niemi & Junn), while also emphasizing the importance of experiential and participatory learning approaches [3,6,9]. However, traditional models of civic instruction centered on abstract institutional knowledge have been increasingly criticized for their limited capacity to promote active engagement and meaningful participation, particularly among younger generations [3]. These

P. Brito et al. (Eds.): IFCS 2026, STUDIES CLASS, pp. 85–94, 2026.
https://doi.org/10.1007/978-3-032-32009-4_11

limitations are especially pronounced in contexts marked by political disengagement, institutional distrust, and social polarization. Research on political socialization further emphasizes that democratic competencies emerge not only through formal instruction but through broader processes combining cognitive learning and experiential engagement, in which schools and universities play a central role [3,4,9]. In response to these challenges, this chapter examines an applied civic education intervention implemented by the Laboratory of Applied Political Research at the Aristotle University of Thessaloniki under the title Teaching Civic Education. Launched in 2017, the intervention adopts an experiential, student-led model in which undergraduate political science students design and deliver interactive civic education workshops to high school students [4]. Rather than emphasizing normative democratic ideals, the program focuses on core democratic processes, representation, electoral choice, deliberation, and participation through structured workshops and role-play activities [4]. Such participatory approaches have been shown to be more effective than purely theoretical instruction in enhancing political interest, civic literacy, and perceived political efficacy among young people [4]. Empirical research also shows that civic education interventions do not produce uniform effects. Instead, outcomes vary according to students' prior orientations, values, and modes of political reasoning, motivating a shift from unidimensional indicators toward multidimensional models capable of capturing heterogeneity in political attitudes and practices [4,8]. Methodologically, this shift has been accompanied by the growing use of multivariate exploratory techniques to identify latent structures, behavioral profiles, and discursive patterns [2,5]. Within this literature, deliberation and representation are increasingly treated as analytically distinct but empirically interconnected dimensions of democratic engagement, encompassing communicative, institutional, participatory, and outcome-oriented elements [4]. Studies of youth electoral behavior further indicate that representative choice often reflects communicative competence and perceived responsibility rather than ideological positioning alone [7]. Building on this body of work, the present chapter does not seek to re-theorize civic education or democratic participation. Instead, it presents an empirical application of a multistage data analysis framework designed to capture the multidimensional effects of a civic education intervention on political attitudes and decision-making [4,5]. Drawing on data collected during the 2024–2025 academic year from high school students in Thessaloniki, the analysis focuses on students' understandings of deliberation and their criteria for selecting representatives in student council elections [4]. By combining Principal Component Analysis, Hierarchical Cluster Analysis, and conjoint analysis, the chapter prioritizes empirical results and methodological demonstration, illustrating how multivariate approaches can be used to map political meanings and interpret democratic preferences in educational settings [2,5].

2 Methodology

The empirical analysis presented in this chapter is based on data collected during the implementation of the Teaching Civic Education intervention in the academic year 2024–2025. The intervention consisted of interactive civic education workshops delivered by undergraduate political science students of the Aristotle University of Thessaloniki to high school students in the city of Thessaloniki, following the didactic framework described in earlier applications of the program [5]. The study sample comprises 162 high school students who participated in the civic education seminars during the winter and spring semesters. Participation was voluntary, and data were collected immediately after completion of the workshops through a structured questionnaire. The research design does not aim at causal inference in the strict experimental sense; rather, it is oriented toward the exploratory identification of attitudinal structures, behavioral profiles, and decision-making logics, which is consistent with the objectives of multivariate data analysis [2]. The questionnaire instrument was designed to capture two analytically distinct but empirically related dimensions of democratic engagement:

1. Students' understanding of deliberation in democracy, operationalized through a ranking task evaluating perceived benefits of deliberative processes.
2. Students' representative choice criteria, operationalized through a conjoint experiment simulating candidate selection in student council elections. This dual structure allows the combination of attitudinal mapping and choice modeling within a unified analytical framework. To capture students' perceptions of deliberation, respondents were asked to rank ten statements (Fig. 1) describing potential benefits arising from deliberative democratic processes. These statements refer to institutional outcomes (e.g. credibility and transparency of institutions), communicative processes (e.g. exchange of views and respect for different opinions), participatory aspects (e.g. citizen involvement in policy-making), and practical policy outcomes (e.g. service improvement and alignment of policies with citizens' needs).

E1	Development of effective, credible and transparent institutions.
E2	Development of a participatory Civil Society.
E3	Exchange of views between policymakers and citizens.
E4	Capture, understand, respect for different opinions.
E5	Improvement of the services provided.
E6	Align planned policies with existing and emerging needs of citizens.
E7	Strengthening democracy.
E8	Informing citizens about the issue under consideration.
E9	Recording of opinions on issues of common interest.
E10	Participation in policy-making.

Fig. 1. Ten (10) statements used to capture students' perceptions of the benefits of deliberation in democratic processes.

The ranking task produces an ordinal dataset in which each respondent assigns a unique rank from 1 (most important) to 10 (least important). Rather than analyzing items individually, the analysis focuses on identifying latent dimensions structuring students' understandings of deliberation. Principal Component Analysis (PCA) was applied to the ranking data as an exploratory dimensionality-reduction technique, enabling the identification of orthogonal components summarizing the main evaluative contrasts [5]. Component selection was guided by eigenvalues and cumulative explained variance. The coordinates of individuals and statements on the retained components were subsequently used for further analysis. Although the ranking task produces ordinal data, the application of Principal Component Analysis (PCA) in this context is consistent with established practices in exploratory multivariate analysis, where ordinal variables are often treated as quasi-metric for the purpose of dimensionality reduction and structure detection [2,5]. The objective of the analysis is not precise measurement of distances, but the identification of latent contrasts and relational patterns among items. In this sense, PCA is employed as an exploratory tool to capture the underlying structure of students' evaluative judgments, rather than to estimate exact metric relationships. This approach is widely used in semantic mapping and attitudinal research, particularly when the focus lies on interpretation and typology construction rather than statistical inference. To identify distinct student profiles, Hierarchical Cluster Analysis (HCA) was applied to the individual factorial coordinates. Clustering on factorial scores is a well-established strategy in multivariate analysis, combining dimensional reduction with typological classification [2,5]. HCA was performed using Ward's linkage criterion and Euclidean distance, and the number of clusters was determined by inspecting the dendrogram and within-cluster inertia, resulting in a five-cluster solution. In a second analytical stage, students' criteria for selecting representatives were examined using Conjoint Analysis, a method well suited to studying multidimensional decision-making where preferences are not directly stated [7]. Participants ranked eight hypothetical candidate profiles defined by five attributes: leadership style, priority setting, school performance, communication skills, and popularity among peers. This ranking-based design reduces social desirability bias and allows estimation of latent utility values for each attribute level. Conjoint analysis was then applied to estimate the relative importance of each attribute and the direction and magnitude of preferences. The resulting utilities complement the deliberation analysis by linking normative understandings of democracy to practical criteria of representative selection. Overall, the methodological strategy prioritizes exploration, structure detection, and interpretability over hypothesis testing. By combining PCA, HCA, and conjoint analysis, the approach captures both how students structure democratic meanings and how these structures translate into representative choice. This multistage workflow is consistent with previous applications of semantic mapping and behavioral profiling in political analysis and is adapted here to the context of civic education interventions [5,7].

3 Results

The first stage of the analysis examined how high school students structure their understanding of the benefits arising from deliberative democratic processes. Principal Component Analysis (PCA) was applied to the ranking data of the ten deliberation-related statements (E1E10), as presented in the questionnaire. The eigenvalue analysis indicates that five components should be retained. As shown in Fig. 2, these five components jointly explain 67.6% of the total variance, confirming the multidimensional nature of students' perceptions of deliberation. No single component dominates the structure, suggesting that deliberation is not understood through a single evaluative lens (Fig. 3).

Components	Eigenvalue	% Variance	Cumulative %
1	1,562906	15,6%	15,63%
2	1,510536	15,1%	30,73%
3	1,437441	14,4%	45,11%
4	1,188444	11,9%	56,99%
5	1,060491	10,6%	67,60%
6	1,020401	10,2%	77,80%
7	0,9178175	9,2%	86,98%
8	0,7111852	7,1%	94,09%
9	0,5907789	5,9%	100,00%
10	1,14409E-30	0,0%	100,00%

Fig. 2. Eigenvalues and Explained Variance of the Principal Component Analysis on Deliberation Items (E1E10).

The coordinates of the deliberation statements on the five (5) retained components are reported in Fig. 4. Examination of these coordinates allows for substantive interpretation of each dimension: • Component 1 (15.6% of variance) contrasts statements emphasizing dialogue, respect for different opinions, and recording of views (E4, E9) with statements emphasizing institutional strengthening and participation in policy-making (E7, E10). • Component 2 (15.1%) opposes practical responsiveness-such as aligning policies with citizens' needs and improving services (E6, E5, E8), to more abstract democratic outcomes. • Component 3 (14.4%) differentiates institutional credibility and exchange with policymakers (E1, E3) from the simple recording of opinions (E9). • Component 4 (11.9%) contrasts participatory civil society (E2) with broader policy and democratic outputs. • Component 5 (10.6%) separates service improvement (E5) from policy alignment (E6).

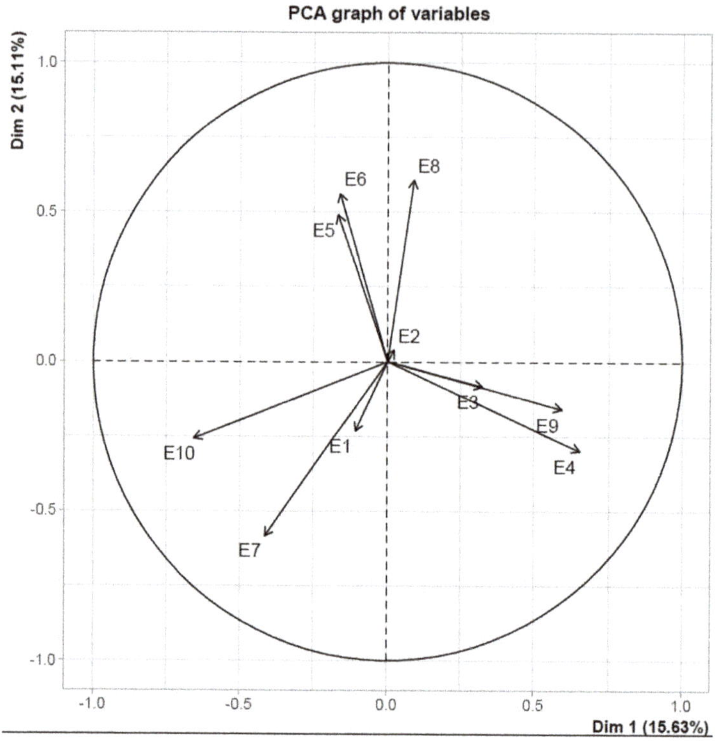

Fig. 3. Component Loadings of Deliberation Items on the Retained Principal Components.

	Dim. 1	Dim. 2	Dim. 3	Dim. 4	Dim. 5
E1	-0,10951	-0,23293	0,679979	-0,19127	0,379736
E2	0,022711	0,038658	0,364648	0,613595	-0,19344
E3	0,323198	-0,08441	0,56269	-0,16291	-0,26532
E4	0,653519	-0,29919	-0,1685	0,016171	-0,04237
E5	-0,1694	0,490946	-0,0302	-0,47771	0,47687
E6	-0,16275	0,561639	-0,09397	-0,27638	-0,67076
E7	-0,41625	-0,58357	-0,41233	-0,31467	-0,15726
E8	0,088303	0,60703	-0,24859	0,396698	0,264258
E9	0,594417	-0,16006	-0,47771	-0,00161	0,165625
E10	-0,65519	-0,25889	-0,1653	0,4331	0,0858

Fig. 4. Coordinates of the deliberation statements on the five retained components.

Together, these components reveal that deliberation is conceptualized by students through multiple, partially competing dimensions, combining communicative, institutional, participatory, and outcome-oriented elements.

To identify distinct patterns among students, Hierarchical Cluster Analysis (HCA) was applied to the individual coordinates obtained from the five retained PCA components. This procedure resulted in a five-cluster solution, corresponding to five interpretable profiles of how students perceive the gains of deliberation. The projection of clusters onto the factorial space is presented in the factorial map shown in the Fig. 5, where both statements and individuals are jointly represented. The spatial proximity between clusters and statements allows for substantive labeling of each profile. The five clusters can be summarized as follows: • Cluster 1: Institutionalists and Democracy Advocates This cluster is located near statements referring to participation in policy-making (E10), strengthening democracy (E7), and institutional credibility (E1). Students in this group understand deliberation primarily as a mechanism for reinforcing democratic institutions and formal participation. • Cluster 2: Abstract Democracy Idealists Positioned away from communicative statements and closer to abstract democratic outcomes, this group perceives deliberation mainly as a symbolic expression of democratic ideals, rather than as a process of dialogue or concrete policy impact. • Cluster 3: Policy Responsiveness Believers This profile is associated with statements emphasizing policy alignment with citizens' needs (E6), improvement of services (E5), information provision (E8), and participatory civil society (E2). Deliberation is interpreted here as a tool for producing effective and responsive policies. • Cluster 4: Dialoguers Students in this cluster are positioned near statements related to respect for different opinions (E4), exchange of views (E3), and recording of opinions (E9). Deliberation is primarily understood as a communicative and inclusive process. • Cluster 5: Integrative Pragmatists This cluster occupies an intermediate position, combining proximity to both dialogue-oriented and policy-oriented statements. Students in this group view deliberation as simultaneously a communicative practice and a means for improving collective outcomes.

In the second analytical stage, students' criteria for selecting representatives in student council elections were examined through conjoint analysis. The estimated utility values for each attribute level are presented in the utilities table shown in Fig. 6.

The results indicate a clear hierarchy of voting criteria. Communication skills emerge as the most influential attribute, with communicative candidates receiving the highest positive utility. Priority setting follows, with students favoring candidates who prioritize their own department over those emphasizing the whole school. Leadership style ranks third, with assertive candidates preferred over consensual ones. By contrast, popularity and school performance have a positive but substantially weaker impact on vote choice. Although these attributes contribute to candidate evaluation, they are secondary compared to communicative capacity and representational focus.

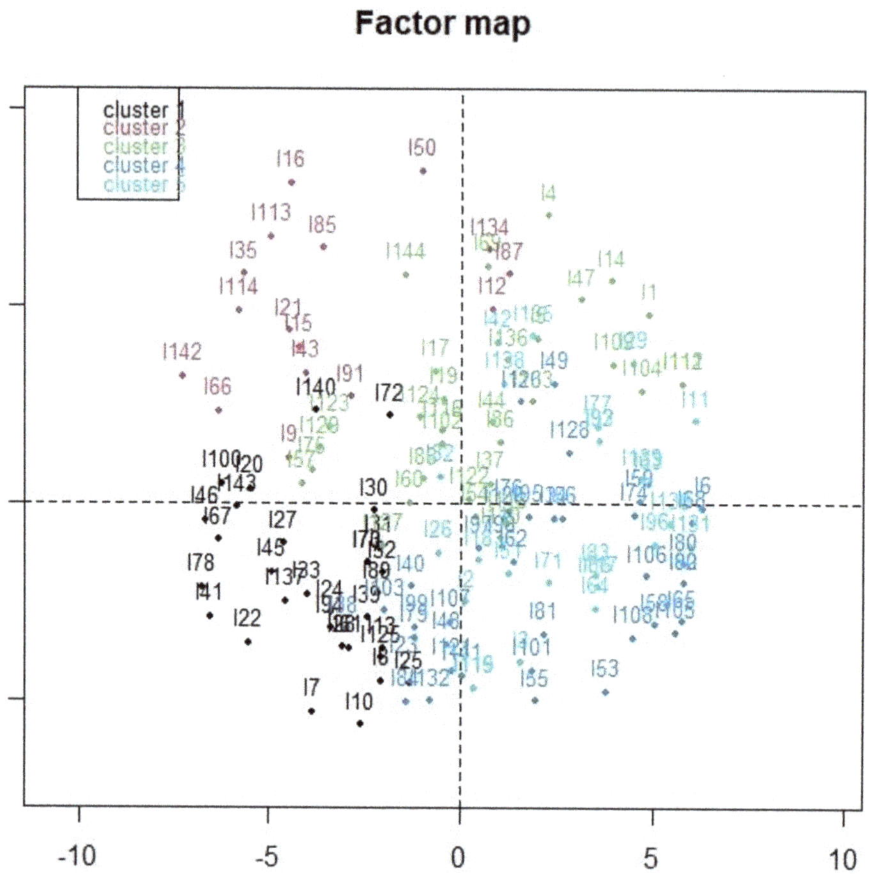

Fig. 5. Cluster Projection of Deliberation Profiles on the Factorial Map.

4 Discussion

The findings of this chapter provide a structured and multidimensional account of how young students conceptualize deliberation and evaluate representatives following a civic education intervention. Rather than converging toward a uniform understanding of democratic processes, students articulate distinct and internally coherent profiles, confirming that democratic attitudes are organized around configurations of meanings and expectations rather than linear scales of participation or interest [1]. The PCA results show that deliberation is perceived as a composite concept, combining communicative norms, institutional outcomes, participatory dimensions, and practical policy effects. This multidimensional structure aligns with research on political socialization and civic education, which emphasizes that democratic competencies emerge through the interaction of normative learning and experiential engagement rather than formal instruction alone [3,4,8]. The identification of five deliberation profiles further illus-

		Utility Estimated	Std. Error
a	Assertive	.325	.092
	Consensual	-.325	.092
b	Set their own department as a matter of priority.	.750	.092
	Support the interest of the whole school as a priority	-.750	.092
c	Good student	.108	.092
	Relatively good student	-.108	.092
d	Communicative	1.149	.092
	Not communicative	-1.149	.092
e	Popular	.402	.092
	Not so popular	-.402	.092
	(Constant)	4.500	.092

Fig. 6. Estimated utility values and relative importance ofaAttributes in representative choice.

trates this heterogeneity. Profiles emphasizing dialogue, policy responsiveness, or integrative reasoning coexist with more abstract orientations toward democracy, indicating that civic education does not automatically translate democratic principles into concrete procedural understandings [4,9]. This finding is consistent with earlier empirical work showing that civic education may reinforce democratic values without homogenizing political reasoning. When combined with the conjoint analysis results, these profiles reveal a meaningful correspondence between students' understandings of democratic processes and their criteria for representative choice. The strong preference for communicative representatives and clarity in representational priorities, alongside the weaker role of popularity and academic performance, points to a shift toward interactional competence and representational responsibility, a pattern also observed in studies of youth political behavior and first-time voters [7]. The findings of this study should be interpreted within the limits of the research design. The sample is not statistically representative, as it consists exclusively of high school students from Thessaloniki who participated in a specific civic education intervention. The results are therefore illustrative rather than generalizable and should be understood as capturing structured patterns of democratic reasoning within this particular context. The analytical emphasis of the chapter lies in identifying relational structures and profiles rather than producing population-level estimates. Methodologically, the chapter demonstrates the analytical advantages of integrating dimensional reduction (PCA), typological classification (HCA), and choice modeling (conjoint analysis). This multistage approach enables the identification of latent structures, interpretable profiles, and decision-making logics that would remain largely invisible using univariate or purely descriptive methods [2,5]. Finally, the results contribute to the evaluation of civic education interventions by showing that their impact lies not in producing uniform outcomes, but in fostering plural and differentiated democratic reasoning. In line with the aims of this volume, the chapter's primary contribution is methodological, illustrating how data-driven, multivariate approaches can be used to systematically analyze complex democratic attitudes and behaviors in educational settings.

References

1. Almond, G.A., Verba, S.: The Civic Culture: Political Attitudes and Democracy in Five Nations. Princeton University Press, Princeton (1963)
2. Benzécri, J.-P.: L'analyse des données. Dunod, Paris (1976)
3. Campbell, D.E.: Voice in the classroom: how an open classroom climate fosters political engagement among adolescents. Polit. Behav. **30**(4), 437–454 (2008)
4. Chadjipadelis, T., Panagiotidou, G.: Students teaching democracy to other students: the effects on political engagement. In: Teaching Civic Engagement Globally. American Political Science Association (2021)
5. Chadjipadelis, T., Panagiotidou, G.: Semantic map: bringing together groups and discourses. In: Data Clustering. IntechOpen (2022)
6. Niemi, R.G., Junn, J.: Civic Education: What Makes Students Learn. Yale University Press, New Haven (1998)
7. Panagiotidou, G., Chadjipadelis, T.: First-time voters in Greece: Views and attitudes of youth on Europe and democracy. Springer, Cham (2020)
8. Parsons, T.: The school class as a social system. Harv. Educ. Rev. **29**(4), 297–318 (1959)
9. Torney-Purta, J., Lehmann, R., Oswald, H., Schulz, W.: Citizenship and Education in Twenty-Eight Countries: Civic Knowledge and Engagement at Age Fourteen. IEA, Amsterdam (2001)

A Landmark Set Learning Algorithm for Scalable Spectral Clustering

Guangliang Chen(✉), Valen Feldmann, and Eli Edwards-Parker

Hope College, Holland, MI 49423, USA
cheng@hope.edu

Abstract. Spectral clustering is an effective method for uncovering nonlinear structures in data sets, but its high computational complexity limits its scalability. Landmark-based spectral clustering alleviates this issue by approximating the similarity graph using a small set of representative points (landmarks), yet the landmark budget is typically fixed *a priori*, without regard to data complexity. We propose an adaptive methodology that learns an effective landmark set by evaluating candidate landmark sets on a bipartite graph using unsupervised criteria. Starting from a small landmark set, the method iteratively expands the landmark set toward underrepresented regions via a batch k-means++ sampling scheme while jointly selecting a compatible similarity scale. Experiments on synthetic and real datasets demonstrate that the proposed approach produces compact, high-quality landmark sets and that the adopted criterion reliably tracks clustering performance.

Keywords: Spectral clustering · landmark learning · k-means++

1 Introduction

Spectral clustering is a powerful technique for uncovering nonlinear manifold structures in data sets by exploiting the eigenvectors of a similarity graph [14]. Despite its strong empirical performance, classical spectral clustering applied to a set of n data points requires $\mathcal{O}(n^3)$ time and $\mathcal{O}(n^2)$ memory, which makes it impractical for large-scale datasets. As a result, considerable effort has been devoted to developing scalable variants of spectral clustering.

Among existing strategies, *landmark-based spectral clustering* has emerged as a particularly effective paradigm [3,7,11,13,16–19]. Instead of constructing the full $n \times n$ similarity matrix, these methods select a set of $m \ll n$ representative points (landmarks) and approximate the original graph using data–landmark affinities. When the landmark set is well chosen, this approach preserves essential geometric and clustering-relevant structure while substantially reducing computational cost. However, the success of landmark-based methods depends critically on the quality of the landmark set.

A variety of landmark selection strategies have been used in practice. Uniform sampling [3,6] is simple and efficient but often produces poorly distributed

P. Brito et al. (Eds.): IFCS 2026, STUDIES CLASS, pp. 95–102, 2026.
https://doi.org/10.1007/978-3-032-32009-4_12

landmarks. k-means-based sampling [3,4] generally yields higher-quality landmark sets and can generate synthetic representatives, but is computationally slow. More advanced approaches include divide-and-conquer strategies [10], lightweight prototype-based methods such as LiteWSC [20], and score-based sampling schemes using leverage scores, determinantal point processes, or graph-based importance measures [2,6,9,12,15]. Despite differences in motivation and implementation, all these methods require the number of landmarks to be specified *a priori* and decouple landmark selection from clustering quality evaluation.

A related but often overlooked issue is the choice of a scale parameter in the similarity function used. For Gaussian kernels, the scale determines the resolution at which clusters are identified and has a major impact on clustering results. Recent work has provided effective unsupervised criteria for scale selection in spectral clustering [5]. In landmark-based settings, this challenge has not been systematically addressed. In the current literature, the scale is typically chosen independently of the landmark set, even though increasing the number of landmarks naturally refines the effective resolution of the data representation. This suggests that landmark selection and scale learning are inherently coupled problems and thus should be handled jointly.

In this paper, we propose an efficient algorithm for *landmark set learning* for scalable spectral clustering. Building on the landmark-based bipartite diffusion maps (LBDM) [11], we evaluate the quality of a candidate landmark set through unsupervised loss functions. Starting from an initial landmark set, the method iteratively expands a landmark set towards underrepresented regions of the data by using a batch implementation of the k-means++ sampling strategy [1], while jointly selecting a compatible similarity scale. The process terminates automatically when the chosen loss stabilizes, thereby determining both the landmark locations and an appropriate landmark budget.

Contributions. The main contributions of this work are:

- A batch k-means++ landmark expansion strategy that continuously targets underrepresented regions of the data.
- An adaptive framework for jointly learning landmark locations, landmark cardinality, and similarity scale for landmark-based spectral clustering.

2 Methodology

Let $X = \{\mathbf{x}_1, \ldots, \mathbf{x}_n\} \subset \mathbb{R}^d$ be a given data set to be partitioned into k clusters. We would like to find a set of landmark points, $Y = \{\mathbf{y}_1, \ldots, \mathbf{y}_m\} \subset \mathbb{R}^d$ and use them to speed up spectral clustering. The size of the landmark set, m, must be set carefully. On one hand, if it is too small, then the landmark set is incapable of capturing the pattern of the data; on the other hand, it needs to be as small as possible (while being still adequate) in order to control the running time.

In this paper, we aim to simultaneously address the following two questions: (1) landmark placement for a given budget (m), and (2) the smallest landmark size m sufficient for accurate clustering. An additional challenge is learning an appropriate similarity scale to match the landmark set. Our proposed algorithm alternates between expanding the landmark set toward underrepresented regions and tuning a compatible scale after each expansion step.

Let $\{m_t\}_{t\geq 0}$ be a monotonically increasing sequence of positive integers, where m_t denotes the size of the landmark set $Y^{(t)}$ constructed at step t. In this work, we use an arithmetic schedule

$$m_t = m_0 + t \cdot b, \qquad 0 \leq t \leq T, \tag{1}$$

where m_0 is the initial landmark set size, b is the fixed increment per step, and T is the total number of steps. We initialize $Y^{(0)}$ by running k-means++ on X for a small number of iterations (10) and taking the resulting m_0 centroids as landmark points. For each subsequent step $t = 1, \ldots, T$, we expand the landmark set $Y^{(t-1)}$ by sampling b additional candidates from X and appending them to $Y^{(t-1)}$. We then refine the enlarged set using k-means (again for 10 iterations), and use the resulting centroids as the landmark set $Y^{(t)}$.

Batch k-means++ Sampling. Let $Y^{(t)} = \{\mathbf{y}_1^{(t)}, \ldots, \mathbf{y}_m^{(t)}\}$ denote the landmark set at step t. In order to expand the landmark set, we introduce a probability distribution over the data points in X. For each $1 \leq i \leq n$, define the (squared) distance from the point $\mathbf{x}_i$ to the current landmark set $Y^{(t)}$ as

$$d_i^{(t)} = \min_{1\leq j\leq m} \|\mathbf{x}_i - \mathbf{y}_j^{(t)}\|^2, \tag{2}$$

We then assign sampling probabilities proportional to those distances:

$$p_i^{(t)} = \frac{d_i^{(t)}}{d_1^{(t)} + \cdots + d_n^{(t)}}, \qquad 1 \leq i \leq n. \tag{3}$$

Using this distribution, we sample a batch of b points i.i.d. from X and append them to $Y^{(t)}$. This procedure is inspired by the k-means++ initialization rule, but differs in one key aspect: instead of selecting one new center at a time and recomputing the probabilities after each selection, we draw b candidates simultaneously, which substantially reduces sampling overhead while still prioritizing underrepresented regions of the data space.

Landmark-Based Clustering. For each landmark set $Y^{(t)}$ formed in the process, we calculate the pairwise similarities between the landmark points in $Y^{(t)}$ and the data in X. Here, we focus on the Gaussian similarity:

$$\mathbf{A}^{(t)} = (a_{ij}^{(t)}) \in \mathbb{R}^{n\times m}, \qquad a_{ij}^{(t)} = \exp\left(-\|\mathbf{x}_i - \mathbf{y}_j^{(t)}\|^2/(2\sigma^2)\right) \tag{4}$$

where $\sigma > 0$ is a fixed parameter.

The LBDM algorithm [11] constructs a bipartite graph between X and $Y^{(t)}$ by using $X \cup Y^{(t)}$ as the node set and

$$\mathbf{W}^{(t)} = \begin{bmatrix} \mathbf{0} & \mathbf{A}^{(t)} \\ (\mathbf{A}^{(t)})^T & \mathbf{0} \end{bmatrix} \in \mathbb{R}^{(n+m)\times(n+m)} \tag{5}$$

as the weight matrix, and then applies the Normalized Cut (NCut) algorithm [14] to co-cluster the points in $X \cup Y^{(t)}$. As a result, the data points in X (and also those in $Y^{(t)}$) are embedded into a low-dimensional space $\mathbb{R}^{k-1}$. Computationally, there is an efficient approach to computing the NCut embedding of $X \cup Y^{(t)}$ through the singular value decomposition (SVD) of a normalized version of $\mathbf{A}^{(t)}$ [11]. Also in [11], the authors proposed three different ways of clustering the data in X by using the embeddings of X, $Y^{(t)}$, and $X \cup Y^{(t)}$, respectively. In this paper, we focus on using the embedding of X alone, denoted $Z^{(t)} = \{\mathbf{z}_1^{(t)}, \ldots, \mathbf{z}_n^{(t)}\}$. Finally, k-means is used to obtain a partition of the embedding, $Z^{(t)} = C_1^{(t)} \cup \cdots \cup C_k^{(t)}$, and correspondingly assign labels to the data in X.

Evaluation of Landmark Sets. To evaluate the quality of a landmark set $Y^{(t)}$ in terms of its resulting clustering performance, we adopt the *relative k-means loss* [5] on the embedding of X. This criterion measures the within-cluster scatter of the k-means partition of $Z^{(t)}$, relative to the overall scatter of the embedded data. Since the choice of the similarity scale σ has a critical impact on the embedding quality, we evaluate $Y^{(t)}$ by optimizing the loss over the scale parameter σ:

$$\ell^{(t)} = \min_{\sigma>0} \frac{\sum_{j=1}^{k} \sum_{\mathbf{z}_i^{(t)} \in C_j^{(t)}} \|\mathbf{z}_i^{(t)} - \mathbf{c}_j^{(t)}\|^2}{\sum_{i=1}^{n} \|\mathbf{z}_i^{(t)} - \mathbf{c}^{(t)}\|^2}, \tag{6}$$

where each $\mathbf{c}_j^{(t)}$ is the centroid of $C_j^{(t)}$ and $\mathbf{c}^{(t)}$ is the global centroid of $Z^{(t)}$. In practice, we solve the minimization by searching a decreasing geometric grid $\sigma \leftarrow \sigma/\beta$ within a finite interval $[\sigma_{\min}^{(t)}, \sigma_{\max}^{(t)}]$, where the bounds are defined as

$$\sigma_{\max}^{(t)} = \frac{1}{nm} \sum_{i=1}^{n} \sum_{j=1}^{m} \|\mathbf{x}_i - \mathbf{y}_j^{(t)}\|, \qquad \sigma_{\min}^{(t)} = \sqrt{\frac{1}{n-m} \sum_{i=1}^{n} d_i^{(t)}}, \tag{7}$$

and $\beta > 1$ is the fixed ratio. See Algorithm 1 for the complete σ-tuning procedure.

Algorithm 1. Scale learning for LBDM [11] with a fixed landmark set

Input: Data X, landmark set $Y^{(t)}$, number of clusters k
Output: Minimal relative k-means loss $\ell^{(t)}$ and corresponding clustering
1: Compute the bounds $\sigma_{\max}^{(t)}, \sigma_{\min}^{(t)}$ by (7), and fix the ratio $\beta > 1$ (e.g., 1.1).
2: Set $\sigma \leftarrow \sigma_{\max}^{(t)}$ and compute $\mathbf{A}^{(t)}$ by (4).
3: **while** $\sigma \geq \sigma_{\min}^{(t)}$ **do**
4: Calculate the embedding $Z^{(t)}$ of the data X by LBDM with affinity matrix $\mathbf{A}^{(t)}$
5: Apply k-means to $Z^{(t)}$ in order to evaluate the relative k-means loss.
6: Update $\mathbf{A}^{(t)}$ by $a_{ij}^{(t)} \leftarrow (a_{ij}^{(t)})^{\beta^2}$ and σ by $\sigma \leftarrow \sigma/\beta$.
7: **end while**
8: **return** Minimal loss $\ell^{(t)}$ and corresponding clustering of X.

The Full Algorithm. Algorithm 2 has the entire process of learning a landmark set for LBDM with the relative k-means loss function.

Algorithm 2. Landmark set learning for LBDM [11] using relative k-means loss

Input: Dataset X, number of clusters k, arithmetic sequence (m_0, b, T)
Output: Optimal landmark set Y and clustering of X

1: Initialize $Y^{(0)}$ by running k-means++ (with 10 iterations) on X to obtain m_0 centroids. Set $t \leftarrow 0$.
2: **repeat**
3: Evaluate $Y^{(t)}$ by Alg. 1 to obtain minimal loss $\ell^{(t)}$.
4: Calculate sampling probabilities (3) and use them to select b points from X to merge with $Y^{(t)}$. Apply k-means for 10 iterations to refine them; the resulting centroids are stored in $Y^{(t+1)}$.
5: Update $t \leftarrow t + 1$.
6: **until** $t = T$, or loss function converges
7: **return** $Y = Y^{(t)}$ and the corresponding clustering of X.

Stopping Criterion. We recommend setting the number of steps to a few dozens, e.g., $T = 30$ (along with reasonable choices of m_0, b). In the meantime, we suggest to continuously monitor the convergence of the loss function. If the loss has already converged within T steps, then we can terminate the process earlier. Otherwise, we can easily extend the process based on the last landmark set $Y^{(T)}$.

3 Experiments

We evaluate Algorithm 2 on both synthetic and real datasets. In the *plain* implementation, we apply 10 k-means refinement updates after each sampling step. We also consider a simplified variant that skips the refinement stage and relies solely on batch k-means++ sampling to grow the landmark set; in this case, the landmark sets produced across steps are strictly nested. We refer to these two versions as k-means++10 and k-means++0, respectively. We compare them against two additional nested landmark selection strategies (each combined with LBDM): (i) greedy farthest-point sampling [21] which always selects the point that is farthest from the current landmark set and (ii) nested uniform sampling.

Synthetic Datasets. We generated three nonconvex patterns in Fig. 1 (top row). For each dataset, we construct a sequence of landmark sets using each of the four sampling methods and record the corresponding LBDM clustering errors for all landmark sizes along the same m-schedule. To account for randomness in the sampling procedures, we repeat each experiment 30 times and report the average clustering error at each value of m; see Fig. 1 (bottom row).

Across all three datasets, k-means++10 (red) converges the fastest and achieves near-zero error with the smallest landmark sets. The k-means++0 variant is consistently worse than k-means++10, demonstrating that even a small number of refinement iterations significantly improves landmark placement. Observe also that nested uniform sampling (black) converges the most slowly, requiring substantially more landmarks to reach comparable accuracy. The greedy farthest-point strategy (blue) performs poorly on the last two datasets (which are noisier). We attribute this behavior to the fact that farthest-point selection tends to over-sample outliers and boundary points. The top row of Fig. 1 also shows representative landmark sets produced by k-means++10 at the smallest m values achieving near-zero error. Although the ground-truth clusters have comparable sizes, the learned landmark allocation is nonuniform: fewer landmarks are assigned to dense regions, while more landmarks are placed in low-density or high-curvature regions where finer resolution is needed.

Finally, since ground-truth labels are unavailable in practice, clustering error cannot be used to select m. Instead, we use the relative k-means loss as an unsupervised criterion. Figure 1 (bottom row) also displays this loss as a function of m for k-means++10, and we observe that its elbow point consistently aligns with the onset of near-zero clustering error.

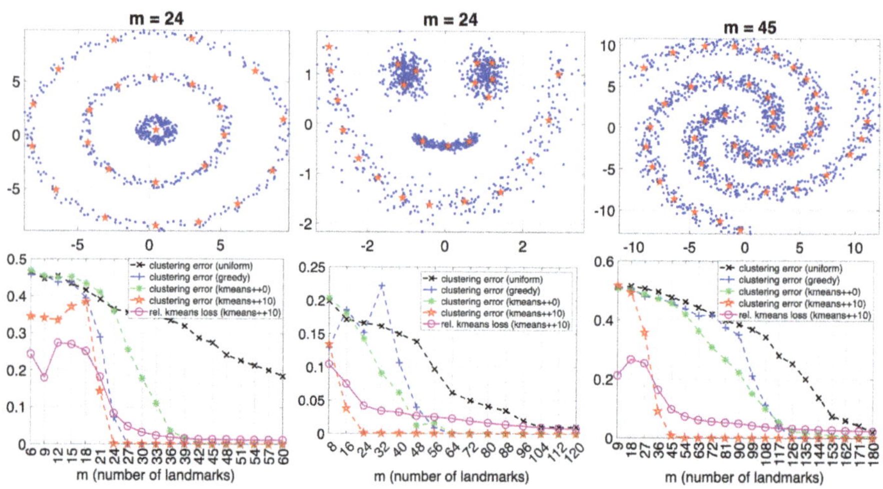

Fig. 1. Top row: synthetic data (blue) and examples of selected landmark sets found by k-means++10 (red); bottom row: clustering errors of LBDM coupled with each of the four sampling schemes. (Color figure online)

USPS Digits. We further evaluate the same landmark selection methods (each coupled with LBDM) on the USPS digits dataset [8], which contains $n = 9298$ grayscale images of size 16×16. The experimental protocol matches that of the synthetic datasets, including 30 independent replications. Figure 2 (left) shows that k-means++10 again consistently achieves the lowest clustering error and saturates after only a few hundred landmarks. Figure 2 (right) visualizes the selected landmark points (reshaped as digit images). The learned landmark set exhibits substantial diversity in writing style, indicating that the sampling strategy captures representative modes of the data. Interestingly, although the ten digit classes have comparable sizes, the learned landmark allocation is highly

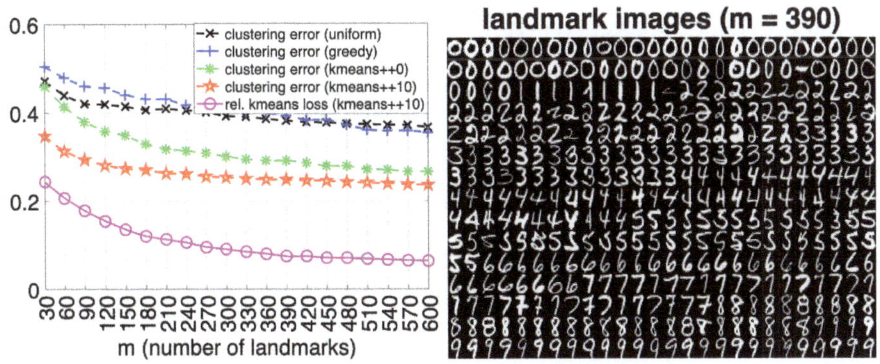

Fig. 2. Results on the USPS digits.

nonuniform: simple digits such as "1" require far fewer landmarks, whereas more variable digits such as "0" require substantially more.

4 Conclusions and Future Work

We presented a k-means++-inspired iterative sampling method for learning an effective landmark set for scalable spectral clustering within the LBDM framework [11]. The proposed algorithm grows the landmark set via a batch k-means++ sampling scheme and evaluates each candidate set using an unsupervised criterion—the relative k-means loss—which has been shown to closely track clustering performance [5]. By monitoring this loss across iterations, Algorithm 2 helps select a compact landmark budget while maintaining clustering accuracy. Experiments on both synthetic and real datasets demonstrate that the procedure converges rapidly and produces well-distributed landmark sets.

This work represents a step toward a general framework for optimal landmark set learning. Although we focused on the combination of LBDM and the relative k-means loss in the setting of Gaussian similarity, the proposed methodology extends naturally to other similarity functions (e.g., cosine similarity), other landmark-based spectral clustering algorithms (e.g., LSC [3]), and alternative unsupervised spectral criteria such as the relative NCut loss [5]. Future work will explore these extensions and further validate the framework.

References

1. Arthur,D., Vassilvitskii, S.: k-means++: the advantages of careful seeding. In: Proceedings of the 18th Annual ACM-SIAM Symposium on Discrete Algorithms, pp. 1027–1035. Society for Industrial and Applied Mathematics. USA (2007)
2. Belabbas, M., Wolfe, P.: On landmark selection and sampling in high-dimensional data analysis. Philos. Trans. R. Soc. A Math. Phys. Eng. Sci. **367**, 4295–4312 (2009)

3. Cai, D., Chen, X.: Large scale spectral clustering via landmark-based sparse representation. IEEE Trans. Cybern. **45**(8), 1669–1680 (2015)
4. Chen, G.: A general framework for scalable spectral clustering based on document models. Pattern Recogn. Lett. **125**, 488–493 (2019)
5. Chen, G., Feldmann, V., Seo, I.: On parameter tuning for spectral clustering: two simple, fast, and effective criteria. IEEE Access **13**, 181686–181703 (2025)
6. Gittens, A., Mahoney, M.: Revisiting the Nyström method for improved large-scale machine learning. J. Mach. Learn. Res. **17**(117), 1–65 (2016)
7. Hong, X., Gao, J., Wei, H., Xiao, J., Mitchell, R.: Two-step scalable spectral clustering algorithm using landmarks and probability density estimation. Neurocomputing **519**, 173–186 (2023)
8. Hull, J.: A database for handwritten text recognition research. IEEE Trans. Pattern Anal. Mach. Intell. **16**(5), 550–554 (1994)
9. Li, C., Jegelka, S., Sra, S.: Fast DPP sampling for Nyström with application to kernel methods. In: Proceedings of the 33rd International Conference on Machine Learning (ICML), vol. 48, pp. 1324–1333. PMLR (2016)
10. Li, H., Ye, X., Imakura, A., Sakurai, T.: Divide-and-conquer based large-scale spectral clustering. Neurocomputing **501**, 664–678 (2022)
11. Pham, K., Chen, G.: Large-scale spectral clustering using diffusion coordinates on landmark-based bipartite graphs. In: Proceedings of the 12th Workshop on Graph-based Natural Language Processing (TextGraphs-12), pp. 28–37. Association for Computational Linguistics, New Orleans (2018)
12. Rafailidis, D., Constantinou, E., Manolopoulos, Y.: Landmark selection for spectral clustering based on weighted pagerank. Futur. Gener. Comput. Syst. **68**, 465–472 (2017)
13. Rouhi, A., Bouyer, A., Arasteh, B., Liu, X.: Two-pronged feature reduction in spectral clustering with optimized landmark selection. Appl. Soft Comput. **161**, 111775 (2024)
14. Shi, J., Malik, J.: Normalized cuts and image segmentation. IEEE Trans. Pattern Anal. Mach. Intell. **22**(8), 888–905 (2000)
15. Wachinger, C., Golland, P.: Sampling from determinantal point processes for scalable manifold learning. In: Ourselin, S., Alexander, D.C., Westin, C.-F., Cardoso, M.J. (eds.) IPMI 2015. LNCS, vol. 9123, pp. 687–698. Springer, Cham (2015). https://doi.org/10.1007/978-3-319-19992-4_54
16. Wang, L., Leckie, C., Kotagiri, R., Bezdek, J.: Approximate pairwise clustering for large datasets via sampling plus extension. Pattern Recognit. **44**, 222–235 (2011)
17. Xue, J., et al.: A comprehensive survey of fast graph clustering. Vicinagearth, **1**(7) (2024)
18. Yan, D., Huang, L., Jordan, M.: Fast approximate spectral clustering. In: Proceedings of the 15th ACM SIGKDD International Conference on Knowledge Discovery and Data Mining, pp. 907–916 (2009)
19. Yang, B., Wu, J., Sun, A., Gao, N., Zhang, X.: Robust landmark graph-based clustering for high-dimensional data. Neurocomputing **496**, 72–84 (2022)
20. Yang, G., Deng, S., Yang, Y., Gong, Z., Chen, X., Hao, Z.: Litewsc: a lightweight framework for web-scale spectral clustering. In: Database Systems for Advanced Applications (DASFAA). LNCS, vol. 13246, pp. 556–573. Springer, Cham (2022)
21. Zhang, X., Zong, L., You, Q., Yong, X.: Sampling for Nyström extension-based spectral clustering: incremental perspective and novel analysis. ACM Trans. Knowl. Discov. Data **11**(1) (2016)

Fast Classification of Textual Richness and Stylistic Beauty

João Cordeiro(✉), Sebastião Pais, and Nuno Pombo

University of Beira Interior, 6201-001 Covilhã, Portugal
{jpcc,sebastiao,ngpombo}@ubi.pt

Abstract. This study examines whether lightweight textual metrics can efficiently characterize textual complexity and stylistic variation across five heterogeneous corpora, including blogs, literary texts, news, and legal documents. One-way ANOVA showed significant differences across all 28 metrics, confirming the descriptive value of low-cost linguistic indicators. In a five-class classification task, XGBoost reached 0.80 accuracy, with clearer separation for EuroLex and greater overlap between the two blog corpora. SHAP-based analyses further indicated that discriminative patterns were class-dependent and that a reduced subset of metrics may be sufficient to capture the most relevant contrasts. Overall, the results support the use of computationally inexpensive metrics for scalable text characterization and motivate future interpretable rule-based classifiers.

Keywords: Text Classification · Linguistic Metrics · Readability · XAI

1 Introduction

The present study examines the performance of computationally efficient metrics for characterizing the *complexity* and *aesthetics* (stylistic quality) across diverse textual corpora. While these two concepts are intuitively recognized by humans, they remain challenging to formalize objectively. What underlying elements render one text more aesthetically pleasing than another? Does a universal pattern exist, and what are the determining factors? The present study does not seek to answer these questions definitively, but rather to offer preliminary insights toward that goal. Drawing upon an established set of textual metrics for evaluating writing complexity, and incorporating novel, promising indicators to total 28 variables, this research evaluates their efficacy in characterizing both textual complexity and aesthetic quality. The dataset comprises corpora across five categories exhibiting varying levels of complexity and aesthetics. Operating on the premise that works by acclaimed authors—including Nobel laureates in Literature—demonstrate a high aesthetic standard, one of the five corpus categories is explicitly dedicated to such literature.

The research aims to reconcile the need for large-scale processing with the accurate assessment of stylistic merit and linguistic depth, focusing on low-cost computational approaches. Although Large Language Models (LLMs) achieve

P. Brito et al. (Eds.): IFCS 2026, STUDIES CLASS, pp. 103–110, 2026.
https://doi.org/10.1007/978-3-032-32009-4_13

state-of-the-art performance in many linguistic tasks, their application is often constrained by substantial computational demands, latency, and operational costs. In real-time or large-scale processing scenarios, lightweight and efficient alternatives are therefore essential. While numerous traditional textual metrics have been proposed, systematic studies comparing and integrating these measures across corpora of varying complexity and aesthetics remain scarce.

2 Related Work

The assessment of text quality and linguistic complexity has long been associated with readability formulas, often developed for specific domains such as health information, where the main concern is accessibility and reading difficulty [13]. Although useful, these approaches offer limited insight into how such metrics behave across markedly different linguistic registers, for example when comparing legal documents with informal online writing.

More recent work has moved beyond classic readability formulas by incorporating richer linguistic features [3]. While these methods often improve predictive performance across diverse textual settings, they also raise interpretability challenges, since the linguistic attributes driving the decisions are not always clear. In parallel, Stajner et al. [12] note that traditional readability measures do not necessarily capture the actual sources of text complexity. Moreover, most studies based on transparent metrics remain centered on accessibility rather than on broader stylistic variation.

Neural approaches have also achieved strong results in tasks such as local coherence [9] and readability assessment [7], but they are typically computationally expensive and less transparent. Our work addresses this gap by prioritizing 28 lightweight and interpretable metrics that can be computed efficiently. By evaluating them across heterogeneous corpora and complementing the analysis with explainability tools, we aim to provide a detailed yet efficient comparison of textual richness spanning formal legal discourse, literary writing, news, and informal blogs.

3 Experimentation

This work evaluates 28 lightweight and computationally efficient metrics for textual analysis, aiming to identify those most suitable for stylistic and complexity-based text categorization. The metrics span four linguistic dimensions: morphological, lexical, syntactic, and semantic. They were tested across five heterogeneous corpora, ranging from news, legal, and literary texts to more informal blog writing. This section describes the corpora, the selected metrics, and the experimental framework illustrated in Fig. 1.

3.1 Corpora

The premise of our analysis is that textual metrics yield varying values across different categories, with certain metrics being more sensitive and suitable for

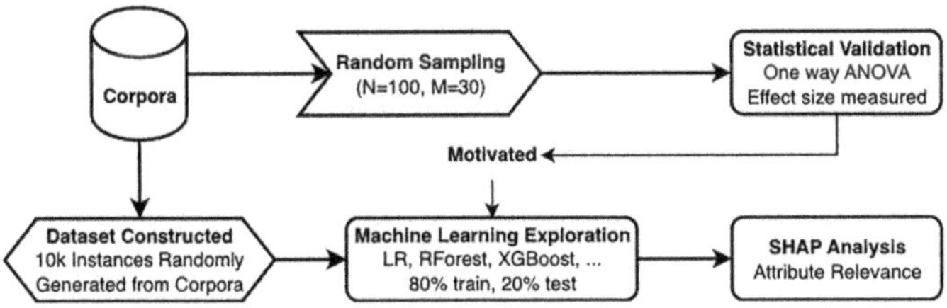

Fig. 1. Workflow of our experimental framework.

measuring specific linguistic phenomena, such as complexity and textual quality. To this end, we curated five heterogeneous corpora to serve as the basis for evaluating the metrics under study and drawing subsequent conclusions.

The texts for **Blogs/13-17** (108 MB) and **Blogs/33-48** (88 MB) were sourced from the study conducted by Schler et Al. [10], representing a less formal and, in the case of **Blogs/13-17**, highly unrefined writing style, serving as proxies for texts of lower quality and complexity. Conversely, the **Books/EN** corpus (16 MB) comprises a collection of literary works, including several authored by Nobel Prize laureates in Literature, such as José Saramago (1998), Orhan Pamuk (2006), Alice Munro (2013), or Kazuo Ishiguro (2017). Consequently, we posit that this corpus represents the highest tier of textual aesthetics. The **News** corpus (70 MB) also exhibits a relatively high linguistic standard, as it is produced by professional journalists. Regarding complexity, it is assumed to be comparable to the **Books/EN** level; however, it lacks the same aesthetic depth, as the latter consists of literary works intentionally crafted for artistic value and widely acclaimed by the general public. Finally, the **EuroLex** corpus (265 MB) consists of legal documents sourced from the European Parliament. This corpus represents the highest level of structural and technical complexity, characterized by highly formal language, specialized terminology, and dense syntactic structures, yet it lacks aesthetic concerns.

3.2 Textual Metrics

The total set of 28 evaluated metrics is organized into four fundamental linguistic categories: morphological, lexical, syntactic, and semantic, as organized in Table 1. This multidimensional approach aims to ensure comprehensive coverage of diverse linguistic phenomena, enabling the identification of quantitative patterns that distinguish structural complexity and aesthetic quality levels across texts of different natures. Each metric is defined herein as a quantitative function that takes a text block as input and returns a numerical value, facilitating the automated and objective categorization of the corpora under analysis.

The **basic** category comprises elemental textual metrics typically utilized in analysis: number of chars, tokens, and sentences, alongside vocabulary size, vocabulary-to-token ratio, and proportion of alphabetic chars. The **readability** category quantifies the cognitive effort and educational level required for com-

Table 1. Classification of the 28 textual metrics across linguistic categories. The range is provided, near each metric.

Basic	Readability	Syntactical	Semantical
Chars, Tokens, Sentences (all in $\mathbb{N}$),	ColemanLiauIndex $\{0...20\}$, FleshKincaidGrade $\{0...18\}$, FleshKincaidEase $\{0...100\}$,	pADJ, pADV, pDET, pNOUN, pPRON, pPROPN, pSCONJ, pVERB (all p* in $[0, 1]$)	Coherence1 $[0, 1]$, Coherence2 $[0, 1]$
Vocab (in $\mathbb{N}$),	GunningFog $\{0...20\}$,	LexicalDensity $[0, 1]$,	SentimentAverage $[-1, 1]$
AlphaRatio $[0, 1]$,	SMOGA $\{3...18\}$, ARI $\{1...14\}$	DepDistMean (in $\mathbb{R}^+$),	
LexicalDiversity $[0, 1]$	LIX, RIX (both in $\mathbb{R}_0^+$)	DepthAverage (in $\mathbb{R}^+$)	

prehension. We evaluate traditional syllable-based formulas [5,6,8] and include efficient character-based metrics [1,2,11] that reduce processing overhead while maintaining a high correlation with human-graded difficulty. Regarding **syntactic** metrics, we utilize *part-of-speech tagging* to calculate the proportion of specific morphosyntactic categories (denoted by the "p" prefix, e.g., pVERB) and lexical density (% content words). We also assess structural complexity through DepthAverage (maximum syntactic tree depth) and DepDistMean (mean distance between dependent nodes), which employ slightly more demanding parsing resources. The **semantic** category captures the information flow and emotional tone. Coherence1 and Coherence2 assess local semantic cohesion between consecutive sentences. Finally, SentimentAverage characterizes the document's overall affective valence by averaging the sentence-level sentiment scores.

3.3 Measurement Protocol

In our study, the process of measuring the value of metric f within a specific corpus C is achieved through a bootstrap resampling strategy, where we randomly sampled $N = 100$ text blocks (each comprising $M = 30$ sentences) with replacement from the corpus' texts, calculating N measurements of f. The final value $\bar{f}$ for a given metric f within C is simply the mean of these N bootstrap iterations. This resampling approach ensures statistical robustness and mitigates the influence of local outliers, providing a more reliable estimation of the underlying textual complexity. The selection of $M = 30$ sentences (approximately 600 tokens) was established to provide an optimal balance: it represents a context window large enough to capture stable syntactic and semantic patterns, yet sufficiently constrained to maintain computational efficiency and prevent the over-smoothing of local stylistic variations. This segment size strongly aligns with established empirical practices in computational stylometry and text classification [4].

3.4 Mining the Metrics

The results from the ANOVA test (Sect. 4.1) motivated us to conduct additional experiments with machine learning (ML) algorithms to better understand which combination of features (metrics) is more promising for characterizing text complexity and aesthetics. Therefore, in this ML experiment, we have generated data instances based on the values of the 28 text metrics, measured in text blocks randomly sampled as described in Sect. 3.3. Each text block yields a data instance from the class where the block was taken (News, Books/EN, EuroLex, etc.), as illustrated here:

Corpus/Class	SMOGA	ARI	DepthAverage	GunFog	Coherence1	pPRON	pVERB	...
EuroLex	18.135	15.680	8.500	15.134	0.525	0.029	0.079	...
Blogs/13-17	8.936	3.795	5.500	5.278	0.623	0.087	0.121	...
Books/EN	8.131	11.309	7.400	11.774	0.929	0.179	0.123	...
Blogs/33-48	11.004	9.274	7.833	8.984	0.802	0.147	0.101	...
News	12.101	7.938	5.867	11.114	0.795	0.072	0.097	...

We have generated a dataset of this kind comprising 10,000 instances, with 2,000 samples taken from each class. These are new random blocks, different from the 500 used in the ANOVA. With this data, we evaluated different classifiers to identify which attributes (metrics) contribute most significantly to the classification across our five classes, which represent corpora from different categories. Section 4 provides a detailed account of the classification process taken and the subsequent analysis of metrics relevancy.

4 Results

4.1 The One-Way ANOVA Test on Textual Metrics

In an initial attempt to assess the importance of each metric in identifying the text categories, we recorded the mean and standard deviation of 100 measurements. Accordingly, we conducted a one-way ANOVA test to compare the means for the 28 metrics considered. For each metric, a one-way ANOVA was performed to test differences across corpora (N = 100 per corpus). The *p-values* were adjusted using the *Benjamini–Hochberg* procedure, and the effect sizes were quantified using η^2. All metrics show statistically significant differences after *Benjamini–Hochberg* correction ($p_{adj} < 0.05$). Effect sizes, measured by η^2, are consistently large, with values ranging from 0.03 (pADJ) to 0.67 (pPRON) (mean $\eta^2 = 0.28$), indicating that corpus type explains a substantial proportion of the variance in most metrics. Given the large sample size, effect sizes were prioritized over *p-values* in the interpretation of results.

4.2 Metrics Mining Experiments

Using the dataset of 10,000 instances, generated as described in Sect. 3.4, we tested several data mining algorithms, including *Logistic Regression*, *Random Forest*, *K-Means Clustering*, and XGBoost, to explore the classification potential of the metrics. The dataset was split into 80% training and 20% test sets using stratified sampling, so that the class proportions were preserved in both subsets. No hyperparameter tuning was performed. XGBoost was trained using the default settings, and no additional validation strategy was used for hyper-parameter selection.

We have observed that XGBoost achieves a high accuracy (80%) in this five-class task, with equally high F1-score values (70% in Blogs/13-17, 57% in Blogs/33-48, 86% in Books/EN, 85% in News, and 98% in Books/EN), with Macro-averaged F1 of 79%, thus demonstrating that the combination of metrics is promising for text classification. Its performance lowers in blog text, especially on the Blogs/33-48 class. To better understand this behavior, the corresponding *Confusion Matrix* is presented in Fig. 2. We can observe that the primary source of error lies in misclassifying Blogs/33-48 as other categories, most notably as Blogs/13-17, likely due to their shared stylistic features as blog-based corpora. Conversely, the lowest misclassification rate occurs with the EuroLex class, which is consistent with its significantly different textual nature.

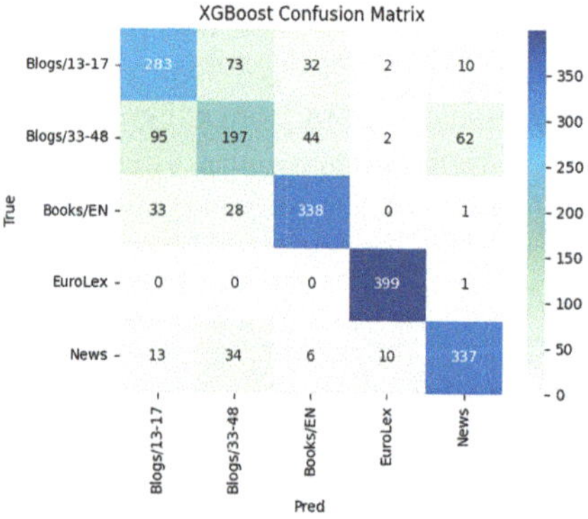

Fig. 2. XGBoost Confusion Matrix.

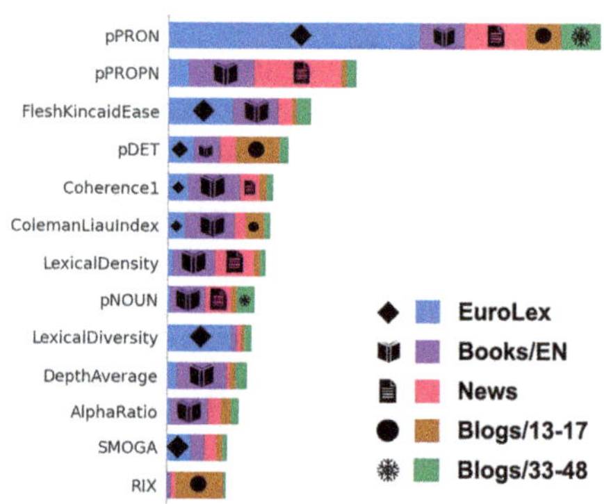

Fig. 3. Class-specific SHAP feature importance across relevant textual metrics.

Furthermore, we leveraged our subsequent analysis on SHAP (*SHapley Additive exPlanations*), a model-interpretability method that quantifies how much each input feature (metrics) contributes to an individual prediction and to the model's overall behavior. Thus, the stacked bar chart, shown in Fig. 3, summarizes the contribution of the most influential features to the discrimination of the five corpus classes by the XGBoost model. Each bar corresponds to one feature, and each colored segment represents its relative contribution to a specific class (EuroLex, Books/EN, News, Blogs/13-17, and Blogs/33-48). The plot highlights that some metrics, particularly `pPRON`, contribute unevenly across classes, indicating stronger class-specific discriminative effects.

As a compact alternative to multiple instance-level waterfall plots, we computed a heatmap (Fig. 4) that condenses the signed SHAP contribution profiles for the five classes using the most relevant features. The class-specific signed SHAP heatmap revealed marked asymmetries in how textual metrics contributed to the prediction of each corpus category. Among all features, pPRON showed the strongest class-specific effect, with a particularly large negative contribution for EuroLex, indicating that this variable systematically influenced the model's discrimination of highly formal legal texts. More moderate negative contributions of pPRON were observed across the remaining corpora, suggesting that pronoun-related usage is broadly informative, but especially distinctive for EuroLex.

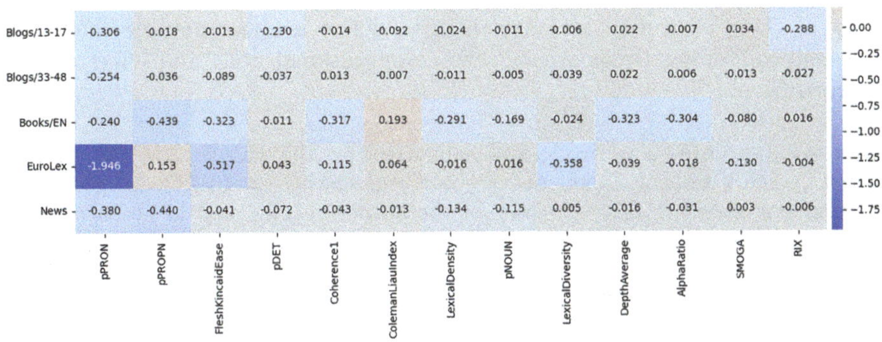

Fig. 4. Mean signed SHAP contribution for relevant metrics.

Books/EN displayed the most distributed pattern of contributions across features, with consistently negative effects from pPROPN, FleshKincaidEase, Coherence1, LexicalDensity, pNOUN, DepthAverage, and AlphaRatio, while ColemanLiauIndex showed a positive contribution. This suggests that the classification of literary texts depends on a broader combination of lexical, structural, and readability-related features, rather than on a single dominant metric.

News also showed a relatively coherent pattern, with negative contributions from pPRON and pPROPN standing out, alongside smaller negative effects from LexicalDensity and pNOUN. By contrast, Blogs/33–48 exhibited generally weaker and more diffuse contributions, with most feature effects remaining close to zero, suggesting less sharply defined separation in the feature space. Blogs/13–17 showed stronger effects than Blogs/33–48, particularly for pPRON, pDET, and RIX, indicating that these metrics may better capture the stylistic profile of younger-blog texts.

5 Conclusion

This study investigated lightweight textual metrics as efficient indicators of textual complexity and stylistic quality across five heterogeneous corpora. Statistical analysis showed that all 28 metrics differed significantly across corpus categories, confirming that computationally inexpensive linguistic descriptors can capture meaningful stylistic variation. However, the results also suggest that a smaller subset of features may characterize these differences more effectively, particularly those related to pronoun and proper-noun usage, readability, lexical density, and structural coherence. The SHAP analysis further showed that the contribution of the most relevant metrics varies across corpora. EuroLex was influenced most strongly by `pPRON`, while Books/EN exhibited a more distributed contribution pattern involving readability, lexical, and structural features. In News, the relevant metrics contributed in a more balanced and consistent way, whereas the two blog corpora were associated with weaker and more diffuse metric contributions.

In future work, we plan to focus on a reduced subset of approximately five or six of these indicators and use them to define an interpretable rule-based classifier, combining transparency, low computational cost, and effective text categorization.

Acknowledgments. This work was supported by the COOPERHATE project, funded by the European Union under Grant Agreement No. 101213938. Views and opinions expressed are however those of the author(s) only and do not necessarily reflect those of the European Union.

References

1. Anderson, J.: Lix and rix: variations on a readability index. J. Read. **26**(6), 490–496 (1983)
2. Coleman, M., Liau, T.L.: A computer readability formula designed for machine scoring. J. Appl. Psychol. **60**(2), 283 (1975)
3. Crossley, S.A., Skalicky, S., Dascalu, M.: Moving beyond classic readability formulas: new methods and new models. J. Res. Read. **42**(3–4), 541–561 (2019)
4. Eder, M.: Does size matter? Authorship attribution, small samples, big problem. Digital Sch. Humanit. **30**(2), 167–182 (2015)
5. Flesch, R.: A new readability yardstick. J. Appl. Psychol. **32**(3), 221 (1948)
6. Gunning, R.: The Technique of Clear Writing. McGraw-Hill, New York (1952)
7. Martinc, M., Pollak, S., Robnik-Šikonja, M.: Supervised and unsupervised neural approaches to text readability. Comput. Linguist. **47**(1), 141–179 (2021)
8. McLaughlin, G.H.: Smog grading-a new readability formula. J. Read. **12**(8), 639–646 (1969)
9. Mesgar, M., Strube, M.: A neural local coherence model for text quality assessment. In: Riloff, E., Chiang, D., Hockenmaier, J., Tsujii, J. (eds.) Proceedings of the 2018 Conference on Empirical Methods in Natural Language Processing, pp. 4328–4339. Association for Computational Linguistics, Brussels (2018). https://aclanthology.org/D18-1464/
10. Schler, J., Koppel, M., Argamon, S., Pennebaker, J.W.: Effects of age and gender on blogging. In: AAAI Spring Symposium: Computational Approaches to Analyzing Weblogs, vol. 6, pp. 199–205. Stanford, CA, USA (2006)
11. Senter, R.J., Smith, E.A.: Automated readability index. Technical report, Aerospace Medical Research Laboratories (1967)
12. Štajner, S., Evans, R., Orasan, C., Mitkov, R.: What can readability measures really tell us about text complexity. In: Proceedings of Workshop on Natural Language Processing for Improving Textual Accessibility, pp. 14–22 (2012)
13. Wang, L.W., Miller, M.J., Schmitt, M.R., Wen, F.K.: Assessing readability formula differences with written health information materials: application, results, and recommendations. Res. Soc. Admin. Pharm. **9**(5), 503–516 (2013)

Sparse Clustering via the Deterministic Information Bottleneck Algorithm

Efthymios Costa[1(✉)], Ioanna Papatsouma[1], and Angelos Markos[2]

[1] Department of Mathematics, Imperial College London, London SW7 2AZ, UK
{efthymios.costa17,i.papatsouma}@imperial.ac.uk

[2] School of Education, Department of Primary Education, Democritus University of Thrace, 68132 Nea Hili, Alexandroupoli, Greece
amarkos@eled.duth.gr

Abstract. Cluster analysis relates to the task of assigning objects into groups which ideally present some desirable characteristics. When a cluster structure is confined to a subset of the feature space, traditional clustering techniques face unprecedented challenges. We present an information theoretic framework that overcomes the problems associated with sparse data, allowing for joint feature weighting and clustering. Our proposal constitutes a competitive alternative to existing clustering algorithms for sparse data, as demonstrated through simulations on synthetic data. The effectiveness of our method is established by an application on a real-world genomics data set.

Keywords: cluster analysis · information theory · information bottleneck · sparse data · high-dimensional data

1 Introduction

Data exhibiting feature-level sparsity, where the relevant signal resides in a small subset of variables, is nowadays commonly encountered in several research areas. For instance in bioinformatics, gene expression data is inherently high-dimensional and sparse, or in chemometrics, the analysis of spectra for material characterisation or product quality control typically involves very few samples and a large number of features. Clustering algorithms can be very useful for assigning objects into groups based on their similarities. However, they are typically accounting for all the variables included in a data set, thus undermining the interpretability of the obtained partition when just a subset of the features is informative about an existing lower-dimensional cluster structure. This inclusion of uninformative variables can obscure the underlying signal and mislead the clustering process, resulting in incorrect partitions. Moreover, distance-based clustering algorithms suffer from the curse of dimensionality for large enough numbers of variables, whereas model-based techniques struggle with singularity issues when the number of features exceeds that of samples.

P. Brito et al. (Eds.): IFCS 2026, STUDIES CLASS, pp. 111–119, 2026.
https://doi.org/10.1007/978-3-032-32009-4_14

In this paper, we present an information-theoretic algorithm for clustering sparse and potentially high-dimensional data. The proposed approach is based on the Information Bottleneck algorithm (IB) [1] and extends its application to cluster analysis that was presented in [2]. The rest of the paper is organised as follows; in Sect. 2 we present the deterministic variant of IB and how it can be used for clustering. Section 3 extends the clustering framework to account for feature-level sparsity, allowing for feature weighting to be performed simultaneously with the clustering. Simulations on synthetic data are presented in Sect. 4, where the proposed method is compared to six algorithms specifically designed for sparse clustering. We illustrate the applicability of our approach on a genomics data set in Sect. 5. We finally give some concluding remarks and future study recommendations in Sect. 6.

2 Deterministic Information Bottleneck Clustering

Traditional clustering methods rely on geometric assumptions or arbitrary distance metrics. Information-theoretic alternatives frame clustering as an optimisation problem, seeking compressed representations that retain maximal information. A characterisation of this compression-relevance tradeoff that produces a crisp partition can be obtained using the deterministic variant of the Information Bottleneck algorithm (DIB) [3]. This Section summarises these ideas for continuous data.

In the clustering paradigm, the random variables Y, T, and X are used to model the observed feature values, the cluster assignment, and the observation index (treated as uninformative). The DIB algorithm assumes that T and Y are conditionally independent given X and seeks a deterministic encoder $q^*(t \mid x)$ satisfying:

$$q^*(t \mid x) = \underset{q(t|x)}{\arg\min}\, H(T) - \beta I(Y;T), \tag{1}$$

where $H(T)$ is the entropy of T (cluster compression measure), $I(Y;T)$ the mutual information (cluster relevance measure), and $\beta > 0$ controls the compression-relevance tradeoff. Solving this optimisation problem yields a partition where each cluster is defined as a group of points whose feature distributions are well approximated by a common prototype $q(\boldsymbol{y} \mid t)$, with assignments driven by information preservation rather than geometric distance. Following [2], we search for a partition maximising $I(Y;T)$ with β updated dynamically to prevent empty clusters, essentially treating $H(T)$ as a regularisation term for cluster imbalance.

The core of DIB is the perturbed similarity matrix $\mathbf{P}'_{Y|X} \in \mathbb{R}^{n\times n}$, with entries $[\mathbf{P}'_{Y|X}]_{i,j} = p(Y = \boldsymbol{y}_i \mid X = x_j)$ representing probabilistic proximities. Density estimation uses product kernels [4]:

$$p(Y = \boldsymbol{y}_i \mid X = x_j) \propto \prod_{m=1}^{p} K_m(y_{i,m}, y_{j,m}; \lambda_m),$$

where $K_m(\cdot, \cdot; \lambda_m)$ is a kernel function (we use the Gaussian kernel) with bandwidth λ_m, chosen via the criterion from [2] or set manually. The algorithm iteratively updates cluster distributions and allocates observations to clusters maximising $\log[q(t)] - \beta D_{KL}[p(\boldsymbol{y}_i \mid x) || q(\boldsymbol{y}_i \mid t)]$, where $D_{KL}(\cdot||\cdot)$ is the Kullback-Leibler (KL) divergence. The term $q(t) = \sum_j q(t \mid x_j)p(x_j)$ is the mass function of the random variable describing the cluster masses, whereas $q(\boldsymbol{y}_i \mid t) = \sum_j q(t \mid x_j)p(\boldsymbol{y}_i \mid x_j)p(x_j)/q(t)$ quantifies the expected feature distribution of points in cluster t. The latter is optimised when the cluster-conditional density estimator matches the feature distribution of points in that cluster, rendering a higher cluster relevance term $I(Y;T)$, or equivalently, a lower KL divergence. Multiple random initialisations are used and the solution yielding the highest mutual information value is selected.

3 An Extension for Sparse Clustering

In sparse clustering, only a small proportion of variables is informative about the underlying cluster structure. Standard techniques like K-Means or DIB assume all variables are equally informative, which is rarely true in practice. We extend DIB to incorporate feature weighting; feature selection is a special case, with zero weights corresponding to unselected variables.

The Sparse DIB optimisation problem becomes:

$$q_W^*(t \mid x) = \underset{q_W(t|x),\mathbf{w}}{\arg\min}\ H(T) - \beta I(Y_W;T), \tag{2}$$
$$\text{subject to } \|\mathbf{w}\|_2 \leq 1,\ \|\mathbf{w}\|_1 \leq u,\ w_j \geq 0\ \forall 1 \leq j \leq p,$$

where Y_W is the weighted random variable and $\mathbf{w} \in \mathbb{R}^p$ are feature weights constrained to the first orthant of the unit L_2-ball with an L_1 constraint controlled by sparsity parameter $u > 0$. The weighted perturbed similarity matrix is:

$$\left[\mathbf{P}'^{W}_{Y|X}\right]_{i,j} \propto \prod_{m=1}^{p} K_m(y_{i,m}, y_{j,m}; \lambda_m)^{w_m},$$

with weights introduced exponentially to control each variable's contribution. Note that for some kernel functions such as the Gaussian, this reduces to bandwidth rescaling ($\lambda_m \leftarrow \lambda_m/\sqrt{w_m}$) but we keep the exponentiated form for broader applicability. The algorithm (summarised in Algorithm 1) alternates between obtaining cluster assignments via DIB and updating weights until these converge (we set a convergence tolerance threshold $\epsilon = 10^{-5}$). Weights are updated using Dykstra's projection algorithm [5] by projecting onto the feasible set $\mathcal{C} = \{\mathbf{w} : \|\mathbf{w}\|_2 \leq 1, \|\mathbf{w}\|_1 \leq u, \mathbf{w} > \mathbf{0}\}$, giving us a vector of weights $\mathbf{w}$ that satisfies both L_1 and L_2 constraints.

Weights can be initialised uniformly ($1/\sqrt{p}$) or via warm start from an initial K-Means solution. To tune u, we run the algorithm across multiple values and plot the normalised entropy of weights against u; a plateau indicates the range of suitable values for which the same features are selected as informative.

4 Simulation Study

We conduct a simulation study in which Sparse DIB is benchmarked against six alternative clustering methods for sparse data, namely Sparse K-Means [6], Random Projection Ensemble Clustering (RPEClust)[7], model-based clustering with variable selection (VarSelLCM) [8], Clustering On Subsets of Attributes followed by Partitioning Around Medoids (COSA/PAM) [9,10] and Principal Component Analysis (PCA) and its sparse extension [11], followed by K-Means. We experimented with several hyperparameter values for each method and selected solutions that maximised the ARI. Uniform initialisation was used for Sparse DIB. Code for reproducing the simulations is publicly available in https://github.com/EfthymiosCosta/SparseDIB_Sims.

We generated synthetic data following a Gaussian mixture model on $\rho = \lfloor p \times \mathrm{q} \rfloor$ informative features (q being the proportion of informative features), with the remaining $p - \rho$ features drawn independently from standard Gaussians. We fixed $n = 200$ and varied $p \in \{100, 200, 400, 1000\}$, $\mathrm{q} \in \{0.05, 0.10, 0.20, 0.50\}$, and $K \in \{3, 5, 8\}$, with balanced/unbalanced and spherical/elliptical configurations (192 settings, 50 replicates each). Performance was assessed using the Adjusted Rand Index (ARI) [12] and Adjusted Mutual Information (AMI) [13].

According to Fig. 1, the most successful methods in recovering cluster structure are VarSelLCM and dimension reduction approaches, with Sparse PCA performing best overall. Sparse DIB performs similarly to Sparse K-Means (mean ARI/AMI: 0.88/0.89 vs 0.91/0.92), highlighting their methodological similarities. COSA/PAM and RPEClust perform suboptimally, with the former never achieving perfect recovery. Notably, Sparse DIB outperforms competitors when ρ is very small (e.g., $p = 100$, $\mathrm{q} = 0.05$); as q increases, most methods converge to near-perfect recovery. We omit the results for K, cluster imbalance, and sphericity for brevity, as performance variations were dominated by changes in p and q.

We finally investigate the effect of the sparsity constraint value u to the performance of Sparse DIB. Recalling that a cluster structure was deliberately defined on a fixed number of variables, we assess whether these features are correctly assigned non-zero weights. Figure 2 shows that the heuristic method described in Sect. 3 identifies the true number of relevant variables in most scenarios. The only exception is probably $p = 1000$ and $\mathrm{q} = 0.5$, where values of u greater than ten would be needed for a clearer picture, hinting that a large number of simulations may be needed to be confident about the choice of u.

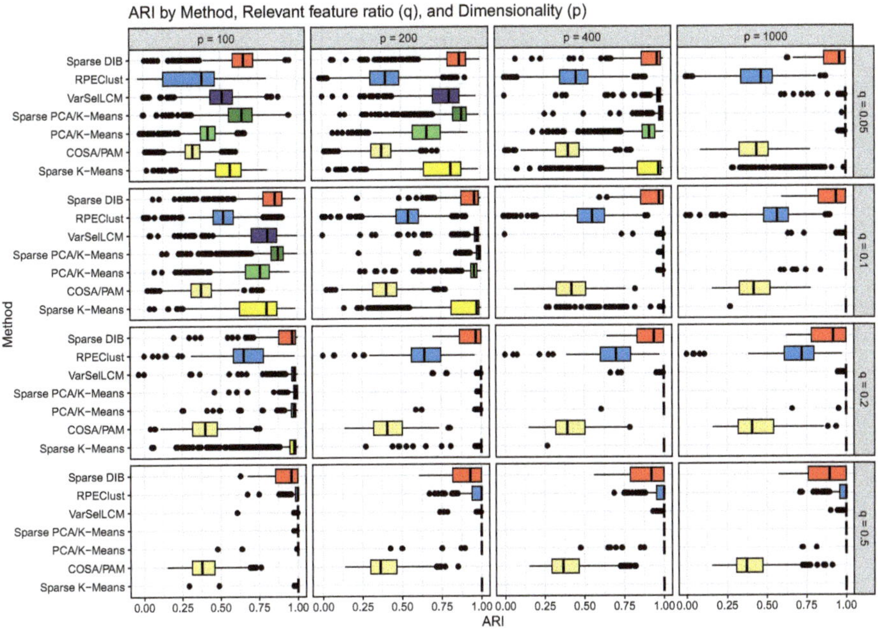

Fig. 1. Box plots of ARI values across different dimensionalities (p) and ratios of relevant features (q) for each method.

Algorithm 1. Sparse DIB clustering

Input: Data $\mathbf{X} \in \mathbb{R}^{n \times p}$, sparsity constraint u, max iterations $m^{\max}$, threshold ϵ
Output: Cluster assignment $q^*_W(t \mid x)$, weights $\mathbf{w}^*$

Initialise $\mathbf{w}^{(0)}$ (uniform or warm start) and set $m \leftarrow 1$
while not converged and $m \leq m^{\max}$ **do**
    Obtain $q^{(m)}_W(t \mid x)$ using DIB with weights $\mathbf{w}^{(m-1)}$
    Update $\mathbf{w}^{(m)}$ by setting $w_j \propto I(Y_j; T)$ using Dykstra's algorithm
    Check convergence; $\sum_j |w_j^{(m)} - w_j^{(m-1)}| / \sum_j |w_j^{(m-1)}| < \epsilon$ and set $m \leftarrow m + 1$
end while

---

5 Application on Bladder Cancer Data

Biomedical data are often sparse and high-dimensional, particularly in cancer genomics where molecular features far outnumber available samples. To demonstrate our approach under such conditions, we analysed RNA-seq gene expression data from The Cancer Genome Atlas (TCGA) bladder cancer cohort (BLCA), accessed via the Genomic Data Commons portal [14]. Expression values (transcripts per million, TPM) were $\log_2(\text{TPM}+1)$ transformed. We retained samples with molecular subtype annotations and filtered to protein-coding genes with variance above 0.01, yielding 412 samples and 18193 genes. The five molecular subtypes were aggregated into three classes: Basal (144 samples), Luminal (248 samples), and Neuronal (20 samples). Sparse DIB was run using a warm start (K-Means with 1000 random initialisations) for sparsity values u from 0.1 to 100. Results are summarised in Table 1.

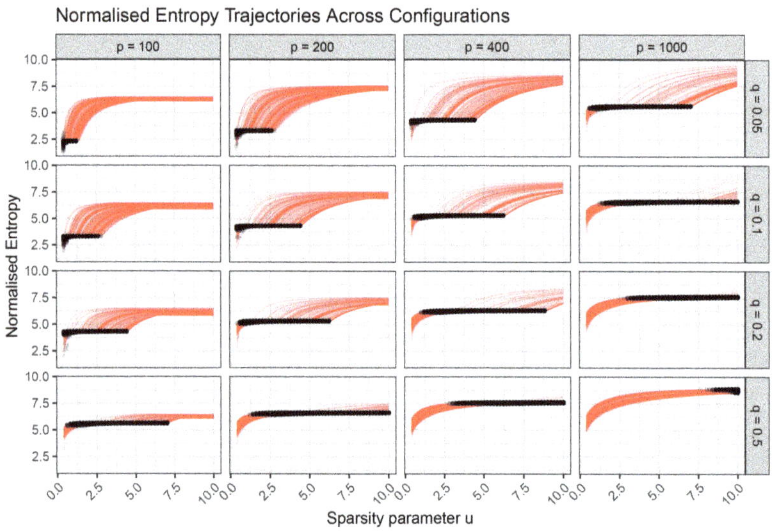

Fig. 2. Trajectories of normalised entropy for sparsity parameter $u \in [0.4, 10]$. The black points mark the value of u and the corresponding normalised entropy where the number of non-zero weights is closest to $\rho = \lfloor p \times \mathrm{q} \rfloor$.

The RPEClust algorithm yields the highest ARI with 0.73, followed by Sparse DIB and COSA/PAM. It is important to point out that the clustering problem on the bladder cancer data set is more of a subspace clustering problem, with clusters being defined on different subsets of features; this justifies the relative success of COSA. The main strengths of Sparse DIB are that it produces the second highest ARI score and also selects a subset of features, unlike the top-performing RPEClust. This enables an interpretation of the obtained partition. Sparse K-Means fails to select a subset of features with a higher ARI compared to when all variables are included, whereas VarSelLCM seems to identify all variables as redundant and returns a random partition. Figure 3 illustrates the 94 genes selected by Sparse DIB using a sparsity constraint value of $u = 3.3$. Of these 94 genes, twelve are known luminal markers, two are basal markers, and one is a neuronal marker, with the remaining 79 genes representing novel discriminative features or genes involved in supporting biological processes [15]. The weight distribution reflects the optimisation objective; Sparse DIB prioritises the largest and most heterogeneous class, namely luminal subtypes (approximately 60% of the samples) by assigning substantially higher weights to luminal-discriminative features. The four uroplakins (UPK1A, UPK2, UPK3A, UPK3B), which are

Table 1. ARI scores and number of selected features for sparse clustering algorithms applied to the cleaned bladder cancer data set.

Algorithm	ARI	Selected Features
Sparse DIB	0.64	94
RPEClust	0.73	18193
VarSelLCM	0.14	0
Sparse PCA/K-Means	0.33	316
PCA/K-Means	0.23	18193
COSA/PAM	0.50	18193
Sparse K-Means	0.46	18193

bladder-specific urothelial differentiation markers, collectively account for nearly 40% of the total weight budget (combined weight 0.395), with UPK2 ranking third overall. Additional high-weight luminal markers include key transcription factors (GATA3, FOXA1, GRHL3, ELF3) and PPARG, a central regulator of luminal differentiation. In contrast, basal markers (S100P, TBX2) and the neuronal marker (SNCG) receive substantially lower weights which is still sufficient for discriminating these smaller and more molecularly homogeneous classes. This weight allocation is consistent with the information-theoretic objective of the algorithm, features that reduce uncertainty about the dominant luminal class contribute more to overall clustering quality. Notably, the canonical luminal marker KRT20 distinguishes luminal-papillary from other luminal subtypes and was not selected, demonstrating that the method correctly avoids features that introduce within-class heterogeneity in the aggregated three-class task.

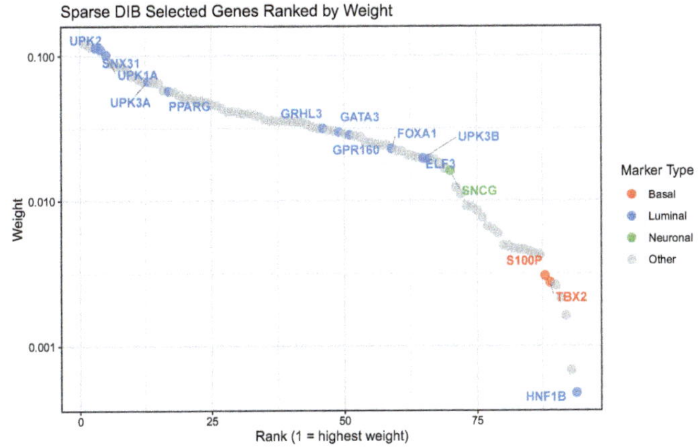

Fig. 3. Genes with non-zero weights selected by Sparse DIB marked by type.

6 Conclusion

This paper introduced Sparse DIB, an extension of the DIB clustering framework to high-dimensional sparse data. Our simulation study demonstrated that Sparse DIB is a competitive alternative to high-dimensional clustering methods, including Sparse K-Means and latent class modelling with variable selection. This competitiveness was further confirmed on a real-world genomics data set of bladder cancer patients, where Sparse DIB performed reasonably well in differentiating between cancer subtypes, while selecting a small and interpretable subset of features. Notably, several of the selected genes have known clinical relevance, underscoring the practical utility of our approach.

Despite these promising results, the use of mutual information for simultaneous clustering and feature weighting warrants deeper theoretical investigation. Future work includes extending the proposed framework to enable sparse hierarchical agglomerative clustering, building on information-theoretic merging criteria [16]. Additionally, developing a version that incorporates cluster-specific feature weights would allow structures to be defined on different feature subsets, thus enhancing flexibility. Finally, a promising direction is to develop an extension for high-dimensional mixed-type data (e.g. combining genetic information with clinical variables), thus creating a unified tool for extracting coherent clusters from complex datasets with heterogeneous features.

References

1. Tishby, N., Pereira, F.C., Bialek, W.: The information bottleneck method In: Proceedings of the 37th Annual Allerton Conference on Communication, Control and Computing, pp. 368–377 (1999)
2. Costa, E., Papatsouma, I., Markos, A.: A deterministic information bottleneck method for clustering mixed-type data. Pattern Recogn. (2026)
3. Strouse, D.J., Schwab, D.J.: The deterministic information bottleneck. Neural Comput. **29**(6), 1611–1630 (2017)
4. Li, Q., Racine, J.: Nonparametric estimation of distributions with categorical and continuous data. J. Multivar. Anal. **86**(2), 266–292 (2003)
5. Dykstra, R.L.: An iterative procedure for obtaining i-projections onto the intersection of convex sets. Ann. Probab. 975–984 (1985)
6. Witten, D.M., Tibshirani, R.: A framework for feature selection in clustering. J. Am. Stat. Assoc. **105**(490), 713–726 (2010)
7. Anderlucci, L., Fortunato, F., Montanari, A.: High-dimensional clustering via random projections. J. Classif. **39**(1), 191–216 (2022)
8. Marbac, M., Sedki, M.: Variable selection for model-based clustering using the integrated complete-data likelihood. Stat. Comput. **27**(4), 1049–1063 (2017)
9. Friedman, J.H., Meulman, J.J.: Clustering objects on subsets of attributes (with discussion). J. R. Stat. Soc. Ser. B Stat Methodol. **66**(4), 815–849 (2004)
10. Kaufman, L., Rousseeuw, P.J.: Finding Groups in Data: an Introduction to Cluster Analysis. Wiley, Hoboken (1990)
11. Zou, H., Hastie, T., Tibshirani, R.: Sparse principal component analysis. J. Comput. Graph. Stat. **15**(2), 265–286 (2006)

12. Hubert, L., Arabie, P.: Comparing partitions. J. Classif. **2**(2), 193–218 (1985)
13. Vinh, N.X., Epps, J., Bailey, J.: Information theoretic measures for clusterings comparison: variants, properties, normalization and correction for chance. J. Mach. Learn. Res. **11**(95), 2837–2854 (2010)
14. Cancer Genome Atlas Research Network et al.: Comprehensive molecular characterization of urothelial bladder carcinoma. Nature **507**(7492), 315 (2014)
15. Robertson, A.G., et al.: Comprehensive molecular characterization of muscle-invasive bladder cancer. Cell **171**(3), 540–556 (2017)
16. Slonim, N., Tishby, N.: Agglomerative information bottleneck. Adv. Neural. Inf. Process. Syst. **12** (1999)

Robust Clustering of Cylindrical Objects

Houyem Demni[1,2](✉)

[1] University of Cassino and Southern Lazio, 03043 Cassino, Italy
houyem.demni@unicas.it
[2] EUt+ Data Science Research Group, European University of Technology, European Union, Cassino, Italy

Abstract. Cylindrical data arise in many scientific domains and consist of observations combining one periodic angular variable with one or more linear variables. Clustering methods designed for Euclidean spaces fail to respect the periodic nature of angular variables, often resulting in artificial boundaries and distorted cluster structures. Although geometry-aware extensions of centroid-based algorithms have been developed for cylindrical data, medoid-based approaches remain comparatively under-explored despite their advantages in interpretability and robustness. This work proposes a robust medoid-based clustering framework that integrates appropriate circular-linear dissimilarities with robust dispersion measures, yielding an effective method for clustering cylindrical data in the presence of outliers.

Keywords: Angular and linear variables · Circular Median Absolute Deviation · Directional statistics · Partitioning Around Medoids

1 Introduction

Cylindrical data arise when observations combine an angular variable $\theta \in \mathbb{S}^1$ with one or more real-valued components, yielding a sample space of the form $\mathbb{S}^1 \times \mathbb{R}^p$. Such representations arise naturally in a variety of applications, including metrology, environmental sciences, medical imaging, engineering, and color spaces (see e.g. [1,12]).

Clustering methods partition observations into groups with high within-cluster similarity and high between-cluster dissimilarity, as defined by a suitable distance or dissimilarity. A widely used approach is the k-means algorithm [4], which minimizes squared Euclidean distances to cluster centroids. A common alternative, adopted here, represents clusters by medoids, i.e., observed data points selected as cluster representatives, with a well-known method in this class being Partitioning Around Medoids (PAM) [14].

Clustering methods are typically designed for Euclidean data. If applied to cylindrical data as they are (i.e., by embedding the cylinder in $\mathbb{R}^{p+1}$), they ignore the periodic nature of the angular component. Since angles differing by integer multiples of 2π represent the same point, Euclidean distance may artificially separate nearby observations on the circle and distort the resulting cluster

P. Brito et al. (Eds.): IFCS 2026, STUDIES CLASS, pp. 120–127, 2026.
https://doi.org/10.1007/978-3-032-32009-4_15

structure. This issue is further amplified by the presence of outliers, which may affect both angular and linear components. In the circular setting, several robust methods have been proposed, including weighted likelihood and minimum disparity estimators for circular location and scale [2], extensions to multivariate circular data on the torus [7], and, more recently, robust measures of circular dispersion for outlier detection [5].

Approaches to cluster cylindrical objects are available. Model-based methods combine circular and Gaussian components [16], while distance-based methods construct geometry-aware dissimilarities, leading to extensions such as cylindrical k-means [6] and related embedding approaches [17]. Within this latter framework, medoid-based clustering has also been proposed (although briefly mentioned and not fully detailed [8]). By selecting medoids as representative data points, PAM is typically less sensitive to outliers and noise than the classical k-means algorithm.

However, in the cylindrical setting, an additional issue arises. The classical PAM must be adapted by introducing an appropriate form of *scaling* (although this concept cannot be applied to circular components in the same manner as to linear variables). If this aspect is not carefully considered, as will be shown later, the robustness of PAM may decrease.

For such a reason, and with the aim of keeping a certain degree of robustness of the cylindrical PAM, this work suggests considering robust forms of scaling. More particularly, the focus will be on a newly introduced robust measure of dispersion for circular data [5].

The paper is organized as follows. Section 2 reviews dissimilarities on the cylinder. Section 3 describes the proposed medoid-based clustering method with the robustly scaled dissimilarity. Section 4 presents a simulation study, while Sect. 5 illustrates the method through a real-data application. Finally, Sect. 6 provides some concluding remarks.

2 A Dissimilarity on the Cylinder

Let $\mathbf{x} = (\theta, r, z)$ be a cylindrical observation, where θ represents the angular component, and r and z are two linear components, hereafter referred to as the radial and axial directions, respectively. To obtain geometry-aware solutions, the available cylindrical k-means method [6] adopts a probabilistic framework and models the angular component $\theta \in [0, 2\pi)$ by a von Mises distribution, $\theta \sim \mathrm{vM}(\mu_\theta, \kappa)$, and the linear components by Gaussian distributions, $r \sim \mathcal{N}(\mu_r, \sigma_r^2)$ and $z \sim \mathcal{N}(\mu_z, \sigma_z^2)$.

Assuming mutual independence between the angular and linear components, the joint probability density function of $\mathbf{x}$ is given by

$$f(\mathbf{x}) = \frac{1}{2\pi I_0(\kappa)} \exp\big(\kappa \cos(\theta - \mu_\theta)\big) \cdot \frac{1}{\sqrt{2\pi\sigma_r^2}} \exp\Big(-\frac{(r-\mu_r)^2}{2\sigma_r^2}\Big) \cdot \frac{1}{\sqrt{2\pi\sigma_z^2}} \exp\Big(-\frac{(z-\mu_z)^2}{2\sigma_z^2}\Big).$$

where $I_0(\cdot)$ denotes the modified Bessel function of the first kind and order zero.

Based on this model, a pairwise dissimilarity between two observations $\mathbf{x}_i = (\theta_i, r_i, z_i)$ and $\mathbf{x}_j = (\theta_j, r_j, z_j)$ is defined by considering the negative log-likelihood of their relative displacement. The resulting cylindrical dissimilarity is

$$D^c(\mathbf{x}_i, \mathbf{x}_j) = -\kappa \cos(\theta_i - \theta_j) + \frac{(r_i - r_j)^2}{2\sigma_r^2} + \frac{(z_i - z_j)^2}{2\sigma_z^2}, \quad (1)$$

where the first term quantifies angular proximity via the von Mises distribution, while the remaining terms capture linear differences along the radial and axial directions. Smaller values of $D^c(\mathbf{x}_i, \mathbf{x}_j)$ correspond to a higher likelihood that observations i and j belong to the same cluster.

3 Robust Scaling for a Cylindrical PAM

In practical applications, data often contain noise, measurement errors, or outliers that can strongly distort results obtained with classical statistical tools. In clustering, distance-based methods such as PAM [14] offer some robustness by relying on medoids rather than means.

According to Hennig [9,10], PAM admits a clustering functional that is robust at the cluster-location level, meaning that the selected medoids remain stable under small perturbations of the data.

For a fixed number of clusters c, PAM seeks medoids $\{m_1, \ldots, m_c\}$ that minimize the total within-cluster dissimilarity:

$$\min_{\{m_1,\ldots,m_c\}} \sum_{i=1}^{n} \min_{j=1,\ldots,c} D(\mathbf{x}_i, m_j), \quad (2)$$

where $\mathbf{x}_i = (\theta_i, r_i, z_i)$ and $D(\mathbf{x}_i, \mathbf{x}_j)$ is a dissimilarity measure.

The resulting medoids $\{m_1, \ldots, m_c\}$ represent the most central observations in each cluster, noting that each m_j must coincide with one of the observed data points.

When cylindrical data are considered, if the dissimilarity in (1) is used (as suggested in [8]), sensitivity to outliers remains. Although $D^c(\mathbf{x}_i, \mathbf{x}_j)$ arises naturally as the negative log-likelihood under independent von Mises and Gaussian components, it remains sensitive to outliers. In particular, the squared linear terms can cause large radial or axial differences to dominate the dissimilarity, while the parameters κ, σ_r^2, and σ_z^2 may themselves be distorted by extreme observations.

For this reason, this work suggests adopting a more robust version of the dissimilarity in (1), defined as

$$D^{c^*}(\mathbf{x}_i, \mathbf{x}_j) = \psi\, d(\theta_i, \theta_j) + \frac{|r_i - r_j|}{S_r} + \frac{|z_i - z_j|}{S_z}. \quad (3)$$

Here ψ is an angular concentration parameter, S_r and S_z are robust scale parameters for the radial and axial components, and $d(\theta_i, \theta_j)$ is the shortest arc distance between the angles θ_i and θ_j (i.e., $d(\theta_i, \theta_j) = \pi - \big|\pi - |\theta_i - \theta_j|\big| \in [0, \pi]$).

Unlike quadratic dissimilarities, which rely on second-moment properties of the distribution, the proposed dissimilarity is geometrically defined and otherwise distribution-free apart from the scale normalization.

Furthermore, to obtain reliable estimates in the presence of outliers, robust estimators are used for both angular concentration and linear scale parameters. Specifically, the linear scales S_r and S_z are estimated using the Median Absolute Deviation (MAD) [15], while the angular concentration is estimated using the recently introduced robust estimator of κ, which is a nonlinear function of the Circular Median Absolute Deviation (CMAD) [5]. These estimators have bounded influence functions and high breakdown points, ensuring stable scaling of the cylindrical dissimilarity.

The resulting scaled cylindrical dissimilarity takes the form

$$\hat{D}^{c^*}(\mathbf{x}_i, \mathbf{x}_j) = \hat{\kappa}_{\text{CMAD}}\, d(\theta_i, \theta_j) + \frac{|r_i - r_j|}{\hat{\sigma}_{r,\text{MAD}}} + \frac{|z_i - z_j|}{\hat{\sigma}_{z,\text{MAD}}}. \tag{4}$$

Here $\hat{\kappa}_{\text{CMAD}}$ denotes the robust estimate of the circular concentration obtained through the circular MAD, while $\hat{\sigma}_{r,\text{MAD}}$ and $\hat{\sigma}_{z,\text{MAD}}$ denote the robust estimates of the radial and axial dispersions, respectively.

The associated cylindrical median functional is conjectured to possess a bounded influence function under suitable regularity conditions. A detailed investigation of this robustness property is omitted here for reasons of space and will be reported elsewhere.

4 A Simulation Study

The performance of the proposed robustly scaled cylindrical PAM clustering method was evaluated through a simulation study. The objectives were to compare the proposed method with the cylindrical PAM with no robust scaling [8], and classical linear PAM, and to assess robustness under contamination.

Three simulation scenarios were considered. Scenario 1 (no contamination). Three clusters with sizes $N = (200, 300, 400)$ were generated. The angular component followed von Mises distributions with means $\mu_\theta = (0, -2, 2)$ and concentrations $\kappa = (1, 2, 3)$. Linear components were generated from Gaussian distributions with means $\mu_r = (0, 0.5, -0.5)$, $\mu_z = (3, 1, 0.5)$ and standard deviations $\sigma_r = \sigma_z = 0.1$. Scenario 2 (Mild contamination). Scenario 1 was reproduced. However, five percent of the obtained observations were replaced in their linear components by Gaussian noise $\epsilon_i \sim \mathcal{N}(0, 1.5^2)$. Scenario 3 (Moderate contamination). Fifteen percent of points belonging to group 1 were shifted by $\pi/2$ in their angular component, and shifted by $(+8, -8)$ in their linear components with $\sigma = 2$.

The three scenarios are visualized in Fig. 1. While the underlying cylindrical cluster structure is preserved, contaminated observations introduce substantial dispersion in the linear components and distort angular concentration.

For each scenario, three clustering methods were compared: classical PAM applied to triples (θ, r, z) treating θ as a linear variable, cylindrical PAM [8], and

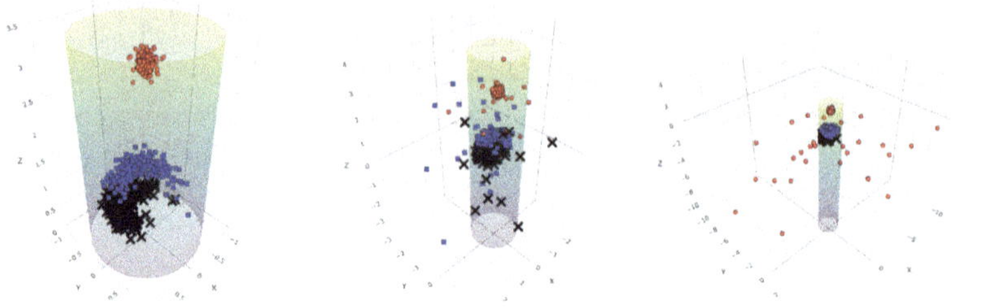

No contamination Mild contamination Moderate contamination

Fig. 1. Simulated cylindrical datasets under the three simulated scenarios. Cluster 1: circles, cluster 2: squares, cluster 3: crosses.

the proposed cylindrical PAM with the robustly scaled dissimilarity. Each scenario was replicated 100 times. Clustering performance was evaluated using the Adjusted Rand Index (ARI), which measures the agreement between estimated and true labels [11]. The average ARI values are reported in Table 1.

Table 1. Average Adjusted Rand Index (ARI) across simulations.

Method	Contamination		
	Absent	Mild (5%)	Moderate (15%)
Linear PAM	0.69	0.63	0.57
Cylindrical PAM	**0.99**	0.88	0.82
Robust cylindrical PAM	**0.99**	**0.95**	**0.92**

In the absence of contamination, cylindrical clustering methods accurately recover the underlying cluster structure, whereas linear PAM performs substantially worse, highlighting the importance of respecting the circular nature of the angular variable. Under mild contamination, performance decreases for all methods, although the robustly scaled cylindrical PAM consistently achieves higher accuracy than the standard cylindrical PAM. Under moderate contamination, the advantage of the robust approach becomes more evident, with the robustly scaled cylindrical PAM maintaining higher ARI values than the non-robust cylindrical method. Overall, the results indicate that accounting for cylindrical geometry is the primary driver of clustering accuracy, while robust scaling further improves resilience to noisy or contaminated observations.

5 Application to Image Segmentation

Image segmentation partitions an image into homogeneous regions by grouping pixels according to selected features. In color images, clustering is commonly used to identify regions of similar color, and its effectiveness depends on the color space and the similarity or dissimilarity measure. The HSV model has a natural cylindrical structure: Hue (H) is the angular coordinate, Saturation (S) the radial coordinate, and Value (V) the vertical coordinate.

This study considers the church test image from the Berkeley Segmentation Dataset (BSD) [13], a standard benchmark in image processing. To obtain a mid-level segmentation, the number of clusters was fixed at $c = 12$, consistent with common color quantization settings. The original 512×512 image was resized to 64×64 to reduce computational cost and memory requirements. To assess the robustness of the clustering methods, artificial noise was introduced in the hue component. Specifically, a proportion of pixel hue values (15%) was randomly selected and perturbed by adding Gaussian noise. Dissimilarity matrices were then constructed from the HSV features and used for clustering.

Fig. 2. HSV image segmentation with $c = 12$ clusters. Top-left: resized original image (64×64). Top-right: reconstruction obtained with linear PAM. Bottom-left: cylindrical PAM. Bottom-right: robustly scaled cylindrical PAM. Differences between clusters are more clearly visible in the online version of this figure.

Figure 2 shows the segmentation results obtained with three PAM-based clustering strategies. The original resized image (top-left) provides a visual reference for the color distribution and structural details of the scene. The Linear PAM result (top-right), which treats the HSV coordinates as Euclidean variables, exhibits noticeable sensitivity to the injected noise. This sensitivity appears as scattered pixels across otherwise homogeneous regions, particularly in the sky and on the building facade, producing a grainy appearance.

The Cylindrical PAM segmentation (bottom-left) improves the result by explicitly accounting for the circular structure of the hue variable. Pixels with similar hue values are therefore grouped more consistently, leading to smoother regions in the sky and on the white walls of the building.

The robustly scaled cylindrical PAM (bottom-right) further enhances the segmentation by reducing the influence of perturbed hue values through robust scaling of the dissimilarity. In particular, the clock region of the bell tower is more clearly represented, and several isolated dark pixels visible in the cylindrical PAM segmentation disappear in the robust result, producing a cleaner and more homogeneous segmentation.

6 Final Remarks

This work introduced a robust medoid-based clustering framework for cylindrical data based on a geometry-aware scaled dissimilarity. By accounting for angular periodicity and using robust dispersion estimates, the proposed method improves clustering stability under contamination.

Although it can be shown that the suggested medoid estimator under robust scaling possesses a bounded influence function, the underlying dissimilarity remains unbounded, so extreme observations may still influence pairwise distances. On the other hand, the clustering procedure introduced here cannot be considered fully robust in the sense of Hennig [9].

Future planned work will investigate bounded variants of the dissimilarity, such as Huberized or Winsorized forms, in order to further limit the influence of extreme observations. Finally, the current framework assumes independence between the components. In some applications, this may become a serious limitation. Extending the proposed method to incorporate correlation structures represents a promising direction for future research: it may further improve its flexibility and practical performance.

Acknowledgment. The work of Houyem Demni was funded by the project ECS 0000024 "Ecosistema dell'innovazione - Rome Technopole" financed by EU in NextGenerationEU plan through MUR Decree n. 1051 23.06.2022 PNRR Missione 4 Componente 2 Investimento 1.5 - CUP H33C2200042000.

References

1. Abe, T., Shimatani, I.K.: Cylindrical distributions and their applications to biological data. In: Ley, C., Verdebout, T. (eds.) Applied Directional Statistics, pp. 163–185. Chapman and Hall/CRC, Boca Raton (2018)
2. Agostinelli, C.: Robust estimation for circular data. Comput. Stat. Data Anal. **51**(12), 5867–5875 (2007)
3. Agostinelli, C., Greco, L., Saraceno, G.: Weighted likelihood methods for robust fitting of wrapped models for p-torus data. AStA Adv. Stat. Anal. **108**(4), 853–888 (2024)
4. Bezdek, J.C.: Pattern Recognition with Fuzzy Objective Function Algorithms. Plenum Press, New York (1981)

5. Demni, H., Porzio, G.C., Hubert, M., Rousseeuw, P.: Robust measures of dispersion for circular data with an anomaly detection rule. arXiv preprint arXiv:2603.01237 (2026)
6. Fujita, A., Kamo, H., Fujita, K., Sugimoto, T.: A clustering method for data in cylindrical coordinates. Comput. Intell. Neurosci. (2017)
7. Greco, L., Saraceno, G., Agostinelli, C.: Robust fitting of a wrapped normal model to multivariate circular data and outlier detection. Stats **4**(2), 454–471 (2021)
8. Hammami, Y., Demni, H., Messaoud, A., Porzio, G.C.: Partitioning around medoids on product spaces: a clustering approach for cylindrical data. In: SAML-25 Workshop on Statistical and Machine Learning, vol. 16 (2025). https://arrow.tudublin.ie/saml/16
9. Hennig, C.: Dissolution point and isolation robustness: robustness criteria for general cluster analysis methods. J. Multivar. Anal. **99**(6), 1154–1176 (2008)
10. Hennig, C.: Breakdown points for maximum likelihood estimators of location-scale mixtures. Ann. Stat. **32**(4), 1313–1340 (2004)
11. Hubert, L., Arabie, P.: Comparing partitions. J. Classif. **2**(1), 193–218 (1985)
12. Mardia, K.V., Sutton, T.W.: A model for cylindrical variables with applications. J. R. Stat. Soc. Ser. B **40**(2), 229–233 (1978)
13. Martin, D., Fowlkes, C., Tal, D., Malik, J.: A database of human segmented natural images and its application to evaluating segmentation algorithms and measuring ecological statistics. In: Proceedings of the Eighth IEEE International Conference on Computer Vision, pp. 416–423. IEEE Press, New York (2001)
14. Kaufman, L., Rousseeuw, P.J.: Finding Groups in Data: An Introduction to Cluster Analysis. Wiley, New York (1990)
15. Rousseeuw, P.J., Croux, C.: Alternatives to the median absolute deviation. J. Am. Stat. Assoc. **88**(424), 1273–1283 (1993)
16. SenGupta, A., Roy, M., Chattopadhyay, A.K.: Model-based clustering for cylindrical data. In: Advances in Statistics - Theory and Applications: Honoring the Contributions of Barry C. Arnold in Statistical Science, pp. 347–363. Springer, Cham (2020)
17. Sun, X., Sajda, P.: Circular clustering with polar coordinate reconstruction. IEEE/ACM Trans. Comput. Biol. Bioinform. **21**(5), 1591–1600 (2024)

Exploring Feature-Graph Alignment in Graph Neural Networks for Spatio-Temporal Forecasting

Abhishek Dey[1], Julia Handl[1(✉)], Margherita Battistotti[1], Arnd Hamburg[2], and Nikolay Mehandjiev[3]

[1] University of Manchester, Manchester, UK
julia.handl@manchester.ac.uk
[2] Airbus, Manching, Germany
[3] University of Birmingham Dubai, Dubai, UAE

Abstract. We explore mechanisms to understand and explain the performance of spatio-temporal graph neural networks (GNNs) for forecasting. Using the England Covid benchmark, we illustrate how a combination of established statistical techniques, from the community detection and clustering literature, can help understand GNN forecasting performance in terms of the interactions between the temporal (time series) and spatial (graph) information in a given problem. Our work contributes towards improved transparency and explainability in the design and application of spatio-temporal GNNs for forecasting.

Keywords: Graph neural network · Spatio-temporal forecasting · Community detection · Clustering

1 Introduction

Many real world systems are naturally described as graphs, where entities (modeled as nodes) are connected by certain relationships (modeled as edges). For example, social networks reflect the connections between people, e.g. based on their interactions in physical or virtual environments. By encoding these relations, graph structures explicitly capture departures from the standard assumption in machine learning that the data is independent and identically distributed (i.i.d.). In contrast, in Graph Neural Networks (GNNs) the state of each node can directly influence and be influenced by its neighbors. GNNs are specifically designed to leverage this relational structure: they aggregate information from a node's neighborhood so that each node can be represented by an embedding that reflects both individual attributes and the broader connectivity pattern.

Although GNNs have seen increasing interest in the machine learning community [2], there is a dearth of studies on the interaction between graph structure

Supported by Royal Academy of Engineering through the Google Deep Mind Research Ready Scheme.

P. Brito et al. (Eds.): IFCS 2026, STUDIES CLASS, pp. 128–136, 2026.
https://doi.org/10.1007/978-3-032-32009-4_16

and feature information in explaining the learning performance of different GNN approaches. In previous work [3], we have started to set out how existing methodologies from the community detection and cluster validation literature may be utilized for the development of explainable AI (XAI) techniques in a node classification context. Here, we expand this work to consider the more complex case of GNNs for spatio-temporal forecasting.

Spatio-temporal GNN models typically combine a spatial GNN, which encodes the graph structure at each time step to generate node embeddings, with a temporal component that models the sequence of these embeddings across time. For each node, the combined model produces a forecast based on inputs that collate both its own history and its neighbors' histories. Depending on the problem, the graph structure in these models can be dynamic or static over time. For example, in the context of traffic forecasting, the spatial graph structure may reflect existing road networks and is likely to remain static over short time horizons, whilst potentially becoming dynamic in the long term. In addition to the usual hyperparameter choices for neural networks (including learning rate, optimizer, model complexity) and forecasting models more generally (number of lags to consider), spatio-temporal graph neural networks require designing the underpinning graph structure, whilst considering the temporal features of the data set.

The different interacting elements and design decisions in spatio-temporal graph neural networks make it particularly challenging to identify key drivers of forecasting performance for these models and to correctly pinpoint situations in which their use will be beneficial. A better understanding of these interactions is instrumental for informing design decisions, supporting the expert- or data-driven construction of graph structures, and facilitating decisions around model choice.

2 Theoretical Background

2.1 Graph Representation Learning

Graph Representation. A graph is formally an ordered pair $\mathcal{G} = (\mathcal{V}, \mathcal{E})$, where $\mathcal{V}$ is a set of nodes, and $\mathcal{E} \subset |\mathcal{V}| \times |\mathcal{V}|$ is the set of edges. An edge between two nodes is defined as $(u, v) \in \mathcal{E}$. A graph can be described by an adjacency matrix $\mathbf{A} \in \mathbb{R}^{|\mathcal{V}| \times |\mathcal{V}|}$. This matrix is indexed such that:

$$A_{uv} = \begin{cases} 1 & \text{if } (u, v) \in \mathcal{E}, \\ 0 & \text{otherwise.} \end{cases} \tag{1}$$

If we allow for the existence of weighted edges, then A_{uv} can take a wider range of values, reflecting the "strength" of each connection.

Message Passing. In a GNN learning context, we further assume that each node may be characterized by a feature vector $\mathbf{x}$, providing its original representation. A GNN produces node embeddings by iteratively passing messages across the graph. At each layer ℓ, node u applies a learnable message function a_θ to each neighbor's embedding $\mathbf{h}_v^{(\ell)}$, then combines the resulting vectors with a permutationinvariant aggregation (e.g. sum, mean, or max). This aggregated neighbor information is then merged with $\mathbf{h}_u^{(\ell)}$ via an update function $\sigma(\cdot)$, producing the nextlayer embedding:

$$\mathbf{h}_u^{(\ell+1)} = \sigma\Big(\mathbf{h}_u^{(\ell)}, \bigoplus_{v \in \mathcal{N}(u)} a_\theta\big(\mathbf{h}_u^{(\ell)}, \mathbf{h}_v^{(\ell)}\big)\Big). \tag{2}$$

In this equation, the quantity $\mathbf{m}_u^{(\ell)} = \bigoplus_{v \in \mathcal{N}(u)} a_\theta\big(\mathbf{h}_u^{(\ell)}, \mathbf{h}_v^{(\ell)}\big)$ represents the aggregated message sent to u from its neighbors, and σ then combines the node's previous state $\mathbf{h}_u^{(\ell)}$ with $\mathbf{m}_u^{(\ell)}$.

2.2 Time Series Forecasting on Graphs

Sequence modelling, in the autoregressive sense, predicts the most likely future observation given the previous T observations:

$$\hat{x}_{t+1} = \mathbb{E}[X_{t+1} \mid x_{t-T+1}, \dots, x_t]. \tag{3}$$

where $\{x_t\}_{t=1}^T$ is some sequence of observations, each of which is a vector $x_t \in \mathbb{R}^{d_x}$. In a spatio-temporal GNN, each node $v \in \mathcal{V}$ of a fixed (or, potentially, dynamic) graph $\mathcal{G} = (\mathcal{V}, \mathcal{E})$ carries its own time series. Stacking the node-level observations across all T time steps yields a spatio-temporal feature tensor $\mathcal{X} \in \mathbb{R}^{|\mathcal{V}| \times T \times d_x}$, where the first index runs over nodes, the second over time, and the third over the d_x-dimensional feature vector at each (v, t).

To capture spatial dependencies, it is possible to apply a message passing layer at each time step and layer, producing an observation at layer ℓ and time t:

$$\mathbf{H}_t^{(\ell)} = \mathrm{MessagePass}^{(\ell)}\big(\mathbf{H}_t^{(\ell-1)}\big) \in \mathbb{R}^{|\mathcal{V}| \times d_h^{(\ell)}}, \tag{4}$$

where we set $\mathbf{H}_t^{(0)} = \mathbf{X}_t$ and each $\mathrm{MessagePass}^{(\ell)}$ is the ℓth graph convolution (or aggregation) operator. The spatial embeddings $\{\mathbf{H}_t^{(\ell)}\}_{t=1}^T$ at any chosen layer ℓ can then be processed by a recurrent model to capture temporal dependencies.

While a simple Recurrent Neural Network (RNN) could be used, such models demonstrate vanishing gradients, which limits the ability of them to capture long range information. To mediate this, [8] uses a Long Short Term Memory (LSTM) architecture, which augments the RNN with gates that regulate the flow of information.

3 Understanding Drivers of GNN Forecasting Performance for Spatio-Temporal Forecasting

Spatio-temporal GNNs for forecasting are largely deployed as black-boxes, with the graph-structure commonly assumed as given and the remaining design choices implemented using automated hyper-parameter tuning or prior settings. Fundamentally though, graph neural networks can be seen as a learning paradigm bringing together (at least) *two distinct types of modalities:* one in the form of edges, i.e. relations between input nodes, and the other in the form of feature-based information within each node. Their performance is likely to depend strongly on the alignment between these modalities (e.g. complementarity versus redundancy versus conflict), but this is an aspect of GNNs that is poorly understood. In the following, we describe mechanisms from community detection and clustering suitable for exploring this relationship, and proceed to experimentally demonstrate the impact of their alignment on forecasting performance.

3.1 Measuring Structure Within the Graph

Established community detection approaches (e.g. the Louvain method [1]) may be used to identify structure within different candidate graphs, assessing the characteristics of the structure using the following distinct measurements:

- The number of communities within the graph, e.g. measured using the number of communities determined by the Louvain method [1] (self-loops removed), potentially for different resolutions (number of communities).
- The quality of the resulting partition of the graph, measured using Modularity [5].

Having established the nature and strength of the community structure, we may then assess the extent to which it aligns with node-level inputs (time series) or (hidden-layer) node representations. This alignment may be measured using, e.g. the Average Silhouette Width (ASW [6]), using the clusters defined by the graph communities, but assessing distance in terms of the node feature space. Two distinct types of time-dependent trajectories can potentially be analyzed here: (i) how this alignment changes between a dynamically changing graph and the training portion of the full time series; (ii) how this alignment changes between a static graph and the node representations considered at each time step, considering either the initial time series information or the learned embedding for each node. This allows us to query the initial agreement between input modalities, as well as the model's success in learning representations that adequately capture both types of input.

3.2 Measuring Structure Within the Time Series

Alternatively, a combination of clustering and cluster validation may be used for the reverse analysis, i.e. starting from structure in the node-level inputs (time series) or (hidden-layer) node representations and relating this back to the graph structure. For example, k-means [4] may be used to identify clusters of nodes, for a range of resolutions, using as input either the training portion of the full time series for each node; the initial node features; or the learned embeddings. A suitable partition can then be selected based on an appropriate cluster validation techniques such as the ASW, or the desired resolution of the partition, supporting the following useful measurements:

- The level of structure inherent to the feature information, measured using the ASW.
- The number of clusters inherent to the time series data, if determined automatically, e.g. using the ASW.

The measure of modularity may then be used to investigate the clusters' alignment with the original graph structure. Specifically, the node labels resulting from cluster analysis are tested with respect to the graph structure, evaluating the extent to which they define a partition with high edge density within classes and low edge density between classes.

Using the combination of the approaches described in Sects. 3.1 and 3.2 above, we can comprehensively examine the alignment between modalities (i.e. between graph structure and node representations), in concrete quantitative terms, and we can monitor how this alignment changes during model training. Note that this quantitative assessment could be further complemented by visual tools, e.g. the use of graph visualization [9], and dimensionality reduction techniques (e.g. t-SNE, [10]), potentially over time, using color (heat maps) to visually assess the consistency of node-level forecasts.

3.3 Experimental Setup

Dataset. The experiments are conducted on the England Covid dataset [7], accessed through PyTorch Geometric Temporal. The England Covid dataset provides daily incidence counts within England NUTS3 regions from March until 12th of May 2020. The graph structure is dynamic and captures mobility patterns across the 129 regions. The graphs are directed and weighted. At each time t, the node feature matrix $\mathbf{X}_t \in \mathbb{R}^{129\times 8}$ holds the counts from the previous $lags = 8$ days, where the targets are the case loads for the next day $t+1$.

Graphs. In addition to the original graph (Original), we consider an alternative data-driven graph structure: a static, distance-based graph, connecting each node to its 10 nearest neighbors on the training data, adding additional edges to ensure symmetry (KNN10). We additionally consider two baselines for model performance: (i) Random, a baseline containing random spatial information, constructed as a static, random graph (unweighted, undirected, symmetric) with an edge sampling probability $p = 0.1$; (ii) Self, a baseline containing only self-loops and no spatial information.

Distance Measure. Given the (pre-)standardized nature of the data, the Euclidean distance function is adopted in this paper, for simplicity.

Models. We deploy GConvLSTM from PyTorch Geometric Temporal [7], which embeds a truncated Chebyshev spectral filter inside each recurrent step. At time t, the raw node features $\mathbf{x}_t$ are first filtered as

$$\mathbf{z}_t = \mathbf{g}_\theta(\mathbf{\Lambda}) *_\mathcal{G} \mathbf{x}_t = \sum_{k=0}^{K-1} \theta_K T_K(\tilde{\mathbf{L}}) \mathbf{x}_t, \tag{5}$$

where $\tilde{\mathbf{L}} = 2\mathbf{L}/\lambda_{\max}\mathbf{I}$ and T_K is the Chebyshev polynomial of degree K. In general, $\mathbf{g}_\theta(\mathbf{\Lambda})$ defines the filter in the spectral domain. Applying it to the rescaled Laplacian via $\mathbf{g}_\theta(\tilde{\mathbf{L}})$ realises the filter on the graph domain rather than in frequency space, with the Chebyshev expansion providing an efficient approximation. The filtered signal $\mathbf{z}_t$ drives the LSTM update $(\mathbf{h}_t, \mathbf{c}_t) = \mathrm{LSTM}(\mathbf{z}_t, \mathbf{h}_{t1}, \mathbf{c}_{t1})$, and a final linear head maps $\mathbf{h}_t$ to the one step ahead forecast $\hat{\mathbf{y}}_t$.

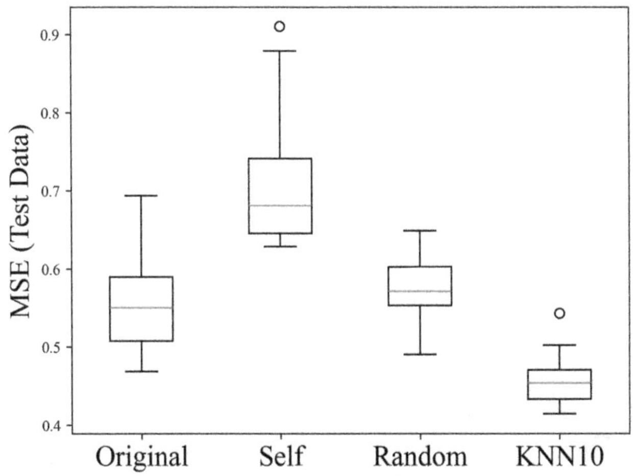

Fig. 1. MSE on the test set over 30 runs. From left to right: Original graph; Graph with self-loops only; Random graph with edge probability p=0.1; 10-nearest-neighbor-graph. The Friedman test is significant (F = 80.0400, p = 3.009 × 10^{-17}). Nemenyi post-hoc analysis indicates significant differences between all four approaches.

We train a onelayer GConvLSTM with hidden dimension $d_h = 64$, drop out probability $p = 0.1$, Chebyshev polynomial order $K = 2$, and onestep horizon, minimising the meansquared error (MSE) via Adam with learning rate $\eta = 10^{-3}$ for up to 500 epochs, with early stopping parameters patience 50 and tolerance 1×10^{-7} (patience sets the maximum number of epochs without improvement larger than tolerance, on the validation set). Use of a sequential training/validation/test split of 40/30/30. Parameter settings rely on default settings and preliminary experimentation, to ensure convergence, with no extensive or automated hyper-parameter tuning performed.

4 Results

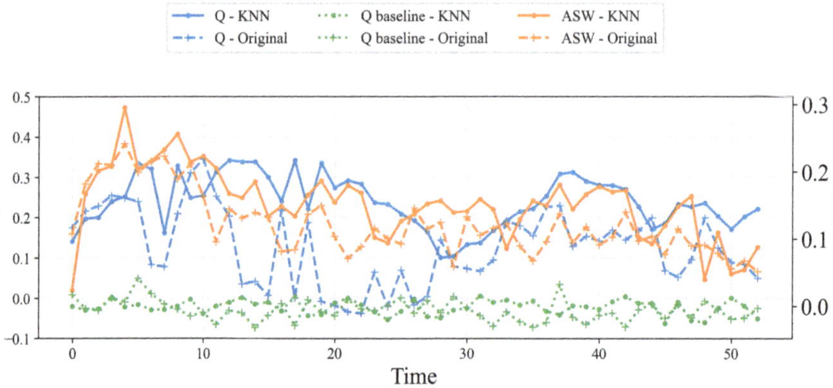

Fig. 2. Comparing the alignment between cluster structure (derived from the hidden-layer node embedding for each time step; k-means solution for 20 clusters) and the two alternative graph structures: Original - Mobility graph for the England Covid dataset (dashed lines); KNN10 - Data-driven (10-nearest-neighbor) graph (full lines). Modularity values Q for the random perturbation of each input graph are included as an additional reference. The paired samples are compared using a Wilcoxon signed-rank test, highlighting statistically significant differences ASW ($W = 213$, $p = 8.647 \times 10^{-6}$) and Q ($W = 92$, $p = 3.396 \times 10^{-8}$), but not in terms of the Q baseline. Results are not included for the two baselines used in Fig. 1 - their graphs contain no meaningful structure, so results are trivial.

In the following, we provide an example of the type of exploration facilitated by the techniques proposed. Figure 1 compares the performance of GConvLSTM across the original and the data-driven graph (as well as the two baselines with random / absent spatial information), for an England Covid data set. The results confirm that an informative spatial structure results in better predictions, compared to the baselines. Perhaps surprisingly though, for the setup considered here, a data-driven graph construction approach significantly outperforms the original graph reflecting mobility data. This emphasises the need to exercise caution in constructing graphs for GNNs, while evaluating the graph's impact on node representation learning and prediction performance.

We further explore how the performance differences between these two graphs can be explained by some of our proposed measures. Figure 2 shows our assessment of structure within the learned (hidden-layer) embeddings, evaluated in terms of the ASW. We then evaluate this structure using modularity, checking directly for the consistency of time-series similarity with the relational information derived from the graph. The analysis reveals clear differences between the results for the Original and the KNN10 graph. First, use of the KNN10 graph results in hidden-layer embeddings that contain a more pronounced cluster structure, as evident from improvements in the ASW, when compared to embeddings learned with the mobility-based graph. This is the case for the majority of time steps considered. Second, use of the KNN10 graph results in hidden-layer embeddings that are more consistent with the information provided by the graph's relational (i.e. spatial) information, as evidenced by improvements to the modularity measure Q. Again, this holds for the large majority of time steps.

These findings highlight that the better MSE results of the KNN10 approach (see Fig. 1) can be explained in terms of a better learned representation. Concretely, this improved representation is characterized by a more pronounced group structure, within the embedding space, as well as the closer integration of the information captured by these node representations and the graph's relational structure.

5 Conclusion

We highlight opportunities for the use of community detection and clustering techniques in understanding the performance of spatio-temporal GNNs for forecasting, addressing the black-box nature of these models. Our particular focus is the analysis of the structure of the graph, the structure of the node representation space, and the alignment between these spaces. The work provides a first step towards designing XAI tools for the design and comparison of data-driven and expert-defined graph structures for forecasting using spatio-temporal GNNs.

References

1. Blondel, V.D., Guillaume, J.-L., Lambiotte, R., Lefebvre, E.: Fast unfolding of communities in large networks. J. Stat. Mech: Theory Exp. **2008**(10), P10008 (2008)
2. Corradini, F., Gerosa, F., Gori, M., Lucheroni, C., Piangerelli, M., Zannotti, M.: A systematic literature review of spatio-temporal graph neural network models for time series forecasting and classification. Neural Netw. **195**, 108269 (2025)
3. Handl, J., Mehandjiev, N.: Towards XAI for the facilitation of human-machine learning of graph structure. In: 2025 IEEE 5th International Conference on Human-Machine Systems (ICHMS), pp. 309–314. IEEE (2025)
4. Hartigan, J.A., Wong, M.A.: Algorithm as 136: a k-means clustering algorithm. J. R. Stat. Soc. Ser. C (Appl. Stat.) **28**(1), 100–108 (1979)
5. Newman, M.E.J., Girvan, M.: Finding and evaluating community structure in networks. Phys. Rev. E **69**(2), 026113 (2004)
6. Rousseeuw, P.J.: Silhouettes: a graphical aid to the interpretation and validation of cluster analysis. J. Comput. Appl. Math. **20**, 53–65 (1987)
7. Rozemberczki, B., et al.: Pytorch geometric temporal: spatiotemporal signal processing with neural machine learning models. In: Proceedings of the 30th ACM International Conference on Information & Knowledge Management, pp. 4564–4573 (2021)
8. Seo, Y., Defferrard, M., Vandergheynst, P., Bresson, X.: Structured sequence modeling with graph convolutional recurrent networks. In: Cheng, L., Leung, A.C.S., Ozawa, S. (eds.) ICONIP 2018. LNCS, vol. 11301, pp. 362–373. Springer, Cham (2018). https://doi.org/10.1007/978-3-030-04167-0_33

9. Tamassia, R. (ed.): Handbook of Graph Drawing and Visualization. CRC Press, Boca Raton (2013)
10. van der Maaten, L., Hinton, G.: Visualizing data using t-sne. J. Mach. Learn. Res. **9**(Nov), 2579–2605 (2008)

Handling Outliers in Compositional Data: A Variance-Inflation Strategy

Agnese Maria Di Brisco(✉)

Department of Studies in Economics and Business (DISEI), University of Piemonte Orientale, Vercelli, Italy
agnese.dibrisco@uniupo.it

Abstract. Outlier detection in compositional data entails a twofold level of complexity, arising from both the multivariate nature of the data and the constraints imposed by the simplex space. To address this issue, a Dirichlet Contaminated Model is proposed, adapting to the compositional setting the variance-inflation rationale commonly used in the Gaussian framework. The approach models the data as a finite two-component mixture, consisting of a regular component and a contaminated component, the latter characterized by inflated variance to account for outlying observations. Parameter estimation is carried out within a Bayesian framework using the Hamiltonian Monte Carlo algorithm. Simulation studies are conducted to assess the model's ability to recover the true generating mechanism, to correctly estimate the model parameters, and to identify outlying and atypical observations. An applicative example is performed to show the behavior of the proposed approach in a real-case scenario.

Keywords: Dirichlet Distribution · Finite Mixture Model · Hamiltonian Monte Carlo · Stan

1 Introduction

Outlier detection represents a central problem in statistics, as atypical observations can severely bias parameter estimation and worsen the fit of statistical models. An outlier is commonly understood as an observation that significantly deviates from the rest of the data [4]. The problem of outlier detection in a multivariate setting is highly involved and cannot be reduced to a straightforward extension of univariate techniques, since multivariate outliers are not necessarily characterized by anomalous values with respect to the marginal coordinates [7].

The difficulty of detecting outliers is further exacerbated when the data are proportions, rates, or parts of some whole. These types of data, known as compositional data, are characterized by strictly positive values summing to one, and lie in a constrained space called the simplex,

$$\mathcal{S}^D = \left\{ (x_1, \ldots, x_D)^\intercal : x_d > 0,\ \sum_{d=1}^{D} x_d = 1 \right\}.$$

P. Brito et al. (Eds.): IFCS 2026, STUDIES CLASS, pp. 137–144, 2026.
https://doi.org/10.1007/978-3-032-32009-4_17

Since the simplex is a non-Euclidean space, the direct application of standard multivariate techniques may lead to misleading results [1]. The detection of outliers in compositional data is therefore particularly challenging, due to the inherently multivariate nature of the problem and the unit-sum constraint [6].

Traditional approaches for robust analysis of compositional data typically rely on log-ratio transformations, such as isometric log-ratio (ilr), to map compositions from $\mathcal{S}^D$ to $\mathbb{R}^{D-1}$. The transformed data can then be analyzed with standard robust multivariate techniques, e.g., the Mahalanobis distance [3]. While these transformation-based methods are effective in many settings, they can pose challenges in terms of interpretability, as the results refer to the transformed coordinates and do not have a clear interpretation in terms of the original compositional parts. Motivated by this, the present work considers a modeling approach that is directly defined on the original compositional space.

The Dirichlet Contaminated Model (DCM) is proposed to address this problem. The model specifies a mixture of two Dirichlet-distributed components: one component models typical observations, while the other is dedicated to capturing outliers. A similar strategy for outlier detection has been proposed in the univariate setting, using a mixture of two beta distributions [2]. From a broader perspective, the DCM can be interpreted as the counterpart on the simplex of contaminated mixture models, which have been extensively studied in the Gaussian setting [9]. The rationale behind these models assumes that data affected by outliers arise from a finite mixture distribution, characterized by a "regular" component capturing the majority of the data and a "contaminated" component with inflated variance accommodating atypical points.

The remainder of the paper is organized as follows. Section 2 introduces the DCM and illustrates a Bayesian estimation approach. Section 3 presents results from three simulated scenarios, and Sect. 4 illustrates the application of the DCM to a real-world compositional dataset.

2 Model Definition and Bayesian Estimation

Before introducing the DCM, it is useful to recall the probability density function (p.d.f.) of a Dirichlet distributed random vector $\mathbf{X}$, expressed in terms of its mean-precision parameterization:

$$f_{DIR}(\mathbf{x}; \boldsymbol{\mu}, \phi) = \frac{\Gamma(\phi)}{\prod_{i=1}^{D} \Gamma(\phi\mu_i)} \prod_{i=1}^{D} x_i^{\phi\mu_i - 1}, \qquad \mathbf{x} \in \mathcal{S}^D,$$

where $\mathbb{E}[\mathbf{X}] = \boldsymbol{\mu} = (\mu_1, \ldots, \mu_D)^\intercal \in \mathcal{S}^D$ is the mean vector, and $\phi > 0$ is a precision parameter, being the marginal variance equal to $Var(X_i) = \frac{\mu_i(1-\mu_i)}{\phi+1}$, for $i = 1, \ldots, D$.

The DCM is defined as a finite mixture model designed to account for extra dispersion and potential outlying values in compositional data. Its p.d.f. is:

$$f(\mathbf{x}; \boldsymbol{\mu}, \phi, p, \delta) = (1-p) f_{DIR}(\mathbf{x}; \boldsymbol{\mu}, \phi) + p f_{DIR}(\mathbf{x}; \boldsymbol{\mu}, \phi/(1+\delta)), \qquad \mathbf{x} \in \mathcal{S}^D,$$

where $\boldsymbol{\mu} \in \mathcal{S}^D$, $\phi > 0$, $\delta > 0$, and $p \in (0,1)$. It is worth noting that the first component models the main data structure, i.e., is the regular one, whereas the second component is the contaminated one with higher dispersion to accommodate for potential outliers. The parameter $\boldsymbol{\mu}$ represents the overall mean of the mixture model and coincides with the mean of both components, implying that the regular and contaminated components share the same center. Since both components have the same mean, the marginal variance of the mixture is $Var(X_i) = \mu_i(1-\mu_i)\left[\frac{1-p}{\phi+1} + \frac{p}{\phi/(1+\delta)+1}\right]$, for $i = 1, \ldots, D$. The parameter ϕ is a precision parameter, the parameter δ is an inflation parameter, which controls the additional variability in the contaminated component, while the parameter p is the mixture weight.

Bayesian inference is performed using the Hamiltonian Monte Carlo (HMC) algorithm, specifically the No-U-Turn Sampler (NUTS), as implemented in the probabilistic programming language Stan [8]. It is worth noting that the DCM is not affected by label switching because its two components, although sharing the same mean vector, are uniquely identified by their precision. Indeed, by construction, the first component has higher precision, while the second component has increased dispersion induced by $\delta > 0$.

Prior distributions are assumed independent and weakly informative. The compositional mean vector $\boldsymbol{\mu}$ is parameterized through the vector $\boldsymbol{\beta} \in \mathbb{R}^{D-1}$, by using the inverse logit link:

$$\mu_i = \frac{\exp(\beta_i)}{\sum_{j=1}^{D} \exp(\beta_j)}, \qquad i = 1, \ldots, D,$$

with $\beta_D = 0$, to improve the efficiency and stability of the HMC sampler. Specifically, a multivariate normal prior with zero mean vector and a diagonal covariance matrix with "large" values is chosen for the vector parameter $\boldsymbol{\beta}$. As for the remaining parameters, the precision $\phi \sim Gamma(0.001, 0.001)$, the mixing weight $p \sim Uniform(0,1)$, and the inflation parameter $\delta \sim Cauchy^+(0, sd)$, with $sd \sim Exponential(1)$. Posterior inference is robust to these prior choices: in particular, alternative specifications on ϕ, such as $Gamma(0.01, 0.01)$ or $Cauchy^+(0,5)$, yield essentially identical posterior summaries for all parameters of interest.

Posterior inference includes the computation of the posterior membership probability for each observation i, and it represents the posterior probability that the i-th observation arises from the contaminated component. The membership probability is computed for each iteration of the MCMC chain, $s = 1, \ldots, S$, as:

$$\pi_i^{(s)} = \frac{p^{(s)} f_{DIR}\left(\mathbf{x}_i; \boldsymbol{\mu}^{(s)}, \frac{\phi^{(s)}}{1+\delta^{(s)}}\right)}{\left(1-p^{(s)}\right) f_{DIR}\left(\mathbf{x}_i; \boldsymbol{\mu}^{(s)}, \phi^{(s)}\right) + p^{(s)} f_{DIR}\left(\mathbf{x}_i; \boldsymbol{\mu}^{(s)}, \frac{\phi^{(s)}}{1+\delta^{(s)}}\right)}.$$

These membership probabilities can be used for model-based clustering, allowing observations with high $\hat{\pi}_i = \frac{1}{S}\sum_{s=1}^{S} \pi_i^{(s)}$, usually higher than 0.5, to be flagged as potential outliers.

3 Simulation Studies

To evaluate the performance of the DCM, three simulation studies were conducted. In each study, 200 Monte Carlo (MC) replicates were generated, each consisting of a sample of $n = 200$ compositional vectors of dimension $D = 3$.

The first two scenarios, i.e., S1 and S2, are fitting studies, in which the data are generated according to the same mixture mechanism assumed by the DCM. The true parameter values are set as follows: the mean vector is $\boldsymbol{\mu} = (0.2, 0.3, 0.5)$, the precision parameter is $\phi = 100$, and the mixture weight is $p = 0.2$ so that 20% of data come from the contaminated component. To investigate the effect of the variance inflation, the first study sets $\delta = 9$, corresponding to a larger inflation, whereas the second study sets $\delta = 3$, corresponding to a smaller one. These studies allow the evaluation of the accuracy and reliability of the Bayesian estimation procedure.

The third study is designed to evaluate the ability of the DCM to detect atypical or overdispersed observations. In study S3 the data-generating mechanism works as follows. First, data are generated from a bivariate normal distribution with mean vector $(0, 0)^\intercal$ and a diagonal covariance matrix with variances equal to 0.05. Next, a fraction $p = 0.2$ of observations is randomly selected, and their original values y_i are replaced with:

$$\tilde{y_i} = \hat{\mu} + 1.25 \times \sqrt{\frac{\chi^2_{0.975;D-1}}{d_i^2}}(y_i - \hat{\mu}) + \epsilon,$$

where $\hat{\mu}$ is the sample mean, the cutoff $\chi^2_{0.975;D-1}$ is the 97.5th quantile of a chi-square distribution with $D - 1 = 2$ degrees of freedom, d_i is the Mahalanobis distance of observation i, and ϵ is small random noise term, $\epsilon \sim Normal(0, 0.1)$. This transformation shifts the selected observations away from the center so that their Mahalanobis distances exceed the chi-square cutoff by an additional 25%. In standard practice, observations beyond this cutoff are considered potential outliers, and here we artificially move them further to ensure they represent clear outlying values. Finally, data are transformed back to the simplex space by using the inverse ilr transformation [1].

Figure 1 displays the ternary plots corresponding to three randomly selected MC replicates, one for each simulation scenario.

Table 1 contains the posterior means, together with 95% credible intervals for the three simulated scenarios. Interestingly, in the first two fitting studies, S1 and S2, the DCM provides precise estimates of all parameters, both with moderate and high levels of variance inflation. In scenario S3, the model overestimates the amount of artificial outliers, and it assumes a regular component correctly centered around the center of the simplex, with a relatively high precision and a moderate level of variance inflation.

For each scenario, posterior membership probabilities were computed for all observations and MC replicates. Based on these probabilities, observations were classified as potential outliers if $\hat{\pi}_i > 0.5$. Model performance was then evaluated in terms of sensitivity and precision. Sensitivity was defined as the proportion

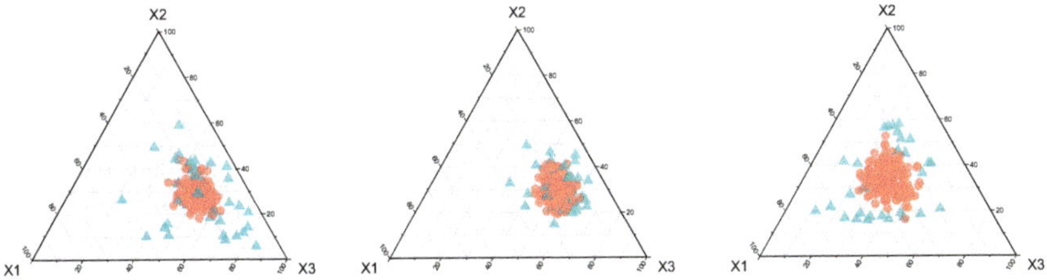

Fig. 1. Ternary plots corresponding to three randomly selected acMC replicates from scenarios S1 (left), S2 (center), and S3 (right). Observations identified as potential outliers are highlighted in light blue triangles.

of truly outlying observations correctly identified by the model, that is, the ratio between true positives and the sum of true positives and false negatives. Precision was defined as the proportion of observations classified as atypical that were truly atypical, corresponding to the ratio between true positives and the sum of true positives and false positives.

Table 1. Posterior means with 95% credible intervals (in parentheses) under the three simulated scenarios.

Parameter	S1	S2	S3
μ_1	0.199 (0.193; 0.207)	0.200 (0.193; 0.207)	0.333 (0.326; 0.341)
μ_2	0.300 (0.293; 0.308)	0.300 (0.293; 0.307)	0.333 (0.325; 0.341)
μ_3	0.500 (0.492; 0.508)	0.499 (0.492; 0.508)	0.334 (0.325; 0.342)
ϕ	102.607 (83.719; 129.500)	103.704 (81.424; 132.890)	74.402 (59.753; 92.962)
p	0.215 (0.138; 0.302)	0.243 (0.117; 0.337)	0.365 (0.294; 0.441)
δ	8.885 (5.261; 13.023)	2.968 (1.462; 4.579)	3.158 (1.942; 4.253)

Table 2. Posterior means with 95% credible intervals (in parentheses) for sensitivity and precision under the three simulated scenarios.

	S1	S2	S3
Sensitivity	0.680 (0.475; 0.850)	0.421 (0.131; 0.667)	0.946 (0.846; 1.000)
Precision	0.889 (0.711; 1.000)	0.778 (0.500; 1.000)	0.610 (0.436; 0.863)

Table 2 reports sensitivity and precision, with corresponding uncertainty intervals, for the three scenarios. In S1, the model exhibits a balanced performance, with moderate sensitivity and high precision, indicating that true positives are detected reliably. S2 represents a more challenging setting. Sensitivity decreases with respect to S1, with a wider interval reflecting unstable detection of true positives, whereas precision remains relatively high. The model is therefore conservative, identifying fewer positives but with high reliability. In S3, the most extreme scenario, sensitivity is the highest and precision is the lowest among the simulated scenarios. In addition, narrower intervals suggest more stable performance compared to previous scenarios.

4 Applicative Example

The coffee dataset [5], available in the R package `robCompositions`, consists of compositional measurements on 30 coffee samples, each described by six chemical components that sum to one. The observations are classified according to three categories: 16 Arabica, 11 Blended, and 3 Robusta. From a compositional perspective, this dataset is known to exhibit a small number of atypical profiles, particularly those corresponding to Robusta coffees. For this reason, it represents a natural benchmark for assessing mixture-based models designed to accommodate heterogeneity and detect compositional outliers.

The DCM is fitted using a HMC algorithm with 40 000 iterations, of which 20 000 are used as warm-up. The posterior chains passed standard convergence tests; however, detailed results are not reported due to space constraints. Posterior summaries are shown in Table 3, in the column labeled "Complete Model".

Table 3. Posterior means with 95% credible intervals (in parentheses) for the complete coffee dataset (Complete Model) and the reduced dataset (Reduced Model).

Parameter	Complete Model	Reduced Model
μ_1	0.199 (0.180; 0.219)	0.207 (0.187; 0.226)
μ_2	0.106 (0.090; 0.123)	0.103 (0.088; 0.119)
μ_3	0.193 (0.173; 0.213)	0.195 (0.175; 0.217)
μ_4	0.263 (0.242; 0.285)	0.260 (0.238; 0.282)
μ_5	0.090 (0.077; 0.105)	0.088 (0.074; 0.103)
μ_6	0.148 (0.131; 0.165)	0.148 (0.132; 0.165)
ϕ	61.792 (45.476; 83.464)	78.707 (50.018; 129.523)
p	0.182 (0.065; 0.347)	0.2770 (0.028; 0.487)
δ	12.173 (6.357; 21.496)	1.343 (0.030; 4.063)
WAIC	-462.3	-453.1

The posterior mean estimates of the common compositional center $\boldsymbol{\mu}$ exhibit narrow CIs, indicating that the dominant compositional profile is estimated with high precision. The estimate of the precision parameter ϕ corresponds to a relatively tight dispersion around the center, reflecting a homogeneous majority

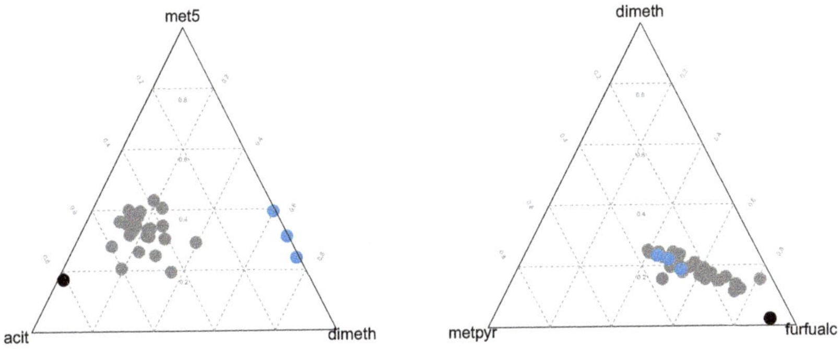

Fig. 2. Ternary plots showing two subsets of three of the six chemical components in the coffee dataset. Robusta observations are colored light blue, the atypical Arabica observation is colored black, and all remaining observations (either Arabica or Blended) are colored gray.

of observations. The estimate of the dispersion inflation parameter δ implies that the second component represents substantially more variable realizations around the same center. The estimate of the mixing weight p is coherent with the presence of a small fraction of atypical observations in the dataset.

To evaluate the model's ability to identify potential outliers, posterior membership probabilities, i.e., $\hat{\pi}_i$, were examined. Results show a clear separation in the data. Four observations exhibit membership probabilities greater than 0.98, and therefore are labeled as outliers. Instead, all remaining observations have membership probabilities below 0.36 (mostly below 0.055), indicating strong assignment to the first component of the mixture. Among the four observations identified as outliers, three correspond to Robusta coffee samples, whereas one is labeled as Arabica. Thus, the model is capable of detecting all the Robusta observations, which are known to be outliers, and additionally identifies one Arabica coffee sample as atypical. An inspection of the ternary plots of all possible subsets of three components among the six chemical components confirms that this Arabica observation deviates from the majority of data along many coordinates. Figure 2 presents only two representative cases for clarity, but similar deviations are observed in other subsets.

To further assess robustness, the DCM is estimated after removing the four atypical observations and results are shown in Table 3, in the column labeled "Reduced Model". Interestingly, the estimated component means remain essentially unchanged with respect to results from the complete model, indicating a high degree of stability of the dominant mixture structure. The model again identifies a contaminated component with a posterior weight of approximately 15%. However, in this case the inflation parameter is small, suggesting only a slight difference in the variability of the two mixture components. Finally, the WAIC [10] values for the complete and the reduced models are very close, indicating similar goodness of fit for both specifications.

5 Discussion

In this work, a DCM was proposed to handle outliers in compositional data by introducing a variance-inflated component within a Bayesian framework. Simulation studies demonstrated that the model provides accurate parameter estimates, effectively identifies atypical observations, and maintains stability even in the presence of moderate/high contamination. The applicative example on the coffee dataset confirmed its practical utility, highlighting the ability to detect both known and previously unrecognized atypical observations. Future extensions may include adaptations to handle compositions with values on the boundary of the simplex, alternative contamination mechanisms beyond variance inflation, and computational improvements for high-dimensional compositional data.

References

1. Aitchison, J.: The Statistical Analysis of Compositional Data. Chapman & Hall, London (1986)
2. Di Brisco, A.M., Migliorati, S., Ongaro, A.: Robustness against outliers: a new variance inflated regression model for proportions. Stat. Model. **20**(3), 274–309 (2020)
3. Filzmoser, P., Hron, K.: Outlier detection for compositional data using robust methods. Math. Geosci. **40**(3), 233–248 (2008)
4. Hawkins, D.M.: Identification of Outliers. Chapman and Hall, London (1980)
5. Korhonová, M., Hron, K., Klimcíková, D., Müller, L., Bednár, P., Barták, P.: Coffee aroma–statistical analysis of compositional data. Talanta **80**(2), 710–715 (2009)
6. Pawlowsky-Glahn, V., Egozcue, J.J., Tolosana-Delgado, R.: Modeling and Analysis of Compositional Data. John Wiley & Sons, Hoboken (2015)
7. Rousseeuw, P.J., Leroy, A.M.: Robust Regression and Outlier Detection. Wiley, New York (1987)
8. Stan Development Team: Stan User's Guide and Reference Manual, Version 2.32 (2025). https://mc-stan.org
9. Tukey, J.W.: A survey of sampling from contaminated distributions. In: Olkin, I. (Ed.), Contributions to Probability and Statistics: Essays in Honor of Harold Hotelling. Stanford Studies in Mathematics and Statistics. Stanford University Press (1960)
10. Vehtari, A., Gelman, A., Gabry, J.: Practical Bayesian model evaluation using leave-one-out cross-validation and WAIC. Stat. Comput. **27**(5), 1413–1432 (2017)

Kernel Deconvolution Regression Estimator for Compositional Data

Marco Di Marzio[1], Stefania Fensore[1(✉)], Agnese Panzera[2], and Chiara Passamonti[3]

[1] G. d'Annunzio University of Chieti-Pescara, Pescara, Italy
stefania.fensore@unich.it
[2] University of Florence, Firenze, Italy
[3] University of Luxembourg, Esch-sur-Alzette, Luxembourg

Abstract. Compositional data arise in many scientific fields and require specialized statistical methods due to the constant-sum constraint. We propose a nonparametric regression approach for compositional responses with either real-valued or compositional predictors subject to measurement error. Specifically, a unified treatment is achieved through a kernel-based approach using a suitable class of kernels on the simplex. A real data application is presented, analyzing the relationship between moss layer composition and O-horizon layer composition in soil.

Keywords: Compositional response · Kola data · Measurement error · Simplicial kernel

1 Introduction

Compositional data are vectors whose entries, called components, are strictly positive and sum up to a constant. The components represent parts of a whole and therefore carry exclusively relative information. For a composition with $D \in Z^+$ components, the associated sample space is the simplex, defined as

$$S^D := \left\{ (x_1, \ldots, x_D)^\top : x_i > 0,\ i = 1, \ldots, D,\ \sum_{i=1}^{D} x_i = k \right\},$$

where k denotes a fixed positive constant, typically equal to 1 in the case of proportions or 100 for percentages. Compositional data arise across a variety of disciplines, including geology, ecology and microbiology, geochemistry, and the study of economic shares, such as the distribution of expenditures across categories within a country. Since classical statistical methods are not suitable for this kind of data due to the constant-sum constraint, methodologies accounting for the geometry of the simplex are required.

While early contributions focused on parametric models, increasing attention has been devoted to nonparametric density estimation for compositional data.

P. Brito et al. (Eds.): IFCS 2026, STUDIES CLASS, pp. 145–152, 2026.
https://doi.org/10.1007/978-3-032-32009-4_18

Concerning kernel methods, [2] proposed the Dirichlet kernel for sparse data situations, and the additive logistic normal kernel based on logarithmic transformation. Then, [3] introduced the isometric log-ratio normal kernel. Recently, [18] exploited the topological equivalence between the simplex and the hypersphere proposing a novel kernel density estimator to address the zero components.

The existing literature highlights a variety of applications of compositional regression analysis, too. From a parametric point of view, [13] introduced experimental design and regression modeling for a real-valued response as a function of a composition, while [14] proposed the simplex-centroid design for studying mixture systems. Later, [7] suggested a linear model using the isometric log-ratio (ILR) coordinates of the response, and [9] studied a partial linear regression for compositional responses with both compositional and Euclidean predictors. For a nonparametric approach, [5] introduced simplicial kernel regression. More recently, [15] proposed α-k-NN and α-kernel approaches tailored to handle compositional data containing zeros, while [16] developed constrained least squares regression for situations where both predictors and responses are compositional.

A largely unaddressed topic in compositional data analysis is the accounting for measurement error. Its impact on compositions was first investigated by [1], while, more recently, [6] discussed a deconvolution approach to density estimation for contaminated compositional data. In this framework, measurement error takes the form of perturbation noise, which is converted into an additive error when compositions are expressed in log-ratio coordinates. Such errors naturally arise, for example, when components do not sum exactly to the constant, or they can remain latent. Typical sources of measurement error include instrument imprecision, subjective estimation procedures, and the intrinsic heterogeneity of the material under study.

In this work, we aim to nonparametrically estimate the regression function when the response is compositional and the predictor, which can be real-valued or compositional, is affected by measurement error. This issue is addressed by adopting a deconvolution kernel strategy.

The paper is organized as follows: in Sect. 2 we provide a theoretical background concerning the Aitchison geometry, along with the nonparametric regression for compositional response. Then, in Sect. 3, we propose the regression model for the errors-in-variables case. Section 4 is dedicated to a real-data application concerning the concentration of chemical elements.

2 Theoretical Background

2.1 Compositional Data and Log-Ratio Transformations

The natural geometry of data on the simplex is the *Aitchison geometry*. Its basic operations are *perturbation* and *powering*, which, for D-part compositions $\boldsymbol{x}$ and $\boldsymbol{y}$, and $a \in \mathbb{R}$, are respectively defined as

$$\boldsymbol{x} \oplus \boldsymbol{y} := \mathcal{C}(x_1 y_1, \ldots, x_D y_D), \qquad a \odot \boldsymbol{x} := \mathcal{C}(x_1^a, \ldots, x_D^a),$$

where $\mathcal{C}(u_1, \ldots, u_D) := \left(u_1/\sum_{i=1}^D u_i, \ldots, u_D/\sum_{i=1}^D u_i\right)^\top$ is the closure operator. The inverse operation of perturbation is given by $\boldsymbol{x} \ominus \boldsymbol{y} := \boldsymbol{x} \oplus ((-1) \odot \boldsymbol{y})$.

Statistical analysis of compositional data is typically conducted through *log-ratio transformations.* Among these, the *centered log-ratio* (CLR) transformation, for $\boldsymbol{x} \in S^D$, is defined as $\operatorname{clr}(\boldsymbol{x}) := \left(\ln\left(x_1/g(\boldsymbol{x})\right), \ldots, \ln\left(x_D/g(\boldsymbol{x})\right)\right)^\top$, where $g(\boldsymbol{x})$ stands for the geometric mean of the components. The CLR embeds compositions into a Euclidean space, but its coordinates are linearly dependent, and it is not subcompositionally coherent, where a subcomposition of $\boldsymbol{x}$ is obtained by selecting a subset of components and renormalizing them to constant sum. These limitations are overcome by the *isometric log-ratio* transformation, which maps compositional data into $\mathbb{R}^{D-1}$. For $\boldsymbol{x} \in S^D$, the ILR is defined as

$$\operatorname{ilr}(\boldsymbol{x}) := \boldsymbol{\Psi}^\top \operatorname{clr}(\boldsymbol{x}),$$

where $\boldsymbol{\Psi}$ is the $D \times (D-1)$ matrix associated with a chosen orthonormal basis of the simplex. The ILR transformation preserves the linear structure of the Aitchison geometry, in the sense that, for all $\boldsymbol{x}, \boldsymbol{y} \in S^D$ and $a \in \mathbb{R}$,

$$\operatorname{ilr}(\boldsymbol{x} \oplus \boldsymbol{y}) = \operatorname{ilr}(\boldsymbol{x}) + \operatorname{ilr}(\boldsymbol{y}), \qquad \operatorname{ilr}(a \odot \boldsymbol{x}) = a \operatorname{ilr}(\boldsymbol{x}),$$

and, as a consequence, $\operatorname{ilr}(\boldsymbol{x} \ominus \boldsymbol{y}) = \operatorname{ilr}(\boldsymbol{x}) - \operatorname{ilr}(\boldsymbol{y})$. Moreover, the *Aitchison norm*,

$$\|\boldsymbol{x}\|_a := \left(\frac{1}{D}\sum_{i \neq j}\left\{\ln\left(\frac{x_i}{x_j}\right)\right\}^2\right)^{1/2},$$

is preserved under the ILR transformation, namely $\|\boldsymbol{x}\|_a = \|\operatorname{ilr}(\boldsymbol{x})\|$, where $\|\cdot\|$ is the usual Euclidean norm. Using the Aitchison geometry, [11] defined the *centre* of a D–part random composition $\boldsymbol{X}$ as

$$\operatorname{Cen}[\boldsymbol{X}] := \arg\min_{\boldsymbol{z} \in S^D} \mathbb{E}\left[\|\boldsymbol{X} \ominus \boldsymbol{z}\|_a^2\right].$$

In virtue of the isometric property of the ILR transformation, the centre admits the representation $\operatorname{Cen}[\boldsymbol{X}] = \operatorname{ilr}^{-1}(\mathbb{E}[\operatorname{ilr}(\boldsymbol{X})])$.

2.2 Nonparametric Regression for Compositional Response

Let $\boldsymbol{Y}$ be a D-part random composition, and let $\boldsymbol{X}$ denote a D-dimensional random variable, which may be either real-valued or compositional. Within the Aitchison geometry, the natural notion of conditional expectation of $\boldsymbol{Y}$ given $\boldsymbol{X} = \boldsymbol{x}$ is provided by the *conditional centre*

$$\boldsymbol{m}(\boldsymbol{x}) := \arg\min_{\boldsymbol{z} \in \mathcal{S}^D} \mathbb{E}\left[\|\boldsymbol{Y} \ominus \boldsymbol{z}\|_a^2 \mid \boldsymbol{X} = \boldsymbol{x}\right] = \operatorname{ilr}^{-1}(\mathbb{E}[\operatorname{ilr}(\boldsymbol{Y}) \mid \boldsymbol{X} = \boldsymbol{x}]).$$

Letting $(\boldsymbol{X}_1, \boldsymbol{Y}_1), \ldots, (\boldsymbol{X}_n, \boldsymbol{Y}_n)$ be independent copies of $(\boldsymbol{X}, \boldsymbol{Y})$, consider the nonparametric regression model

$$\boldsymbol{Y}_i = \boldsymbol{m}(\boldsymbol{X}_i) \oplus \boldsymbol{\varepsilon}_i, \qquad i = 1, \ldots, n, \tag{1}$$

with the $\boldsymbol{\varepsilon}_i$s being compositional error terms with centre equal to the neutral element of the perturbation, $\text{Cen}[\boldsymbol{\varepsilon}_i \mid \boldsymbol{X}_i] = \boldsymbol{s}$, where $\boldsymbol{s} = \mathcal{C}(1, \dots, 1)$, and satisfy $\mathbb{E}\left[\|\boldsymbol{\varepsilon}_i\|_a^2 \mid \boldsymbol{X}_i\right] < \infty$.

Throughout the paper, we use the following notation

$$\boldsymbol{Z}^* := \begin{cases} \text{ilr}(\boldsymbol{Z}), & \text{if } \boldsymbol{Z} \in S^D, \\ \boldsymbol{Z}, & \text{if } \boldsymbol{Z} \in \mathbb{R}^D. \end{cases} \tag{2}$$

Under the above convention, the ILR transformation, when applied to (1), gives the Euclidean representation

$$\boldsymbol{Y}_i^* = \boldsymbol{m}^*(\boldsymbol{X}_i) + \boldsymbol{\varepsilon}_i^*, \tag{3}$$

where $\boldsymbol{m}^*(\boldsymbol{x}) := \text{ilr}(\boldsymbol{m}(\boldsymbol{x})) = \mathbb{E}[\boldsymbol{Y}^* \mid \boldsymbol{X} = \boldsymbol{x}]$, and $\boldsymbol{\varepsilon}_i^* = \text{ilr}(\boldsymbol{\varepsilon}_i)$. Then, a kernel (or local constant) estimator of $\boldsymbol{m}(\boldsymbol{x})$, as introduced by [5], is given by

$$\widehat{\boldsymbol{m}}(\text{ x}) := \text{ilr}^{-1}\left(\widehat{\boldsymbol{m}^*}(\boldsymbol{x})\right),$$

where $\widehat{\boldsymbol{m}^*}(\boldsymbol{x}) = (\widehat{m^*}_1(\boldsymbol{x}), \dots, \widehat{m^*}_{D-1}(\boldsymbol{x}))^\top$, and, for $j = 1, \dots, D-1$,

$$\widehat{m^*}_j(\boldsymbol{x}) := \frac{\sum_{i=1}^n \mathcal{W}(\boldsymbol{x}, \boldsymbol{X}_i) Y_{ij}^*}{\sum_{i=1}^n \mathcal{W}(\boldsymbol{x}, \boldsymbol{X}_i)}, \tag{4}$$

with $\mathcal{W}(\boldsymbol{x}, \boldsymbol{X}_i)$ being a weight defined according to the nature of the predictor.

3 Regression with Measurement Errors

In the classical regression problem with errors-in-variables, the goal is to estimate a regression curve from data in which the predictor variable is observed with error. For the case of a compositional response, we consider model (1) but, instead of $(\boldsymbol{X}_i, \boldsymbol{Y}_i)$, we observe $(\boldsymbol{W}_i, \boldsymbol{Y}_i)$, where, for $i = 1, \dots, n$,

$$\boldsymbol{W}_i = \begin{cases} \boldsymbol{X}_i \oplus \boldsymbol{U}_i, & \text{if } \boldsymbol{X}_i \text{ takes values in } S^D, \\ \boldsymbol{X}_i + \boldsymbol{U}_i, & \text{if } \boldsymbol{X}_i \text{ takes values in } \mathbb{R}^D. \end{cases}$$

The measurement errors $\boldsymbol{U}_i$, as in the classical errors-in-variables problems, are assumed to be independent of $(\boldsymbol{X}_i, \boldsymbol{\varepsilon}_i)$ with a known distribution.

To estimate $\boldsymbol{m}^*(\boldsymbol{x})$, we consider a modified version of (4) with weights based on specific classes of kernels. In particular, for the Euclidean predictor, we consider an Euclidean kernel K, defined on $\mathbb{R}^D$, scaled by a bandwidth matrix H of order D, $K_H(\cdot) := |H|^{-1}K(H^{-1}\cdot)$. In the simplicial predictor case, we start with the class of *simplicial* kernels presented by [5] and defined as

$$\widetilde{K}(\boldsymbol{u}) := \frac{\mathcal{K}(\|\boldsymbol{u}\|_a)}{\int_{S^D} \mathcal{K}(\|\boldsymbol{x}\|_a)\, d\lambda_a(\boldsymbol{x})},$$

where $\mathcal{K}$ is a non–negative, continuous, even function, with maximum at 0, while λ_a denotes the *Aitchison measure* on S^D. Notice that, due to the property of the ILR, these kernels admit the representation

$$\widetilde{K}(\boldsymbol{u}) = K(\boldsymbol{u}^*) := \frac{\mathcal{K}(\|\boldsymbol{u}^*\|)}{\int_{\mathbb{R}^{D-1}} \mathcal{K}(\|\boldsymbol{z}\|)\, d\boldsymbol{z}}.$$

Hence, $\widetilde{K}$ is a density on S^D with respect to λ_a, while K is a density on $\mathbb{R}^{D-1}$ with respect to the Lebesgue measure. In particular, choosing $\mathcal{K}(u) := \exp(-u^2/2)$ yields the isometric log-ratio normal kernel, that is a Gaussian density on the ILR space (see, for example, [3]). Using the above class of simplicial kernels, the notation in (2) allows for a unified formulation of the kernel functions as $K(\boldsymbol{x}^* - \boldsymbol{X}_i^*)$, which corresponds to a D-variate Euclidean kernel in the case of real-valued predictors, or to a simplicial kernel (equivalently, a $(D-1)$-variate Euclidean kernel in the ILR space) in the case of compositional predictors.

Now, let φ_{U^*} be the common characteristic function of the $\boldsymbol{U}_i^*$s, assumed known and such that $\varphi_{U^*}(\boldsymbol{t}) \neq 0$ for all $\boldsymbol{t}$. Setting $d^* := D-1$ in the simplicial case and $d^* := D$ in the Euclidean one, let $H = hI_{d^*}$ be a bandwidth matrix, with $h > 0$ and I_{d^*} being the $d^* \times d^*$ identity matrix. A deconvolution kernel with single bandwidth h can be then defined as

$$K_h^{\text{dec}}(\boldsymbol{u}) = \frac{1}{(2\pi)^{d^*}} \int_{\mathbb{R}^{d^*}} e^{-i\boldsymbol{t}^\top \boldsymbol{u}^*} \frac{\varphi_K(h\boldsymbol{t})}{\varphi_{U^*}(\boldsymbol{t})}\, d\boldsymbol{t},$$

where $\varphi_K(\cdot)$ is the characteristic function of the kernel K discussed above. The deconvolution estimator of $m_j^*(\boldsymbol{x})$ is then given by

$$\widehat{m}_{j,\text{dec}}^*(\boldsymbol{x}) = \frac{\sum_{i=1}^n K_h^{\text{dec}}(\boldsymbol{x}^* - \boldsymbol{W}_i^*)\, Y_{ij}^*}{\sum_{i=1}^n K_h^{\text{dec}}(\boldsymbol{x}^* - \boldsymbol{W}_i^*)}, \tag{5}$$

and, letting $\widehat{\boldsymbol{m}^*}_{\text{dec}}(\boldsymbol{x}) := \left(\widehat{m}_{1,\text{dec}}^*(\boldsymbol{x}), \ldots, \widehat{m}_{D-1,\text{dec}}^*(\boldsymbol{x})\right)^\top$, the deconvolution regression estimator is defined as

$$\widehat{\boldsymbol{m}}_{\text{dec}}(\boldsymbol{x}) := \text{ilr}^{-1}\left(\widehat{\boldsymbol{m}^*}_{\text{dec}}(\boldsymbol{x})\right).$$

The above estimator can be equivalently obtained using the *unbiased score* approach discussed by [4] to estimate the $m^*{}_j$s. Specifically, estimator (4) is the ratio of the quantities $M(\boldsymbol{x}, \boldsymbol{X}_i, T_i) := \mathcal{W}(\boldsymbol{x}, \boldsymbol{X}_i)T_i$, where $T_i = Y_{ij}^*$ at the numerator, and $T_i = 1$ at the denominator. The unbiased score for $n^{-1}\sum_{i=1}^n M(\boldsymbol{x}, \boldsymbol{X}_i, T_i)$ is $n^{-1}\sum_{i=1}^n \mathcal{L}(\boldsymbol{x}, \boldsymbol{W}_i, T_i)$, where $\mathcal{L}(\cdot)$ satisfies

$$\mathbb{E}[\mathcal{L}(\boldsymbol{x}, \boldsymbol{W}_i, T_i) \mid \boldsymbol{X}_i, T_i] = M(\boldsymbol{x}, \boldsymbol{X}_i, T_i).$$

Then, solving the above equation for both $T_i = Y_{ij}^*$ and $T_i = 1$, leads to (5).

4 Application

As an illustrative example, we consider a dataset analyzed for a simplicial–simplicial regression problem in [5], based on the Kola Ecogeochemistry Project (see [12]). In total, approximately 600 soil samples were collected between 1993 and 1998 from four different layers (moss, O-horizon, B-horizon, C-horizon) and analyzed for more than 50 chemical elements. The data are available in the `StatDA` package for `R`. Following [5], we focus on a three-part composition from the O-horizon layer, namely Strontium (Sr), Rubidium (Rb), and Calcium (Ca), used as compositional predictor, and a three-part composition from the moss layer, Magnesium (Mg), Potassium (K), and Phosphorus (P), taken as compositional response. In this regression context, it is reasonable to assume that the O-horizon composition is affected by measurement error. This is because the observed proportions may deviate from the true site-level composition due to several sources of variability. First, O-horizon material is micro-spatially heterogeneous, containing a mix of mineral grains, organic fragments, and inclusions, meaning small sampling variations can cause significant compositional differences (see [19]). Second, errors are introduced during sample preparation, such as sieving and grinding, which may fail to produce perfectly representative aliquots. By contrast, the moss-layer composition is treated as an error-free response, in the sense that it represents the realized bioaccumulation outcome; any remaining variability including biological heterogeneity and potential minor measurement imprecision is assumed to be absorbed by the model's residual term.

We compare the proposed deconvolution estimator with the classical local constant estimator for simplicial-simplicial regression, where the measurement error affecting the covariate is ignored. Let $\boldsymbol{X} = (\mathrm{Sr}, \mathrm{Rb}, \mathrm{Ca}) \in S^3$ denote the O-horizon composition and $\boldsymbol{Y} = (\mathrm{Mg}, \mathrm{K}, \mathrm{P}) \in S^3$ the moss-layer composition. As in [5], we express both compositions through the following isometric log-ratio balances:

$$X_1^* = \sqrt{\frac{2}{3}} \ln\left(\frac{\mathrm{Rb}}{\sqrt{\mathrm{Sr} \cdot \mathrm{Ca}}}\right), \qquad X_2^* = \frac{1}{\sqrt{2}} \ln\left(\frac{\mathrm{Ca}}{\mathrm{Sr}}\right),$$

$$Y_1^* = \sqrt{\frac{2}{3}} \ln\left(\frac{\sqrt{\mathrm{K} \cdot \mathrm{P}}}{\mathrm{Mg}}\right), \qquad Y_2^* = \frac{1}{\sqrt{2}} \ln\left(\frac{\mathrm{K}}{\mathrm{P}}\right).$$

This allows the simplicial regression function to be estimated coordinatewise in the ilr space, by fitting separate regression models on each balance. For the errors-in-variables case, we assume for the compositional error a isometric log-ratio normal density with diagonal covariance matrix and standard deviations 0.4 and 0.25.

For $n = 617$ observations of $(\boldsymbol{X}, \boldsymbol{Y})$, using a diagonal bandwidth matrix with entries h_1 and h_2 for a gaussian kernel K, we consider the corrected Akaike Information Criterion (see [10]) for selecting the smoothing parameters. This criterion leads to the values $h_1 = 0.60$ and $h_2 = 0.38$ for estimating m_1^*, and $h_1 = 0.45$ and $h_2 = 0.29$ for m_2^*. We compare the classical local constant estimator with our deconvoluted version, in terms of the following compositional residual sum of squares

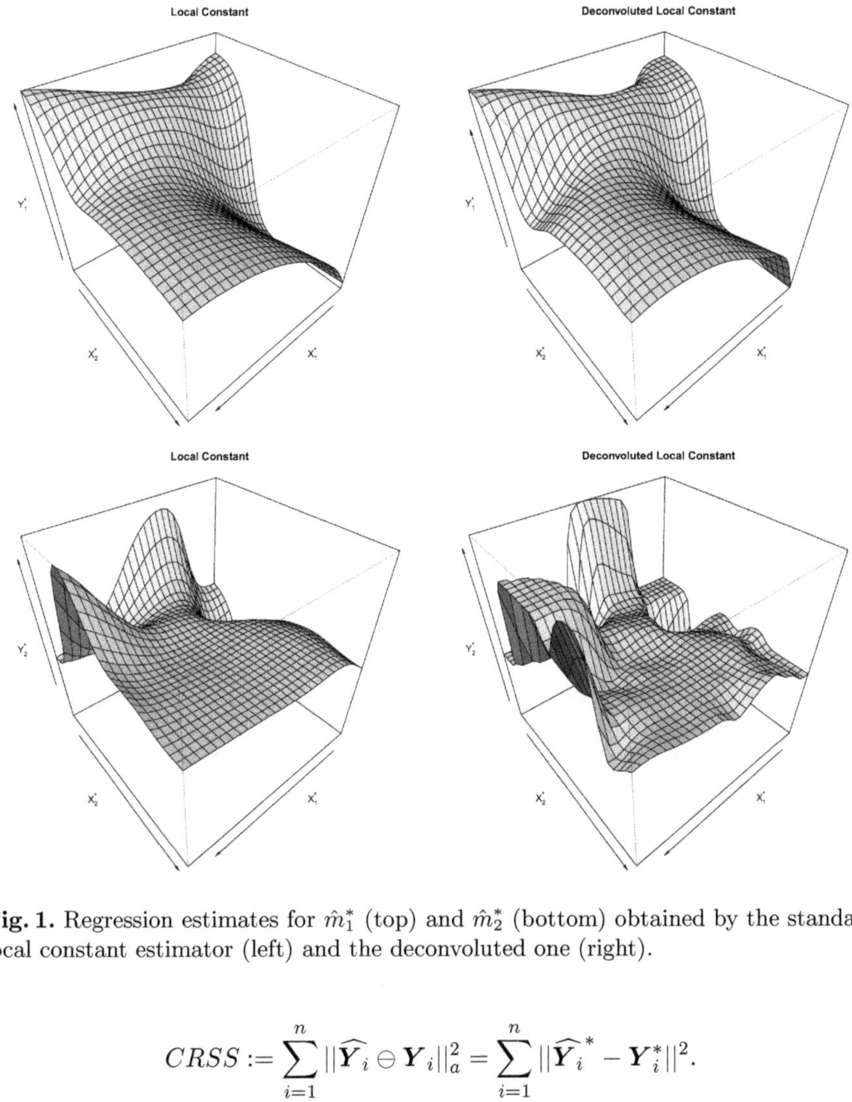

Fig. 1. Regression estimates for $\hat{m}_1^*$ (top) and $\hat{m}_2^*$ (bottom) obtained by the standard local constant estimator (left) and the deconvoluted one (right).

$$CRSS := \sum_{i=1}^{n} ||\widehat{\boldsymbol{Y}}_i \ominus \boldsymbol{Y}_i||_a^2 = \sum_{i=1}^{n} ||\widehat{\boldsymbol{Y}}_i^* - \boldsymbol{Y}_i^*||^2.$$

The resulting estimates are presented in Fig. 1. As expected, the surface obtained using the deconvolution approach is characterized by reduced smoothness relative to the naive one. The prediction accuracy, measured by the compositional residual sum of squares is 93 for the nonparametric errors-in-variables specification, while it decreases to 81.4 for the deconvoluted model.

References

1. Aitchison, J., Shen, S.M.: Measurement error in compositional data. Math. Geol. **16**(6), 637–650 (1984)
2. Aitchison, J., Lauder, I.J.: Kernel density estimation for compositional data. J. R. Stat. Soc.: Ser. C: Appl. Stat. **34**(2), 129–137 (1985)
3. Chacón, J.E., Mateu-Figueras, G., Martin-Fernández, J.A.: Gaussian kernels for density estimation with compositional data. Comput. Geosci. **37**, 702–711 (2011)
4. Delaigle, A.: Nonparametric kernel methods with errors-in-variables: constructing estimators, computing them, and avoiding common mistakes. Aust. N. Z. J. Stat. **56**(2), 105–124 (2014)
5. Di Marzio, M., Panzera, A., Venieri, C.: Non-parametric regression for compositional data. Stat. Mod. **15**(2), 113–133 (2015)
6. Marzio, D., Fensore, M., Panzera, S., Passamonti, A.C.: Deconvolution density estimation for compositional data (2025). Submitted
7. Egozcue, J.J., Daunis-I-Estadella, J., Pawlowsky-Glahn, V., Hron, K., Filzmoser, P.: Simplicial regression. The normal model. J. Appl. Probab. Stat. **6**, 87–108 (2012)
8. Filzmoser, P.: StatDA': Statistical Analysis for Environmental Data. R Package Version, 1.7.11 (2025)
9. Han, H., Yu, K.: Partial linear regression of compositional data. J. Korean Stat. Soc. **51**(4), 1090–1116 (2022)
10. Hurvich, C.M., Simonoff, J.S., Tsai, C.L.: Smoothing parameter selection in nonparametric regression using an improved Akaike information criterion. J. R. Stat. Soc. Ser. B Stat Methodol. **60**(2), 271–293 (1998)
11. Pawlowsky-Glahn, V., Egozcue, J.J.: Geometric approach to statistical analysis on the simplex. SERRA **15**, 384–398 (2001)
12. Reimann, C., Äyräs, M., Chekushin, V.A., Bogatyrev, I.V., Boyd, R., Caritat, P.D., et al.: Environmental geochemical atlas of the central Barents region. Geological Survey of Norway (NGU), Geological Survey of Finland (GTK) and Central Kola Expedition (CKE). Special publication, Trondheim, Espoo, Monchegorsk (2011)
13. Scheffé, H.: Experiments with mixtures. J. R. Stat. Soc. Ser. B Stat Methodol. **20**(2), 344–360 (1958)
14. Scheffé, H.: The simplex-centroid design for experiments with mixtures. J. R. Stat. Soc. Ser. B Stat. Methodol. **25**(2), 235–251 (1963)
15. Tsagris, M., Alenazi, A., Stewart, C.: Flexible non-parametric regression models for compositional response data with zeros. Stat. Comput. **33**(5), 106 (2023)
16. Tsagris, M.: Constrained least squares simplicial-simplicial regression. Stat. Comput. **35**(2), 27 (2025)
17. Wing, M.G., Frank, J.: Vertical measurement accuracy and reliability of mapping-grade GPS receivers. Comput. Electron. Agric. **78**(2), 188–194 (2011)
18. Yoon, C., Choi, H., Ahn, J.: Kernel density estimation for compositional data with zeros via hypersphere mapping. Comput. Stat. Data Anal. 108249 (2025)
19. Reimann, C., et al.: Geochemical gradients in soil O-horizon samples from southern Norway: natural or anthropogenic? Appl. Geochem. **24**(1), 62–76 (2009)

On Hybrid Principal Hessian Directions for Sufficient Dimension Reduction with Binary Response

Yuexiao Dong[1(✉)], Lei Li[2], and Abdul-Nasah Soale[3]

[1] Department of Statistics, Operations, and Data Science, Temple University, Philadelphia, PA 19122, USA
ydong@temple.edu
[2] LunarAI LLC, Dover, DE 19901, USA
[3] Department of Mathematics, Applied Mathematics, and Statistics, Case Western Reserve University, Cleveland, OH 44106, USA

Abstract. We propose a novel hybrid principal Hessian directions method for dimension reduction with binary responses. To achieve data-adaptive weighting between an ordinary least squares component and a principal Hessian directions component, we introduce a working Gaussian index model in which the binary response is governed by two additive index components. The effectiveness of the proposed approach is demonstrated through extensive simulation studies and a real data analysis.

Keywords: central subspace · Gaussian index model · generalized additive model · Stein's identity

1 Introduction

Consider a univariate response Y and a p-dimensional predictor $\mathbf{X} \in \mathbb{R}^p$. In classical sufficient dimension reduction (SDR), we aim to find a matrix $\mathbf{B}$ with the smallest possible column space such that $Y \perp\!\!\!\perp \mathbf{X} \mid \mathbf{B}^\top \mathbf{X}$, where "$\perp\!\!\!\perp$" denotes independence. The corresponding minimal column space is known as the *central subspace* for the regression of Y versus $\mathbf{X}$, and is denoted by $\mathcal{S}_{Y|\mathbf{X}}$. Popular SDR methods, among many others, include ordinary least squares (OLS) [3], sliced inverse regression (SIR) [2], minimum average variance estimation (MAVE) [10], and principal Hessian directions (PHD) [1].

Sufficient dimension reduction with binary responses has gained considerable momentum in recent years. [6] and [7] proposed probability-enhanced SDR methods that utilize the original binary response together with $\Pr(Y = 1 \mid \mathbf{X})$. [8] studied envelope SIR and envelope linear discriminant analysis for dimension reduction and binary classification. When Li [2] first introduced SIR, it involved only the first inverse moment $\mathrm{E}(\mathbf{X} \mid Y)$. In his rejoinder, Li [2] proposed a hybrid estimator, SIR-α, that combines the original SIR with an additional term involving the second inverse moment $\mathrm{Var}(\mathbf{X} \mid Y)$.

P. Brito et al. (Eds.): IFCS 2026, STUDIES CLASS, pp. 153–161, 2026.
https://doi.org/10.1007/978-3-032-32009-4_19

In this paper, we propose a binary response hybrid SDR estimator that combines the first-order cross moment $\mathrm{E}(\mathbf{X}Y)$ and the second-order cross moment $\mathrm{E}\{(\mathbf{X}\mathbf{X}^\top - \mathbf{I}_p)Y\}$. On one hand, we leverage the idea of SIR-α and synthetically combine the central subspace information contained in both the first-order and the second-order moments. On the other hand, similar to probability-enhanced SDR methods designed specifically for binary responses, we employ a working model for $\Pr(Y = 1 \mid \mathbf{X})$ to guide the selection of data-adaptive weights. In the next section, we present the population-level development and the corresponding sample-level algorithm. Simulation studies and a real-world data analysis are reported in Sect. 3.

2 Hybrid PHD

2.1 Population Level Development

We propose a novel Gaussian index model to facilitate SDR with a binary response. For $(Y, \mathbf{X}) \in \{0, 1\} \times \mathbb{R}^p$, consider

$$\begin{aligned} &\log \frac{\pi(\mathbf{X})}{1-\pi(\mathbf{X})} = c_0 + c_1 f_1(U) + c_2 f_2(W^2), \text{ where} \\ &\pi(\mathbf{X}) = \Pr(Y = 1 \mid \mathbf{X}), \mathbf{X} \sim N_p(\mathbf{0}, \mathbf{I}_p), U = \boldsymbol{\beta}_1^\top \mathbf{X}, \text{ and } W = \boldsymbol{\beta}_2^\top \mathbf{X}. \end{aligned} \tag{1}$$

For identifiability, we assume $\mathrm{E}\{f_1(U)\} = \mathrm{E}\{f_2(W^2)\} = 0$, and $\mathrm{Var}\{f_1(U)\} = \mathrm{Var}\{f_2(W^2)\} = 1$.

Our first result connects the OLS estimator $\mathbf{m} = \mathrm{E}(\mathbf{X}\mathbf{Y})$ with $\boldsymbol{\beta}_1$. We implicitly assume that $f_1(\cdot)$ and $f_2(\cdot)$ are continuously differentiable and that all involved moments exist.

Proposition 1. *Assume $\|\boldsymbol{\beta}_1\| = \|\boldsymbol{\beta}_2\| = 1$ and $\boldsymbol{\beta}_1^\top \boldsymbol{\beta}_2 = 0$. Under model (1), we have $\mathbf{m} \propto \boldsymbol{\beta}_1$.*

Proof: First note that $U = \boldsymbol{\beta}_1^\top \mathbf{X}$ and $W = \boldsymbol{\beta}_2^\top \mathbf{X}$ become independent $N(0, 1)$. Denote $\pi(\mathbf{X}) = \sigma\{\eta(\mathbf{X})\}$, where

$$\sigma(t) = (1 + e^{-t})^{-1}, \text{ and } \eta(\mathbf{X}) = c_0 + c_1 f_1(U) + c_2 f_2(W^2).$$

It follows that

$$\sigma'(t) = \sigma(t)\{1 - \sigma(t)\}, \text{ and } \nabla\eta(\mathbf{X}) = c_1 f_1'(U)\,\boldsymbol{\beta}_1 + 2c_2\, W f_2'(W^2)\,\boldsymbol{\beta}_2. \tag{2}$$

From the law of iterated expectations, we have

$$\mathbf{m} = \mathrm{E}(\mathbf{X}Y) = \mathrm{E}\{\mathbf{X}\mathrm{E}(Y \mid \mathbf{X})\} = \mathrm{E}\{\mathbf{X}\pi(\mathbf{X})\}.$$

By Stein's identity and $\mathbf{X} \sim N_p(\mathbf{0}, \mathbf{I}_p)$,

$$\mathbf{m} = \mathrm{E}\{\mathbf{X}\pi(\mathbf{X})\} = \mathrm{E}[\nabla\pi(\mathbf{X})] = \mathrm{E}[\sigma'(\eta)\,\nabla\eta(\mathbf{X})], \tag{3}$$

where $\sigma'(\eta) = \sigma'\{\eta(\mathbf{X})\} = \pi(\mathbf{X})(1 - \pi(\mathbf{X}))$. Plug (2) into (3), and we get

$$\mathbf{m} = c_1 \boldsymbol{\beta}_1 \,\mathrm{E}\{\sigma'(\eta) f_1'(U)\} + 2c_2 \boldsymbol{\beta}_2 \,\mathrm{E}\{\sigma'(\eta)\, W f_2'(W^2)\}\,. \tag{4}$$

Given $U = u$, it is easy to see that $\sigma'(\eta)\, W f_2'(W^2)$ is an odd function of W. Together with $W \sim N(0, 1)$, we have

$$\mathrm{E}\{\sigma'(\eta)\, W f_2'(W^2)\} = \mathrm{E}[\mathrm{E}\{\sigma'(\eta)\, W f_2'(W^2) \mid U\}] = 0.$$

Thus the second term on the right-hand side of (4) becomes 0, and we get the desired result. □

The next result connects the PHD kernel matrix $\mathbf{M} = \mathrm{E}\{(\mathbf{X}\mathbf{X}^\top - \mathbf{I}_p)Y\}$ with $\boldsymbol{\beta}_2$. We implicitly assume that $f_1(\cdot)$ and $f_2(\cdot)$ are twice continuously differentiable and that all involved moments exist.

Proposition 2. *Assume* $\|\boldsymbol{\beta}_1\| = \|\boldsymbol{\beta}_2\| = 1$. *Let* $\mathbf{P} = \mathbf{I}_p - \boldsymbol{\beta}_1\boldsymbol{\beta}_1^\top$. *Under model (1), we have* $\mathbf{PMPMP} = a^2\boldsymbol{\beta}_2\boldsymbol{\beta}_2^\top$ *for some constant scalar* a.

Proof: From the law of iterated expectations, we have

$$\mathbf{M} = \mathrm{E}\{(\mathbf{X}\mathbf{X}^\top - \mathbf{I}_p)\mathrm{E}(Y \mid \mathbf{X})\} = \mathrm{E}\{(\mathbf{X}\mathbf{X}^\top - \mathbf{I}_p)\pi(\mathbf{X})\}.$$

By the second-order Stein's identity for $\mathbf{X} \sim N_p(\mathbf{0}, \mathbf{I}_p)$,

$$\mathrm{E}\{(\mathbf{X}\mathbf{X}^\top - \mathbf{I}_p)\pi(\mathbf{X})\} = \mathrm{E}\{\nabla^2\pi(\mathbf{X})\}\,.$$

Using the chain rule,

$$\nabla^2\pi(\mathbf{X}) = \sigma''(\eta)\, \nabla\eta(\mathbf{X})\nabla\eta(\mathbf{X})^\top + \sigma'(\eta)\, \nabla^2\eta(\mathbf{X}),$$

Here $\sigma''(\eta) = \sigma'(\eta)\{1 - 2\sigma(\eta)\}$, $\nabla\eta(\mathbf{X})$ is defined in (2), and

$$\nabla^2\eta(\mathbf{X}) = c_1 f_1''(U)\, \boldsymbol{\beta}_1\boldsymbol{\beta}_1^\top + c_2\{2f_2'(W^2) + 4W^2 f_2''(W^2)\}\, \boldsymbol{\beta}_2\boldsymbol{\beta}_2^\top.$$

Note that $\mathbf{PMP}$ will drop all the terms that involve $\boldsymbol{\beta}_1$. We get

$$a = \mathrm{E}\big[\sigma''(\eta)\, \{2c_2 W f_2'(W^2)\}^2 + \sigma'(\eta)\, c_2\{2f_2'(W^2) + 4W^2 f_2''(W^2)\}\big]$$

and $\mathbf{PMP} = a\boldsymbol{\beta}_2\boldsymbol{\beta}_2^\top$. It follows that $\mathbf{PMPMP} = a^2\boldsymbol{\beta}_2\boldsymbol{\beta}_2^\top\boldsymbol{\beta}_2\boldsymbol{\beta}_2^\top = a^2\boldsymbol{\beta}_2\boldsymbol{\beta}_2^\top$. □

We propose the following hybrid PHD kernel matrix

$$\mathbf{A} = c_1^2\mathbf{m}\mathbf{m}^\top + c_2^2\mathbf{M}^2. \tag{5}$$

From the proof of Proposition 2, we observe that the PHD kernel matrix $\mathbf{M}$ may possess negative eigenvalues. Since the matrix $c_1^2\,\mathbf{m}\mathbf{m}^\top$ is always nonnegative definite, we define the hybrid kernel as $\mathbf{A} = c_1^2\,\mathbf{m}\mathbf{m}^\top + c_2^2\,\mathbf{M}^2$, thereby avoiding signal cancellation that could arise from negative eigenvalues of $\mathbf{M}$.

The next result establishes that the column space of $\mathbf{A}$ can be used to recover the central subspace. Although the working Gaussian index model (1) is required for Propositions 1 and 2, it is no longer needed for Proposition 3. Even when the weights c_1^2 and c_2^2 are derived from a misspecified working model (1), Proposition 3 guarantees that span($\mathbf{A}$) remains Fisher consistent. Similarly, under the working model (1), we have $Y \perp\!\!\!\perp \mathbf{X} \mid (\boldsymbol{\beta}_1^\top \mathbf{X}, \boldsymbol{\beta}_2^\top \mathbf{X})$, which implies that the dimension of $\mathcal{S}_{Y|\mathbf{X}}$ is at most 2. In contrast, the hybrid PHD estimator defined in (5) does not impose this restriction.

Proposition 3. *Suppose* $\mathbf{X} \sim N_p(\mathbf{0}, \mathbf{I}_p)$. *Then we have* $\text{span}(\mathbf{A}) \subseteq \mathcal{S}_{Y|\mathbf{X}}$.

Proof: Let $\mathbf{B}$ be the orthogonal basis of $\mathcal{S}_{Y|\mathbf{X}}$ such that $Y \perp\!\!\!\perp \mathbf{X} \mid \mathbf{B}^\top \mathbf{X}$ and $\mathbf{B}^\top \mathbf{B} = \mathbf{I}$. The normality of $\mathbf{X}$ guarantees that

$$\mathrm{E}(\mathbf{X} \mid \mathbf{B}^\top \mathbf{X}) = \mathbf{B}\mathbf{B}^\top \mathbf{X} \text{ and } \mathrm{Var}(\mathbf{X} \mid \mathbf{B}^\top \mathbf{X}) = \mathbf{I}_p - \mathbf{B}\mathbf{B}^\top. \tag{6}$$

From the law of iterated expectations, we have

$$\mathbf{m} = \mathrm{E}\{\mathbf{X}\mathrm{E}(Y \mid \mathbf{X})\} = \mathrm{E}\{\mathbf{X}\mathrm{E}(Y \mid \mathbf{B}^\top \mathbf{X})\} = \mathrm{E}\{\mathrm{E}(\mathbf{X} \mid \mathbf{B}^\top \mathbf{X})Y\} = \mathbf{B}\mathbf{B}^\top \mathbf{m}. \tag{7}$$

Similar argument leads to

$$\mathbf{M} = \mathrm{E}\left[\mathrm{E}\{(\mathbf{X}\mathbf{X}^\top - \mathbf{I}_p) \mid \mathbf{B}^\top \mathbf{X}\}Y\right]. \tag{8}$$

Note that $\mathrm{E}(\mathbf{X}\mathbf{X}^\top \mid \mathbf{B}^\top \mathbf{X}) = \mathrm{Var}(\mathbf{X} \mid \mathbf{B}^\top \mathbf{X}) + \mathrm{E}(\mathbf{X} \mid \mathbf{B}^\top \mathbf{X})\mathrm{E}^\top(\mathbf{X} \mid \mathbf{B}^\top \mathbf{X})$. (6), (8) and $\mathbf{B}^\top \mathbf{B} = \mathbf{I}$ together guarantee that

$$\mathbf{M} = \mathrm{E}\{(\mathbf{B}\mathbf{B}^\top \mathbf{X}\mathbf{X}^\top \mathbf{B}\mathbf{B}^\top - \mathbf{B}\mathbf{B}^\top)Y\} = \mathbf{B}\mathbf{B}^\top \mathbf{M}\mathbf{B}\mathbf{B}^\top. \tag{9}$$

Plug (7) and (9) into $\mathbf{A} = c_1^2 \mathbf{m}\mathbf{m}^\top + c_2^2 \mathbf{M}^2$, and we have $\text{span}(\mathbf{A}) \subseteq \text{span}(\mathbf{B}) = \mathcal{S}_{Y|\mathbf{X}}$. □

To fix the idea, we consider two toy examples. Assume $\mathbf{X} = (X_1, X_2)^\top$, $\boldsymbol{\beta}_1 = (1, 0)^\top$, $\boldsymbol{\beta}_2 = (0, 1)^\top$, and model (1) holds. From the proofs of Proposition 1 and Proposition 2, we can calcuate $\mathbf{m} = (c_1 \mathrm{E}\{\sigma'(\eta)\}, 0)^\top$ and

$$\mathbf{M} = \mathrm{diag}(c_1^2 \mathrm{E}\{\sigma''(\eta)\}, 2c_2^2 \mathrm{E}\{X_2^2 \sigma''(\eta)\} + \sqrt{2} c_2 \mathrm{E}\{\sigma'(\eta)\}).$$

In the first example, let $\eta(\mathbf{X}) = c_2 X_2^2$ with $c_2 \neq 0$. Then $\mathcal{S}_{Y|\mathbf{X}} = \text{span}(\boldsymbol{\beta}_2)$. OLS fails in this case as $c_1 = 0$ and $\mathbf{m} = \mathbf{0}$. It is easy to see that $\mathbf{M} = \mathrm{diag}(0, c_3)$ for some $c_3 \neq 0$ as long as $c_2 \neq -\sqrt{2}^{-1} \mathrm{E}^{-1}\{X_2^2 \sigma''(\eta)\}\mathrm{E}\{\sigma'(\eta)\}$. It follows that $\text{span}(\mathbf{M}) = \text{span}(\boldsymbol{\beta}_2) = \mathcal{S}_{Y|\mathbf{X}}$ and PHD works for SDR. Our hybrid PHD also works because $\mathbf{A} = c_2^2 \mathbf{M}^2$ and $\text{span}(\mathbf{A}) = \text{span}(\mathbf{M}) = \mathcal{S}_{Y|\mathbf{X}}$.

In the second example, let $\eta(\mathbf{X}) = c_1 X_1$ with $c_1 \neq 0$. Then $\mathcal{S}_{Y|\mathbf{X}} = \text{span}(\boldsymbol{\beta}_1)$. OLS works because $\mathbf{m} = c_1 \mathrm{E}\{\sigma'(\eta)\}\boldsymbol{\beta}_1$ with $\mathrm{E}\{\sigma'(\eta)\} = \mathrm{E}\{\pi(\mathbf{X})(1 - \pi(\mathbf{X}))\} > 0$ and $c_1 \neq 0$. The second diagonal element of $\mathbf{M}$ is 0 because $c_2 = 0$. Due to the symmetry of the logistic function $\sigma(\cdot)$, we can show $\sigma''(-t) = -\sigma''(t)$. It follows that $\sigma''\{\eta(-\mathbf{X})\} = \sigma''(-c_1 X_1) = -\sigma''(c_1 X_1) = -\sigma''\{\eta(\mathbf{X})\}$. Together with the symmetry of $\mathbf{X}$, the first diagonal element of $\mathbf{M}$ becomes $c_1^2 \mathrm{E}\{\sigma''(\eta)\} = 0$. Thus we have $\mathbf{M} = \mathbf{0}$ and PHD fails. Our hybrid PHD still works because $\mathbf{A} = c_1^2 \mathbf{m}\mathbf{m}^\top = \text{span}(\mathbf{m}) = \mathcal{S}_{Y|\mathbf{X}}$.

We remark that PHD fails in the second example because $\eta(\mathbf{X})$ is an odd function of $\mathbf{X}$. If $\eta(\mathbf{X}) = c_0 + c_1 X_1$ with $c_0 \neq 0$, then $\eta(\mathbf{X})$ is no longer odd and $\mathrm{E}\{\sigma''(\eta)\} \neq 0$ as a result. PHD will be able to recover $\mathcal{S}_{Y|\mathbf{X}} = \text{span}(\boldsymbol{\beta}_1)$ in this modified example.

2.2 Sample Level Algorithm

Denote the dimension of $\mathcal{S}_{Y|\mathbf{X}}$ as d, which we assume to be known. Sequential test is used in SIR [2] and PHD [1] to determine d. Given $\{(Y_i, \mathbf{X}_i)\}_{i=1}^n$ and d, we propose the following hybrid PHD algorithm:

1. Center and standardize $\mathbf{X}$. Let $\bar{\mathbf{X}} = n^{-1}\sum_{i=1}^n \mathbf{X}_i$, $\mathbf{S} = n^{-1}\sum_{i=1}^n(\mathbf{X}_i - \bar{\mathbf{X}})(\mathbf{X}_i - \bar{\mathbf{X}})^\top$, and $\tilde{\mathbf{X}}_i = \mathbf{S}^{-1/2}(\mathbf{X}_i - \bar{\mathbf{X}})$ for $i = 1, \ldots, n$. Calculate OLS estimator $\hat{\mathbf{m}} = n^{-1}\sum_{i=1}^n \tilde{\mathbf{X}}_i Y_i$. Let $\hat{\mathbf{P}} = \mathbf{I}_p - \tilde{\mathbf{m}}\tilde{\mathbf{m}}^\top$, where $\tilde{\mathbf{m}} = \hat{\mathbf{m}}/\sqrt{\hat{\mathbf{m}}^\top\hat{\mathbf{m}}}$.
2. Estimate PHD kernel matrix by $\hat{\mathbf{M}} = n^{-1}\sum_{i=1}^n(\tilde{\mathbf{X}}_i\tilde{\mathbf{X}}_i^\top - \mathbf{I}_p)Y_i$. Let $\hat{\gamma}$ be the eigenvector of $\hat{\mathbf{P}}\hat{\mathbf{M}}\hat{\mathbf{P}}\hat{\mathbf{M}}\hat{\mathbf{P}}$ that corresponds to the largest eigenvalue.
3. Let $\hat{U}_i = \hat{\mathbf{m}}^\top\tilde{\mathbf{X}}_i$ and $\hat{W}_i = \hat{\gamma}^\top\tilde{\mathbf{X}}_i$ for $i = 1, \ldots, n$. Use the R package *mgcv* to fit $\text{logit}\{\Pr(Y_i = 1)\} = f_1^*(\hat{U}_i) + f_2^*(\hat{W}_i^2)$. Denote the estimated additive components as $\hat{f}_1^*(\cdot)$ and $\hat{f}_2^*(\cdot)$.
4. Let $\bar{f}_1^* = n^{-1}\sum_{i=1}^n \hat{f}_1^*(\hat{U}_i)$, $\bar{f}_2^* = n^{-1}\sum_{i=1}^n \hat{f}_2^*(\hat{W}_i^2)$, $\hat{c}_1^2 = n^{-1}\sum_{i=1}^n\{\hat{f}_1^*(\hat{U}_i) - \bar{f}_1^*\}^2$ and $\hat{c}_2^2 = n^{-1}\sum_{i=1}^n\{\hat{f}_2^*(\hat{W}_i^2) - \bar{f}_2^*\}^2$. Estimate hybrid PHD kernel matrix by $\hat{\mathbf{A}} = \hat{c}_1^2\hat{\mathbf{m}}\hat{\mathbf{m}}^\top + \hat{c}_2^2\hat{\mathbf{M}}^2$. Let $\hat{\boldsymbol{\zeta}}_1, \ldots, \hat{\boldsymbol{\zeta}}_d$ be the eigenvectors that correspond to the d largest eigenvalues of $\hat{\mathbf{A}}$.
5. Let $\hat{\mathbf{v}}_j = \mathbf{S}^{-1/2}\hat{\boldsymbol{\zeta}}_j$, $j = 1, \ldots, d$. The final hybrid PHD estimator of $\mathcal{S}_{Y|\mathbf{X}}$ is $\text{span}(\hat{\mathbf{B}})$, where $\hat{\mathbf{B}} = (\hat{\mathbf{v}}_1, \ldots, \hat{\mathbf{v}}_d)$.

Step 1 of the algorithm follows Proposition 1, where we recover the index associated with the first additive component. Step 2 follows Proposition 2, where we recover the index associated with the second additive component. Step 3 is motivated by the proposed working Gaussian index model (1). Because c_1 and f_1 are not separable without the identifiability constraint $\text{Var}\{f_1(U)\} = 1$, the *mgcv* package directly estimates their product $f_1^*(\cdot) = c_1 f_1(\cdot)$. See, for example, equation (8) in Wood [9] about generalized additive models. Under the unit variance constraint, we have $c_1^2 = \text{Var}\{c_1 f_1(U)\} = \text{Var}\{f_1^*(U)\}$, which can be estimated by the sample variance of $\hat{f}_1^*(\hat{U})$. A similar argument is used to estimate c_2^2. Step 4 follows from Proposition 3 and the subsequent discussion. Finally, Step 5 follows from an invariance property of the central subspace; see, for example, Corollary 3.1 of the classical SIR method [2].

As illustrated by the toy examples following Proposition 3, a key advantage of hybrid PHD over the original PHD is its data-adaptive nature. It is well known in the SDR literature that ordinary least squares is more effective than PHD in recovering $\mathcal{S}_{Y|\mathbf{X}}$ when the response Y depends on $\mathbf{X}$ through a monotone link function, whereas PHD tends to outperform OLS when the unknown link function is symmetric. The parameter c_2 in the working model (1) quantifies the signal strength of the symmetric component through c_2^2. The hybrid PHD kernel matrix $\mathbf{A}$ is therefore constructed to adaptively weight the PHD component, assigning greater weight when the symmetric component explains a larger proportion of the variation in $\pi(\mathbf{X})$, and less weight otherwise.

We briefly discuss the consistency of the hybrid PHD estimator. Under the normality assumption of $\mathbf{X}$ and the existence of the required moments, both $\hat{\mathbf{m}}$ and $\hat{\mathbf{M}}$ are consistent estimators of directions in the central subspace $\mathcal{S}_{Y|\mathbf{X}}$

according to classical SDR theory. It follows that, even if $\hat{c}_1^2$ and $\hat{c}_2^2$ are biased estimators of c_1^2 and c_2^2, respectively, the proposed hybrid estimator still consistently recovers a subspace of $\mathcal{S}_{Y|\mathbf{X}}$.

3 Numerical Studies

We compare hybrid PHD (h-PHD) with several existing SDR methods for central subspace recovery through simulations. We include MAVE [10], principal weighted SVM (pw-SVM) [7], probability-enhanced SIR (pre-SIR) [6] and PHD [1] as our competitors. MAVE is implemented through R package *MAVE*, pw-SVM and pre-SIR are implemented through the R functions available on https://www.sjshin.net/software, and PHD is implemented following the algorithm described in Section 3 of [1]. For a fair comparison, we use the true dimension d of $\mathcal{S}_{Y|\mathbf{X}}$ for all estimators. OLS is not included for comparison as it recovers at most one direction in $\mathcal{S}_{Y|\mathbf{X}}$. The performance of $\hat{\mathbf{B}}$ is evaluated by

$$\Delta = \|\mathbf{P}_{\mathbf{B}} - \mathbf{P}_{\hat{\mathbf{B}}}\|_F,$$

where $\mathbf{B}$ is the basis of $\mathcal{S}_{Y|\mathbf{X}}$, $\mathbf{P}_{\mathbf{B}} = \mathbf{B}(\mathbf{B}^\top\mathbf{B})^{-1}\mathbf{B}^\top$, and $\|.\|_F$ is the Frobenius norm. Smaller values of Δ indicate better estimation.

For all models, we generate $\mathbf{X} = (X_1, \ldots, X_p)^\top$ from $\mathbf{X} \sim N_p(\mathbf{0}, \mathbf{I}_p)$. Given $\mathbf{X}$, we generate Y from Bernoulli$\{\pi(\mathbf{X})\}$, where $\pi(\mathbf{X}) = \sigma\{\eta(\mathbf{X})\}$ and $\sigma(t) = (1 + e^{-t})^{-1}$. Let $\boldsymbol{\beta}_1 = (1, 1, \ldots, 1)^\top/\sqrt{p}$, $\boldsymbol{\beta}_2 = (1, -1, 0, \ldots, 0)^\top/\sqrt{2}$, $\boldsymbol{\beta}_3 = (1, 1, 0, \ldots, 0)^\top/\sqrt{2}$, $\boldsymbol{\beta}_4 = (1, 0, \ldots, 0, 1)^\top/\sqrt{2}$, $\boldsymbol{\beta}_5 = (1, 0, \ldots, 0)^\top$, $\boldsymbol{\beta}_6 = (0, 1, 0, \ldots, 0)^\top$, and $\boldsymbol{\beta}_7 = (0, \ldots, 0, 1)^\top$. The $\eta(\cdot)$ of each model is specified as follows.

Model I: $\eta(\mathbf{X}) = \boldsymbol{\beta}_1^\top\mathbf{X}$;
Model II: $\eta(\mathbf{X}) = (0.1 + \boldsymbol{\beta}_2^\top\mathbf{X})^2$;
Model III: $\eta(\mathbf{X}) = \boldsymbol{\beta}_3^\top\mathbf{X} + (\boldsymbol{\beta}_4^\top\mathbf{X})^2$;
Model IV: $\eta(\mathbf{X}) = \boldsymbol{\beta}_5^\top\mathbf{X}/\{0.5 + (1 + \boldsymbol{\beta}_6^\top\mathbf{X})^2\} + |\boldsymbol{\beta}_7^\top\mathbf{X}|$.

Model I and Model II results are summarized in Table 1. The mean and the standard deviation (in parenthesis) of Δ are reported based on 100 replications. The top two performers are boldfaced. Note that $d = 1$ for both models, where $\mathcal{S}_{Y|\mathbf{X}} = \text{span}(\boldsymbol{\beta}_1)$ for Model I and $\mathcal{S}_{Y|\mathbf{X}} = \text{span}(\boldsymbol{\beta}_2)$ for Model II. With linear $\eta(\cdot)$ in Model I, we see that pw-SVM has the best performance and h-PHD is a close second. With mild asymmetry in Model II due to the 0.1 intercept in $\eta(\cdot)$, h-PHD enjoys the best overall performance and PHD is a close second. With fixed $n = 200$ and p increasing from 5 to 15, performances of all methods deteriorate.

The mean and the standard deviation (in parenthesis) of Δ based on 100 replications from Model III and Model IV are reported in Table 2. Note that $d = 2$ for Model III with $\mathcal{S}_{Y|\mathbf{X}} = \text{span}(\boldsymbol{\beta}_3, \boldsymbol{\beta}_4)$, and $d = 3$ for Model IV with $\mathcal{S}_{Y|\mathbf{X}} = \text{span}(\boldsymbol{\beta}_5, \boldsymbol{\beta}_6, \boldsymbol{\beta}_7)$. Note that $\boldsymbol{\beta}_3\boldsymbol{\beta}_4^\top \neq 0$, so the orthogonal indices assumption in Proposition 1 is violated for Model III. Model IV, on the other hand, does not

Table 1. Model I and Model II results based on 100 replications

	$n = 200$	MAVE	pw-SVM	pre-SIR	PHD	h-PHD
Model I	$p = 5$	0.4651	**0.4153**	1.1350	1.3110	**0.4313**
		(0.1961)	(0.1424)	(0.2969)	(0.1338)	(0.1917)
	$p = 10$	0.7524	**0.6314**	1.2643	1.3509	**0.7204**
		(0.2227)	(0.1320)	(0.2206)	(0.0907)	(0.2293)
	$p = 15$	**0.8831**	**0.7381**	1.2627	1.3794	0.9466
		(0.2290)	(0.1393)	(0.2155)	(0.0773)	(0.2860)
Model II	$p = 5$	0.4687	1.1202	1.2419	**0.3633**	**0.3781**
		(0.3297)	(0.3102)	(0.2074)	(0.1907)	(0.2004)
	$p = 10$	0.8867	1.2973	1.3524	**0.7718**	**0.7026**
		(0.3825)	(0.1519)	(0.0737)	(0.3531)	(0.2903)
	$p = 15$	**1.1739**	1.3467	1.3575	1.1748	**1.1251**
		(0.2960)	(0.0957)	(0.0729)	(0.2934)	(0.2868)

follow the additive index structure specified in model (1). MAVE and h-PHD are consistently better than all the other methods in both models. With fixed $p = 5$, performances of PHD, h-PHD and MAVE improve as n increases from 100 to 200. Note that h-PHD consistently outperforms PHD, suggesting that incorporating information from OLS is beneficial in more complex settings. The trade-off, however, is a modest increase in the standard deviation of the h-PHD estimator, driven by the variability in estimating the respective weights of the OLS and PHD components.

For the real data analysis, we compare the performance of h-PHD and PHD using the sonar dataset from the UCI machine learning repository. The goal is to discriminate between sonar signals bounced off a metal cylinder and those bounced off a roughly cylindrical rock. The original data contains $n = 208$ observations with 60 predictors. From the simulation results in Models I and II, we observe that h-PHD deteriorates when the predictor dimension p is moderate to large. Moreover, under the working model (1) and Proposition 3, h-PHD relies on normality of the predictors. These considerations motivate the following preprocessing steps. First, we apply the Kolmogorov filter [5] to retain the top five predictors by testing the equality of the conditional distribution functions $F_{+j}(x)$ and $F_{-j}(x)$, which correspond to the jth predictor given $Y = 1$ and 0, respectively. After screening, we perform rank-based Gaussianization following [4]. The R programming language is used to implement screening and Gaussianization. Denoting the preprocessed predictor as $\mathbf{X}$, we then apply the h-PHD algorithm. The estimates $\hat{c}_1^2 = 3.0886$ and $\hat{c}_2^2 = 1.4883$ imply that the OLS component receives approximately twice the weight of the PHD component in h-PHD. The eigenvalues of $\hat{\mathbf{A}}$ are 0.3332, 0.1614, 0.0426, 0.0135, and 0.0014, suggesting $\hat{d} = 2$ and $\mathcal{S}_{Y|\mathbf{X}}$ is recovered by $\hat{\mathbf{v}}_1$ and $\hat{\mathbf{v}}_2$. We apply k-Nearest Neighbors (kNN) to classify Y based on $\hat{\mathbf{v}}_1^\top \mathbf{X}$ and $\hat{\mathbf{v}}_2^\top \mathbf{X}$. Based on 5-fold cross-validation and 100

Table 2. Model III and Model IV results based on 100 replications

	$p = 5$	MAVE	pw-SVM	pre-SIR	PHD	h-PHD
Model III	$n = 100$	**1.0774**	1.2113	1.5363	1.3436	**1.1447**
		(0.3772)	(0.2964)	(0.2204)	(0.2045)	(0.3097)
	$n = 150$	**0.8875**	1.0935	1.5198	1.2463	**0.8822**
		(0.3354)	(0.3080)	(0.2147)	(0.2179)	(0.3281)
	$n = 200$	**0.7100**	1.1156	1.5281	1.2408	**0.7984**
		(0.2706)	(0.3002)	(0.2318)	(0.2338)	(0.2832)
Model IV	$n = 100$	**1.3444**	1.4192	1.5218	1.3622	**1.3040**
		(0.3067)	(0.2466)	(0.2389)	(0.2761)	(0.2569)
	$n = 150$	**1.2279**	1.4248	1.5419	1.3237	**1.1563**
		(0.3288)	(0.2031)	(0.2177)	(0.2667)	(0.3045)
	$n = 200$	**1.1225**	1.3751	1.5493	1.2546	**1.1224**
		(0.3367)	(0.2777)	(0.2316)	(0.2901)	(0.3007)

random splits, the resulting average error rates are reported in Table 3. We also include the results based on PHD (with $\hat{d} = 2$) as well as kNN applied to $\mathbf{X}$ directly. The results show that h-PHD achieves the two smallest testing error rates and exhibits limited sensitivity to the choice of neighborhood size k.

Table 3. Classification error rates for the sonar data

	PHD		H-PHD		$\mathbf{X}$	
	$k = 5$	$k = 10$	$k = 5$	$k = 10$	$k = 5$	$k = 10$
Training	0.2268	0.2815	0.1824	0.2259	0.1958	0.2328
Testing	0.3576	0.3565	**0.2614**	**0.2506**	0.2673	0.2662

References

1. Cook, R.D.: Principal Hessian directions revisited (with discussion). J. Am. Stat. Assoc. **93**, 84–100 (1998)
2. Li, K.C.: Sliced inverse regression for dimension reduction (with discussion). J. Am. Stat. Assoc. **86**, 316–342 (1991)
3. Li, K.C., Duan, N.: Regression analysis under link violation. Ann. Stat. **17**, 1009–1052 (1989)
4. Mai, Q., He, D., Zou, H.: Coordinatewise gaussianization: theory and applications. J. Am. Stat. Assoc. **118**, 2329–2343 (2023)
5. Mai, Q., Zou, H.: A Kolmogorov filter for variable screening in high-dimensional binary classification. Biometrika **100**, 229–234 (2013)

6. Shin, S., Wu, Y., Zhang, H., Liu, Y.: Probability-enhanced sufficient dimension reduction for binary classification. Biometrics **70**, 546–555 (2014)
7. Shin, S., Wu, Y., Zhang, H., Liu, Y.: Principal weighted support vector machines for sufficient dimension reduction in binary classification. Biometrika **104**, 67–81 (2017)
8. Soale, A.N., Dong, Y.: Envelope dimension reduction with application to binary classification. J. Syst. Sci. Complexity **39**, 79–87 (2026)
9. Wood, S.N.: Generalized additive models. Ann. Rev. Stat. Appl. **12**, 497–526 (2025)
10. Xia, Y., Tong, H., Li, W.K., Zhu, L.-X.: An adaptive estimation of dimension reduction space (with discussion). J. R. Stat. Soc. Ser. B (Stat. Methodol.) **64**, 363–410 (2002)

Approximate Inference for Mixtures of Latent Trait Analyzers

Dalila Failli[1], Maria Francesca Marino[1(✉)], and Francesca Martella[2]

[1] Dipartimento di Statistica, Informatica, Applicazioni, Università degli Studi di Firenze, Florence, Italy
{dalila.failli,mariafrancesca.marino}@unifi.it
[2] Dipartimento di Scienze Statistiche, Sapienza Università di Roma, Rome, Italy

Abstract. The Mixture of Latent Trait Analyzers (MLTA) model combines finite mixture models and latent trait analysis to cluster units while also accounting for within-cluster dependence among variables. Parameter estimation is performed via an EM algorithm, where the intractable integrals in the likelihood can be approximated using either Guass-Hermite quadrature or variational approximation. A simulation study is conducted to compare the two approximation methods under different scenarios, evaluating their performance in terms of parameters' recovery, clustering accuracy, and computational time. The results highlight the trade-off between precision and efficiency, suggesting that the choice of approximation should balance the desired level of accuracy with the available computational resources.

Keywords: Model-based clustering · Latent Variable Models · EM Algorithm · Variational Approximation · GaussHermite Quadrature

1 Introduction

Model-based clustering is widely used to identify hidden patterns in complex data [8]. It provides a tool for grouping units by assuming that observed data arise from a finite mixture of underlying probability distributions. This approach has been applied in diverse fields, including genetics, marketing, and social sciences, where heterogeneity among units is expected. A key development in this framework is the Mixture of Latent Trait Analyzers (MLTA) proposed by [6], which combines finite mixture modeling with latent trait analysis. In detail, in the framework of multivariate binary data, the MLTA provides a clustering of units via a finite mixture specification, while simultaneously capturing within-cluster dependence among variables through a continuous (multidimensional) latent variable (trait). Covariates can be straightforwardly incorporated at the latent level to account for their effect on component membership, enhancing interpretability and model flexibility [2]. [4] proposed a further generalization of

P. Brito et al. (Eds.): IFCS 2026, STUDIES CLASS, pp. 162–169, 2026.
https://doi.org/10.1007/978-3-032-32009-4_20

the model to enable hierarchical clustering of three-way binary data, with variables observed across multiple aggregation levels. On the other side, [5] extended the MLTA model to a biclustering context, focusing on the identification of homogeneous clusters of units with respect to selected subsets of variables.

Parameter estimation for the MLTA model is typically performed via an Expectation-Maximization (EM) algorithm [1]. However, evaluating the model log-likelihood function involves multidimensional integrals that are not available in closed form. Two commonly used approximation strategies to deal with intractable integrals are the Gauss-Hermite (GH) quadrature [9] and the variational approximation [7]. The former yields accurate results at the expense of high computational costs; the latter offers a fast, yet less accurate, approximation to the log-likelihood function. Thus, choosing between these two methods entails a trade-off between computational efficiency and estimation accuracy, motivating a systematic comparison.

The remainder of the paper is organized as follows. In Sect. 2, we introduce the MLTA model. Section 3 describes parameter estimation via the EM algorithm and the two approximation strategies. Section 4 presents a simulation study designed to compare the performance of the two approximation methods under different scenarios. Finally, Sect. 5 concludes with a discussion.

2 The Mixture of Latent Trait Analyzers

Let Y_{ik} denote the k-th ($k = 1, \dots, R$) binary random variable for the i-th unit ($i = 1, \dots, N$) and y_{ik} the corresponding realization. The MLTA assumes that each unit $i = 1, \dots, N$, belongs to one out of G unobserved clusters; these are identified by means of a categorical latent variable $\boldsymbol{z}_i = (z_{i1}, \dots, z_{iG})' \sim \text{Multinomial}(1, (\eta_1, \dots, \eta_G)')'$, with element $z_{ig} = 1$ if unit i belongs to cluster g, and $z_{ig} = 0$ otherwise. The parameter $\eta_g, g = 1, \dots, G$, denotes the probability for a randomly selected unit to belong to cluster g, under the constraints $\sum_{g=1}^{G} \eta_g = 1$ and $\eta_g > 0$. Furthermore, the MLTA assumes the existence of a continuous latent variable (trait) $\boldsymbol{u}_i$ following a D-variate standard Gaussian distribution. Last, MLTA postulates a local independence assumption. Accordingly, conditional on unit i belonging to cluster g (i.e., $z_{ig} = 1$) and the latent trait $\boldsymbol{u}_i$, the elements in the individual response vector $\boldsymbol{Y}_i = (Y_{i1}, \dots, Y_{iR})'$ are assumed to be independent and to follow a (conditional) Bernoulli distribution with success probability modeled according to the following latent trait specification:

$$\pi_{gk}(\boldsymbol{u}_i) = \Pr(Y_{ik} = 1 \mid z_{ig} = 1, \boldsymbol{u}_i) = \frac{1}{1 + \exp[-(b_{gk} + \boldsymbol{w}'_{gk}\boldsymbol{u}_i)]}.$$

In the above equation, b_{gk} represents the baseline propensity for units in the g-th cluster ($g = 1, \dots, G$) towards a positive answer to variable $k = 1, \dots, R$. On the other hand, $\boldsymbol{w}_{gk}$ is a loading parameter capturing heterogeneity between units in cluster g with respect to the baseline level b_{gk}. As clear, when $D = 0$, the MLTA reduces to a standard latent class model. Similarly, when $G = 1$

and $D > 0$, it reduces to a standard latent trait model. In this sense, it can be considered as a more general modeling framework that encompasses both discrete and continuous latent heterogeneity.

3 Model Parameters' Estimation

Let $\boldsymbol{\theta}$ represent the vector of all free model parameters. Starting from the model assumptions detailed in the previous section, the model log-likelihood function can be written as:

$$\ell(\boldsymbol{\theta}) = \sum_{i=1}^{N} \log\Big(\sum_{g=1}^{G} \eta_g \int_{\mathbb{R}^D} f(\boldsymbol{y}_i \mid z_{ig} = 1, \boldsymbol{u}_i) f(\boldsymbol{u}_i) d\boldsymbol{u}_i\Big), \tag{1}$$

where, based on the modeling assumptions detailed above,

$$f(\boldsymbol{y}_i \mid z_{ig} = 1, \boldsymbol{u}_i) = \prod_{k=1}^{R} \left[\pi_{gk}(\boldsymbol{u}_i)\right]^{y_{ik}} \left[1 - \pi_{gk}(\boldsymbol{u}_i)\right]^{1-y_{ik}},$$

while $f(\boldsymbol{u}_i)$ is a D-variate standard Gaussian density.

As usual when dealing with latent variables, an estimate for $\boldsymbol{\theta}$ can be indirectly derived from the log-likelihood function in Eq. (1) by means of an EM algorithm [1]. Starting from the complete-data log-likelihood, this algorithm alternates two steps until convergence. Specifically, under the MLTA specification, the following expression holds:

$$\ell_c(\boldsymbol{\theta}) = \sum_{i=1}^{N}\sum_{g=1}^{G} z_{ig} \log f(\boldsymbol{y}_i \mid z_{ig} = 1, \boldsymbol{u}_i) + \sum_{i=1}^{N}\sum_{g=1}^{G} z_{ig} \log \eta_g + \sum_{i=1}^{N} \log f(\boldsymbol{u}_i). \tag{2}$$

At the generic iteration of the algorithm, in the Expectation-step (E-step), we compute the expected value of Eq. (2), conditional on the observed data and current parameter estimates. That is, we compute

$$\begin{aligned} Q(\boldsymbol{\theta} \mid \boldsymbol{\theta}^{(t)}) = &\sum_{i=1}^{N}\sum_{g=1}^{G} \int_{\mathbb{R}^D} \hat{z}_{ig} f(\boldsymbol{u}_i \mid z_{ig} = 1, \boldsymbol{y}_i) \log f(\boldsymbol{y}_i \mid z_{ig} = 1, \boldsymbol{u}_i)\, d\boldsymbol{u}_i \\ &+ \sum_{i=1}^{N}\sum_{g=1}^{G} \hat{z}_{ig} \log \eta_g + \sum_{i=1}^{N} \int_{\mathbb{R}^D} f(\boldsymbol{u}_i \mid \boldsymbol{y}_i) \log f(\boldsymbol{u}_i)\, d\boldsymbol{u}_i, \end{aligned} \tag{3}$$

where

$$\hat{z}_{ig} = \frac{\eta_g \int_{\mathbb{R}^D} f(\boldsymbol{y}_i \mid z_{ig} = 1, \boldsymbol{u}_i) f(\boldsymbol{u}_i) d\boldsymbol{u}_i}{\sum_{g=1}^{G} \eta_g \int_{\mathbb{R}^D} f(\boldsymbol{y}_i \mid z_{ig} = 1, \boldsymbol{u}_i) f(\boldsymbol{u}_i) d\boldsymbol{u}_i}$$

denotes the the posterior expectation of the indicator variable $z_{ig} \in \boldsymbol{z}_i$.

In the Maximization-step (M-step), Eq. (3) is maximized with respect $\boldsymbol{\theta}$. In detail, model parameters are estimated by finding the zeros of the expected score function, computed with respect to the joint posterior distribution of the indicator variables in $\boldsymbol{z}_i$, and the continuous latent trait, $\boldsymbol{u}_i$. The two steps are alternated until convergence. To avoid local maxima, the EM is run from multiple starting points, and the solution corresponding to the highest log-likelihood is retained as the optimal one. As regards clustering of units, this relies on a MAP rule based on the estimated posterior probabilities, $\hat{z}_{ig}$, obtained at the convergence of the algorithm.

As clear from the above equations, estimation of MLTA parameters requires dealing with a set of multidimensional integrals over the latent trait domain. However, these integrals are not available in closed form, so that approximate inference is needed. In this sense, [2,3,6] propose the use of a variational approximation method. While this represents a computationally efficient strategy and is straightforward to implement, it may compromise estimation accuracy, as relying on the maximization of a lower bound of the log-likelihood function, rather than the log-likelihood itself. A greater accuracy can (at least in principle) be achieved by approximating the integrals via a GH quadrature. This proves to be particularly convenient in the case of low-dimensional latent traits at the expense of higher computational costs. A description of both approximation methods is provided in the following sections.

3.1 Approximation via Gauss-Hermite Quadrature

The main idea of Gaussian quadrature methods is to approximate an integral through a weighted sum of the integrand function, evaluated on a fixed set of abscissas, with given weights. In the MLTA framework, we aim at approximating integrals of the form

$$\Xi_{ig} = \int_{\mathbb{R}^D} f(\boldsymbol{y}_i \mid z_{ig} = 1, \boldsymbol{u}_i) f(\boldsymbol{u}_i) d\boldsymbol{u}_i. \tag{4}$$

These are required to derive the likelihood function, the posterior probabilities $\hat{z}_{ig}$, as well as the expected score functions for model parameters' estimation. Here, given the Gaussian nature of the latent traits $\boldsymbol{u}_i$, $i = 1, \ldots, N$, we can rely on the following GH approximation:

$$\int_{\mathbb{R}^D} h(\boldsymbol{u}_i) e^{-||\boldsymbol{u}_i||^2} d\boldsymbol{u}_i \approx \sum_{m_1 \ldots m_D} h(\boldsymbol{u}_{m_1 \ldots m_D}) \prod_{l=1}^{D} \omega_{m_l},$$

where $\omega_{m_l} = [2^{M-1} M! \sqrt{\pi}]] / [M^2 [H_{M-1}(\boldsymbol{u}_{m_l})]^2]$ are the quadrature weights, $\boldsymbol{u}_{m_1 \ldots m_D} = (u_{m_1}, \ldots, u_{m_D})'$ represents a M-tuple defined by the Cartesian product of the roots of the M-th order GH polynomial $H_M(\boldsymbol{u}_{m_l})$, for $m_l \in \{1, \ldots, M\}$, $l \in \{1, \ldots, D\}$, and $\sum_{m_1 \ldots m_D}$ is used as a shorthand for $\sum_{m_1=1}^{M} \cdots \sum_{m_D=1}^{M}$. To approximate the integral in Eq. (4) via a GH quadrature, we start by rewriting it as a function of $\tilde{\boldsymbol{u}}_i = \boldsymbol{u}_i / \sqrt{2}$:

$$\Xi_{ig} \propto \sqrt{2^D} \int_{\mathbb{R}^D} f(\boldsymbol{y}_i \mid z_{ig} = 1, \sqrt{2} \tilde{\boldsymbol{u}}_i)\, e^{-||\tilde{\boldsymbol{u}}_i||^2} d\tilde{\boldsymbol{u}}_i.$$

Accordingly, the following GH approximation holds:

$$\Xi_{ig} \approx \sqrt{2^D} \sum_{m_1 \dots m_D} f(\boldsymbol{y}_i \mid z_{ig} = 1, \boldsymbol{u}^*_{m_1 \dots m_D}) \prod_{l=1}^{D} \omega_{m_l},$$

where $\boldsymbol{u}^*_{m_1 \dots m_D} = \sqrt{2}\boldsymbol{u}_{m_1 \dots m_D}$. As a result, the log-likelihood function is approximated as

$$\ell(\boldsymbol{\theta}) \approx \sum_{i=1}^{N} \log \left(\sum_{g=1}^{G} \eta_g \sqrt{2^D} \sum_{m_1 \dots m_D} f(\boldsymbol{y}_i \mid z_{ig} = 1, \boldsymbol{u}^*_{m_1 \dots m_D}) \prod_{l=1}^{D} \omega_{m_l} \right).$$

3.2 Variational Approximation

Under a variational approximation framework, intractable integrals are handled by introducing a tractable variational distribution and optimizing a lower bound on the target quantity, yielding an objective that can be evaluated in closed form. In the MLTA framework, it proceeds by replacing the logistic conditional likelihood contribution $f(\boldsymbol{y}_i \mid z_{ig} = 1, \boldsymbol{u}_i)$ with a lower bound that is conjugate to the Gaussian density postulated for $\boldsymbol{u}_i$. Following [7], the conditional likelihood contribution admits the lower bound

$$f(\boldsymbol{y}_i \mid z_{ig} = 1, \boldsymbol{u}_i) \geq \tilde{f}(\boldsymbol{y}_i \mid z_{ig} = 1, \boldsymbol{u}_i, \boldsymbol{\xi}_{ig}),$$

where $\boldsymbol{\xi}_{ig} = (\xi_{ig1}, \dots, \xi_{igR})'$ are auxiliary variational parameters, with $\xi_{igk} \neq 0$, $\forall\, k = 1, \dots, R$, and

$$\tilde{f}(\boldsymbol{y}_i \mid z_{ig} = 1, \boldsymbol{u}_i, \boldsymbol{\xi}_{ig}) = \prod_{k=1}^{R} \sigma(\xi_{igk}) \exp\left\{ \frac{A_{igk} - \xi_{igk}}{2} + \lambda(\xi_{igk})(A^2_{igk} - \xi^2_{igk}) \right\}.$$

In the above expression,

$$\begin{aligned} A_{igk} &= (2y_{ik} - 1)(b_{gk} + \boldsymbol{w}'_{gk}\boldsymbol{u}_i), \\ \sigma(\xi_{igk}) &= (1 + e^{-\xi_{igk}})^{-1}, \\ \lambda(\xi_{igk}) &= \frac{1/2 - \sigma(\xi_{igk})}{2\xi_{igk}}, \end{aligned}$$

respectively. This approximation yields a closed-form expression for Ξ_{ig} in Eq. (4), thereby allowing the latent trait $\boldsymbol{u}_i$ to be integrated out analytically. In particular, the following approximation holds:

$$\begin{aligned} \Xi_{ig} \approx \Xi_{ig}(\boldsymbol{\xi}) = \prod_{k=1}^{R} \sigma(\xi_{igk}) + \exp\left\{ \left(y_{ik} - \frac{1}{2}\right) b_{gk} - \frac{\xi_{igk}}{2} + \lambda(\xi_{igk}) b^2_{gk} + \right. \\ \left. - \lambda(\xi_{igk})\xi^2_{igk} + \frac{\boldsymbol{\mu}'_{ig}\boldsymbol{\Sigma}^{-1}_{ig}\boldsymbol{\mu}_{ig}}{2} + \frac{\log \mid \boldsymbol{\Sigma}_{ig} \mid}{2} \right\}, \end{aligned}$$

with

$$\boldsymbol{\mu}_{ig} = \boldsymbol{\Sigma}_{ig}\Big[\sum_{k=1}^{R} y_{ik} - \frac{1}{2} + 2\lambda(\xi_{igk})b_{gk}\Big]\boldsymbol{w}_{gk} \tag{5}$$

and

$$\boldsymbol{\Sigma}_{ig} = \Big[\boldsymbol{I} - 2\sum_{k=1}^{R}\lambda(\xi_{igk})\boldsymbol{w}_{gk}\boldsymbol{w}'_{gk}\Big]^{-1} \tag{6}$$

denoting the posterior mean vector and covariance matrix of the latent trait $\boldsymbol{u}_i$, conditional on unit i being in the g-th cluster, respectively. In this framework, the EM algorithm described in Sect. 3 needs to be slightly modified. In this respect, additional steps are required to update $\boldsymbol{\mu}_{ig}$, $\boldsymbol{\Sigma}_{ig}$, and ξ_{igk}, together with the remaining model parameters. In particular, we update $\boldsymbol{\mu}_{ig}$ and $\boldsymbol{\Sigma}_{ig}$, conditional on $\boldsymbol{y}_i$, $\hat{z}_{ig}$, and the variational parameters $\boldsymbol{\xi}_{ig}$, according to Eq. (5) and (6), respectively. Variational parameters, ξ_{igk}, and the latent trait parameters in the vector $\boldsymbol{\zeta}_{gk} = (\boldsymbol{w}'_{gk}, b_{gk})'$ are updated as

$$\xi^2_{igk} = b^2_{gk} + 2b_{gk}\boldsymbol{w}'_{gk}\boldsymbol{\mu}_{ig} + \boldsymbol{w}'_{gk}(\boldsymbol{\Sigma}_{ig} + \boldsymbol{\mu}_{ig}\boldsymbol{\mu}'_{ig})\boldsymbol{w}_{gk},$$

and

$$\boldsymbol{\zeta}_{gk} = -\left[2\sum_{i=1}^{N} z_{ig}\lambda(\xi_{igk})E[\boldsymbol{u}_i\boldsymbol{u}'_i]_g\right]^{-1} \times \left[\sum_{i=1}^{N} z_{ig}\Big(y_{ik} - \tfrac{1}{2}\Big)\boldsymbol{\alpha}_{ig}\right],$$

respectively, where $\boldsymbol{\alpha}_{ig} = (\boldsymbol{\mu}'_{ig}, 1)'$ and $E[\boldsymbol{u}_i\boldsymbol{u}'_i]_g = \begin{bmatrix}\boldsymbol{\Sigma}_{ig} + \boldsymbol{\mu}_{ig}\boldsymbol{\mu}'_{ig} & \boldsymbol{\mu}_{ig} \\ \boldsymbol{\mu}'_{ig} & 1\end{bmatrix}$.

4 Simulation Study

In this section, the two approximation methods detailed above are compared through a simulation study, based on different scenarios. The benchmark setting (Scenario 1) considers $N = 100$, $R = 10$, $G = 2$, and $D = 1$. Each factor is then varied in turn – $N = 500$, $R = 20$, $G = 3$, and $D = 2$ (Scenarios 2–5) – to evaluate the impact on clustering performance, computational time, and parameter estimation. For each scenario, 100 datasets are generated and a multi-start strategy, based on 20 random starting values, is considered. Model parameters to simulate the data are set as follows. Prior component probabilities are uniformly distributed ($\eta_g = 1/G, g = 1, \ldots, G$), while the baseline and loading parameters are chosen to ensure adequate separation between components, while controlling within-class variability. Specifically, when $G = 2$, the b_{gk} parameters are drawn from a multivariate Gaussian distribution, with mean vector $(-2, 2)'$ and identity covariance matrix, while the w_{gk} parameters are constrained to be equal across clusters and are drawn from a Gaussian distribution with mean -1 and variance 1. For the scenario based on $G = 3$ components, the b_{gk}'s are drawn from a multivariate Gaussian distribution with mean vector $(-4, 0, 4)'$ and covariance matrix $0.2^2\boldsymbol{I}$, while the w_{gk}'s are drawn from a Gaussian distribution with mean −0.5 and variance 0.2. Finally, when $D = 2$, loading parameters $\boldsymbol{w}_{gk}$ are drawn from a multivariate Gaussian distribution with mean vector $(-0.5, 0.5)'$ and covariance matrix $0.5^2\boldsymbol{I}$.

Table 1 compares the two approximation methods across the simulation scenarios. Both methods perform very well in simpler settings (Scenarios 1–3), with ARI values close to 1 and only minor differences. As model complexity increases (Scenarios 4–5), clustering accuracy declines for both methods. Computational differences are substantial. The variational method requires negligible time in all scenarios, whereas GH becomes increasingly time-consuming as complexity grows, reaching over 175 min on average in Scenario 5 versus 0.020 min for the variational approach. This reflects the curse of dimensionality affecting quadrature-based integration. Regarding parameter estimation, GH generally attains lower MSE, particularly for the mixing proportions η_g, with smaller advantages for the baseline and loading parameters, b_{gk} and $\boldsymbol{w}_{gk}$. As expected, increasing the sample size improves MSE for both methods. Overall, the variational approximation provides comparable or superior clustering performance at a negligible computational cost, while GH offers slightly more accurate parameter estimates at the expense of substantially higher computation time.

Table 1. Simulation study: ARI median, average time in minutes, and average MSE over clusters, variables, and latent dimensions, across scenarios.

		Scenario				
		1	2	3	4	5
		Benchmark	$N = 500$	$R = 20$	$G = 3$	$D = 2$
ARI	Variational	0.96	0.98	1.00	0.84	0.81
	GH	0.96	0.98	1.00	0.84	0.84
Time	Variational	0.002	0.009	0.003	0.008	0.020
	GH	2.587	18.887	3.047	52.542	175.083
MSE(η_g)	Variational	0.053	0.024	0.056	0.091	0.074
	GH	0.002	0.001	0.009	0.030	0.003
MSE(b_{gk})	Variational	0.289	0.099	0.272	0.342	0.728
	GH	0.230	0.033	0.231	0.283	0.605
MSE($\boldsymbol{w}_{gk}$)	Variational	0.212	0.213	0.187	0.245	0.540
	GH	0.223	0.031	0.195	0.222	0.432

5 Conclusions

In this work, we compare two approximation approaches for the estimation of the MLTA model, which combines finite mixture modeling with latent trait analysis to cluster units while accounting for within-cluster dependencies. Within this modeling framework, parameter estimation is performed via an EM algorithm, whose implementation requires the evaluation of multidimensional integrals that are not available in closed form. To address this computational challenge, we consider the use of either GH quadrature or a variational approximation, and assess their practical implications in terms of estimation accuracy and computational efficiency.

A simulation study is conducted to evaluate the performance of the two approaches under different data-generating scenarios, focusing on parameter recovery, clustering accuracy, and computational time. The variational approach is computationally efficient and straightforward to implement, but relies on a lower-bound approximation of the log-likelihood, ensuring that the resulting estimate does not exceed the true log-likelihood. In contrast, GH quadrature directly approximates the required integrals and proves particularly well suited to models with low-dimensional latent traits, yielding more accurate parameter estimates. Overall, GH quadrature provides higher estimation accuracy, whereas the variational approximation attains comparable inferential performance at a substantially lower computational cost. Consequently, the choice between the two approximation strategies should be guided by the trade-off between estimation precision and computational efficiency, especially in high-dimensional or large-scale applications.

References

1. Dempster, A.P., Laird, N.M., Rubin, D.B.: Maximum likelihood from incomplete data via the EM algorithm. J. Roy. Stat. Soc. Ser. B Stat. Methodol. **39**(1) 1–38 (1977)
2. Failli, D., Marino, M.F., Martella, F.: Finite mixtures of latent trait analyzers with concomitant variables for bipartite networks: an analysis of COVID-19 data. Multivar. Behav. Res. **59**(4), 801–817 (2024)
3. Failli, D., Arpino, B., Marino, M.F.: A finite mixture approach for the analysis of digital skills in Bulgaria, Finland and Italy: the role of socio-economic factors. Stat. Methods Appl. **33**, 1483–1511 (2024)
4. Failli, D., Marino, M.F., Arpino, B.: Hierarchical mixtures of latent trait analyzers with concomitant variables for multivariate binary data. Stat. Comput. **35**, 177 (2025)
5. Failli, D., Marino, M.F., Martella, F.: A novel approach for biclustering bipartite networks: an extension of finite mixtures of latent trait analyzers. J. Classif. (2025)
6. Gollini, I., Murphy, T.B.: Mixture of latent trait analyzers for model-based clustering of categorical data. Stat. Comput. **24**, 569–588 (2014)
7. Jaakkola, T.S., Jordan, M.I.: Bayesian logistic regression: a variational approach. In: Madigan, D., Smyth, P. (eds.) Proceedings of the 1997 Conference on Artificial Intelligence and Statistics, Ft.Lauderdale, FL (1997)
8. McLachlan, G.J., Peel, D.: Finite Mixture Models. Wiley Series in Probability and Statistics (2000)
9. Pinheiro, J.C., Bates, D.M.: Approximations to the log-likelihood function in the nonlinear mixedeffects model. J. Comput. Graph. Stat. **4**(1), 12–35 (1995)

Mixture-Based Latent Trait Modeling for Biclustering Overdispersed Counts

Dalila Failli[1(✉)], Maria Francesca Marino[1], and Francesca Martella[2]

[1] Dipartimento di Statistica, Informatica, Applicazioni, Università degli Studi di Firenze, Florence, Italy
dalila.failli@unifi.it

[2] Dipartimento di Scienze Statistiche, Sapienza Università di Roma, Rome, Italy

Abstract. We propose a mixture of latent trait models for biclustering units and variables in the presence of multivariate, overdispersed count data. Units are grouped into homogeneous clusters called components through a finite mixture model. Simultaneously, within each component, variables are grouped into segments using a flexible specification of the linear predictor. Covariates are incorporated at the latent level of the model to account for their effect on component formation, while residual dependence among variables is captured by a multidimensional latent trait. A simulation study based on a varying number of units and variables is conducted to assess both clustering performance and the ability to correctly estimate model parameters.

Keywords: Co-clustering · Latent variable models · Negative Binomial · EM algorithm · Gauss-Hermite quadrature

1 Introduction

The joint partitioning of rows (i.e., units) and columns (i.e., variables) of a data matrix into homogeneous blocks is commonly known in the literature as biclustering, co-clustering, or block clustering (see, e.g., [5]). This approach is used to identify subsets of units that behave similarly across specific subsets of variables. Typical applications include genetics, where biclustering helps detect groups of genes showing similar expression patterns under certain experimental conditions, and market analysis, where the approach is used to segment customers according to their preferences for particular groups of products.

We propose a biclustering approach for multivariate, overdispersed count data based on a mixture of latent trait analyzers (MLTA; [9]) model. Units are partitioned into subsets (components) via a finite mixture of latent variable models, with component probabilities modeled as a function of observed covariates as in [7]. A multidimensional latent variable (trait) is also introduced to capture residual dependence among variables. Within each component, variables are partitioned into subsets (segments) via a flexible specification of the linear predictor, following an approach similar to that firstly introduced by [12], and

P. Brito et al. (Eds.): IFCS 2026, STUDIES CLASS, pp. 170–177, 2026.
https://doi.org/10.1007/978-3-032-32009-4_21

subsequently extended to the MLTA framework by [8] to deal with multivariate binary data.

In this work, we focus the attention on multivariate count responses. While the Poisson distribution is a standard choice for the modeling of count data, it often fails to accommodate the extra variability commonly observed in practice. To address this issue, we assume that, conditional on the latent trait and component membership, the responses of interest are independent of each other (local independence assumption) and follow a Negative Binomial (NB) distribution. Such a distribution introduces an additional dispersion parameter that allows the conditional variance to exceed the conditional mean, thus offering greater flexibility in the modeling of heterogeneous count processes.

A closely related approach is the Latent Block Model (LBM) for contingency tables proposed by [10], which provides a well-established framework for simultaneously clustering rows and columns by assuming block-wise homogeneity of the data matrix. While both approaches perform a joint clustering of units and variables, the proposed model introduces several important extensions compared to the LBM. First, as mentioned above, our proposal explicitly accommodates overdispersion in the data. Second, the model accounts for residual heterogeneity not fully captured by the block configuration. Third, the proposed model allows for cluster-specific partitions of the variables. In other words, for each row cluster, a potentially different grouping of variables is identified, enabling a more flexible representation of the data structure compared to the single global column partition assumed in the standard LBM framework. In addition, unlike the LBM, the proposed model allows for the inclusion of covariates to inform the clustering structure. Another related contribution is the model-based biclustering framework for overdispersed count data proposed in [2], which represents a special case of LBM that explicitly addresses overdispersion within a biclustering setting. Although this approach effectively relaxes the equi-dispersion assumption, it does not explicitly model residual heterogeneity beyond the bicluster structure. Moreover, it does not allow for cluster-specific partitions of variables or for clustering informed by covariates, thereby limiting its flexibility in more complex data scenarios.

The paper is organized as follows. In Sect. 2, we describe the model. Section 3 shows the results of the simulation study, while Sect. 4 contains concluding remarks and details further extensions of the approach.

2 The Proposed Model

Let Y_{ik} denote the k-th random variable and y_{ik} the corresponding observed value for the i-th unit, with $i = 1, \ldots, N$, and $k = 1, \ldots, R$. Each unit is assigned to one out of G unobserved clusters (components), identified by a discrete latent variable

$$\boldsymbol{z}_i = (z_{i1}, \ldots, z_{iG})' \overset{iid}{\sim} \text{Multinomial}(1, (\eta_{i1}, \ldots, \eta_{iG})').$$

The generic element of this vector, z_{ig}, is equal to 1 if the i-th unit belongs to the g-th component (with probability η_{ig}), and is equal to 0 otherwise, for $g =$

$1, \dots, G$, and $i = 1, \dots, N$. Following the proposal by [7], we model component probabilities η_{ig} via the following multinomial logit specification

$$\eta_{ig} = \Pr(z_{ig} = 1 \mid \boldsymbol{x}_i; \boldsymbol{\beta}_g) = \frac{\exp\{\boldsymbol{x}_i' \boldsymbol{\beta}_g\}}{1 + \sum_{g'=2}^{G} \exp\{\boldsymbol{x}_i' \boldsymbol{\beta}_{g'}\}}, \quad g = 2, \dots, G,$$

where $\boldsymbol{\beta}_g = (\beta_{0g}, \beta_{1g}, \dots, \beta_{Jg})'$ is a $(J+1)$-dimensional vector of coefficients measuring the impact of covariates $\boldsymbol{x}_i = (1, x_{i1}, \dots, x_{iJ})'$ on η_{ig}.

For each unit $i = 1, \dots, N$,, we also assume the existence of a D-variate, Gaussian latent variable, $\boldsymbol{u}_i$, with mean vector and covariance matrix equal to $\boldsymbol{0}$ and $\boldsymbol{I}_D$, respectively, to ensure identifiability.

The proposed model relies on a local independence assumption. Specifically, to deal with overdispersed counts, we assume that, conditional on $\boldsymbol{z}_i$ and $\boldsymbol{u}_i$, responses in the vector $\boldsymbol{Y}_i = (Y_{i1}, \dots, Y_{iR})'$ are independent of each other and follow a NB distribution. Therefore, the joint (conditional) density of the vector $\boldsymbol{y}_i$ (a realization of $\boldsymbol{Y}_i$) may be expressed as

$$\begin{aligned} f(\boldsymbol{y}_i \mid \boldsymbol{u}_i, z_{ig} = 1) &= \prod_{k=1}^{R} f(y_{ik} \mid \boldsymbol{u}_i, z_{ig} = 1) \\ &= \prod_{k=1}^{R} \binom{y_{ik} + \alpha_{gk} - 1}{\alpha_{gk} - 1} \left(\frac{\gamma_{gk}}{\gamma_{gk} + \alpha_{gk}} \right)^{y_{ik}} \left(\frac{\alpha_{gk}}{\gamma_{gk} + \alpha_{gk}} \right)^{\alpha_{gk}}, \end{aligned}$$

with location and overdispersion parameters denoted by γ_{gk} and α_{gk}, respectively.

To achieve a joint clustering of units and variables, we follow the proposal by [1,12], and [8] and assume that, within each component, the location parameter γ_{gk} is modeled according to the following specification:

$$\log\left[\gamma_{gk}(\boldsymbol{u}_i)\right] = b_g + \boldsymbol{a}_{gk}'(\boldsymbol{\zeta} + \boldsymbol{u}_i).$$

Here, b_g is a component-specific effect, while $\boldsymbol{\zeta}$, $\boldsymbol{u}_i$, and $\boldsymbol{a}_{gk}$ identify a set of D-dimensional vectors allowing for a partitioning of variables into one out of $D \leq R$ clusters (hereafter, segments). In particular, $\boldsymbol{a}_{gk} = (a_{gk1}, \dots, a_{gkD})'$ is a D-dimensional row stochastic (parameter) vector, whose generic element, a_{gkd}, is equal to 1 if, within component g, the k-th variable belongs to segment d, and 0 otherwise. Furthermore, the parameter b_g represents the baseline (log-transformed) mean count for units in the g-th component, while $\boldsymbol{\zeta} = (\zeta_1, \dots, \zeta_D)'$ represents a vector of fixed model parameters assumed to be constant across components. Therefore, the fixed effect ζ_d captures systematic deviations from the baseline values implied by b_g for those variables belonging to the d-th segment. Last, $u_{id} \in \boldsymbol{u}_i$ is a unit- and segment-specific latent trait that characterizes individual profiles by capturing residual heterogeneity in the transformed mean counts of variables belonging to segment d.

Denoting with $\boldsymbol{\Phi}$ the vector of all free model parameters and considering the modeling assumptions described above, the likelihood function for the proposed model can be written as:

$$L(\mathbf{\Phi}) = \prod_{i=1}^{N} L_i(\mathbf{\Phi}) = \prod_{i=1}^{N} \left(\sum_{g=1}^{G} \eta_{ig} \int_{\mathbb{R}^D} f(\boldsymbol{y}_i \mid \boldsymbol{u}_i, z_{ig} = 1) f(\boldsymbol{u}_i) d\boldsymbol{u}_i \right). \quad (1)$$

Maximum likelihood estimation of $\mathbf{\Phi}$ can be performed via an Expectation-Maximization (EM) algorithm [6]. However, the multidimensional integral appearing in Eq. (1) does not admit a closed-form solution and approximation methods are thus needed. Here, we opt for a Gauss-Hermite quadrature approach [13]. Accordingly, the integral is approximated via a weighted summation over a pre-specified set of quadrature locations (say, Q), with given weights; see [8] for further details.

Note that within the Gauss-Hermite quadrature framework, the computational cost grows in the order of Q^D. However, in our framework, the latent dimension D directly corresponds to the number of segments and is therefore small, ensuring that the approximation remains computationally feasible and numerically stable. However, the exploration of alternative and more convenient approximation methods for higher-dimensional latent traits remains a direction for future work.

Regarding model selection, the proposal is estimated over a range of values for components G and segments D; the optimal specification is selected by minimizing a chosen information criterion, such as the Bayesian Information Criterion (BIC; [14]).

3 Simulation Study

To assess the performance of the proposed approach in terms of parameter estimation and clustering, we conduct a simulation study based on $S = 100$ replicates, each initialized with 100 random starting values to avoid the EM algorithm remains trapped in local maxima solutions, and different data sizes. Specifically, we simulate data from the proposed specification, by considering $G = 3$ components and $D = 2$ segments, along with a varying number of units ($N = 100, 500, 1000$) and variables ($R = 10, 20$). A single covariate drawn from a Gaussian distribution with unit mean and variance is included in the model for prior component probabilities – i.e., $x_{i1} \sim \mathcal{N}(1, 1)$. Model parameters are set as $\beta_{0g} = 0$, $\beta_{12} = -0.4$, $\beta_{13} = -0.9$, $\boldsymbol{b} = (-1.7, 0, 1.7)'$, and $\boldsymbol{\zeta} = (-2, 0.5)'$. These values are chosen to have moderate separation between clusters and segments, ensuring an identifiable block structure. For simplicity, the dispersion parameters are constrained to be constant across variables, so that they only depend on mixture components (i.e., $\alpha_{gk} = \alpha_g$, for all $k = 1, \ldots, R$) and these are set as follows: $\alpha_1 = 0.5, \alpha_2 = 1$, and $\alpha_3 = 1.5$. These values allows for varying overdispersion of counts across clusters.

3.1 Clustering Performance

The ability of the proposal to correctly partition units and variables is evaluated through the Adjusted Rand Index (ARI; [11]); this represents a standard measure to evaluate the agreement between the true and the estimated clustering. Results of the simulation study are shown in Table 1 and demonstrate that, as the number of units and, above all, variables, increases, clustering performance improves.

Table 1. ARI mean and standard error (SE) across samples, for varying N and R.

		$R = 10$		$R = 20$	
		Unit	Variable	Unit	Variable
	100	0.64 (0.11)	0.69 (0.25)	0.79 (0.09)	0.83 (0.24)
N	500	0.66 (0.09)	0.78 (0.17)	0.83 (0.06)	0.97 (0.11)
	1000	0.69 (0.08)	0.79 (0.18)	0.84 (0.04)	0.99 (0.06)

3.2 Parameter Recovery

Table 2 reports the Relative Root Mean Squared Error (RRMSE) values for the model parameters $\boldsymbol{b}$, $\boldsymbol{\zeta}$, $\boldsymbol{\beta}$, and $\boldsymbol{\alpha}$ across different sample sizes (N) and numbers of variables (R). Values equal to zero indicate parameters that were perfectly estimated, while non-zero values reflect the typical relative error. Overall, the table shows that parameter estimates are generally accurate, with RRMSEs decreasing as either the number of units or the number of variables increases. For instance, the RRMSE for $\boldsymbol{\beta}$ decreases from $[0.48, 0.40]$ at $N = 100$ and $R = 10$ to $[0.09, 0.09]$ at $N = 1000$ and $R = 20$, demonstrating improved estimation precision with larger samples and a larger number of observed variables.

Table 2. RRMSE values across samples for $\boldsymbol{b}$, $\boldsymbol{\zeta}$, $\boldsymbol{\beta}$ and $\boldsymbol{\alpha}$, for varying N and R.

		$R = 10$			
		$\boldsymbol{b}$	$\boldsymbol{\zeta}$	$\boldsymbol{\beta}$	α
	100	$[0.38, 0, 0.22]$	$[0.13, 0]$	$[0.48, 0.40]$	$[0.47, 0, 0.35]$
N	500	$[0.21, 0, 0.11]$	$[0.06, 0]$	$[0.33, 0.17]$	$[0.40, 0, 0.16]$
	1000	$[0.18, 0, 0.13]$	$[0.06, 0]$	$[0.27, 0.11]$	$[0.41, 0, 0.15]$
		$R = 20$			
		$\boldsymbol{b}$	$\boldsymbol{\zeta}$	$\boldsymbol{\beta}$	α
	100	$[0.31, 0, 0.24]$	$[0.12, 0]$	$[0.37, 0.32]$	$[0.29, 0, 0.27]$
N	500	$[0.12, 0, 0.15]$	$[0.07, 0]$	$[0.14, 0.14]$	$[0.12, 0, 0.22]$
	1000	$[0.08, 0, 0.08]$	$[0.05, 0]$	$[0.09, 0.09]$	$[0.01, 0, 0.02]$

3.3 A Comparison with Competitive Models

In this section, we compare the performance of the proposed model in terms of clustering accuracy with that of the LBM [10] and the overdispersed LBM (OD-LBM) [2]. Specifically, we considered two scenarios, based on $G = 3$ components, $D = 2$ segments, $N = 500$ units, and $R = 20$ variables. In the first scenario, data are generated according to the proposed model under the parameter configuration described in Sect. 3; in the second scenario, data are generated from an LBM specification. This study allows us to assess (i) the performance of the proposed model under correct specification and (ii) its behavior when the underlying structure originates from a different mechanism. In both scenarios, $S = 100$ datasets are generated and, for each method, clustering ability is evaluated via the ARI. The LBM is estimated using the `blockcluster` R package [4], while the OD-LBM is fitted using the `cobiclust` R package [3].

As shown in Table 3, under Scenario 1, both LBM and OD-LBM perform poorly, failing to capture the biclustering structure of the data. Note that, in the absence of overdispersion, the OD-LBM reduces to the LBM, which explains the nearly identical results observed between the two models. In Scenario 2, the LBM and OD-LBM achieve almost perfect ARI, reflecting that these models are correctly specified. On the other hand, despite the misspecification, the proposed model remains competitive, recovering most of the biclustering structure and thus indicating robustness of the results even in less ideal settings.

Table 3. ARI mean and standard error (SE) for the proposed model (PROP), the LBM, and the overdispersed LBM (OD-LBM) across the two simulation scenarios.

Scenario	Model	Unit ARI	Variable ARI
	PROP*	0.83 (0.06)	0.97 (0.11)
1	LBM	0.23 (0.05)	0.00 (0.02)
	OD-LBM	0.21 (0.05)	0.00 (0.02)
	PROP	0.82 (0.21)	0.93 (0.25)
2	LBM	0.97 (0.12)	1.00 (0.01)
	OD-LBM	0.99 (0.01)	1.00 (0.01)

* Results coincide with those in Table 1 for $N = 500$ and $R = 20$.

4 Conclusions

The proposed model allows for a joint partitioning of units and variables in the context of overdispersed counts, which arise quite often in real data applications, such as genomics, social sciences, and epidemiology. In detail, units are partitioned into components, and, in each of them, variables are partitioned into segments, thanks to a very flexible finite mixture model specification. Furthermore, a continuous latent trait captures residual heterogeneity in individual profiles that is not explained by the joint partitioning of units and variables, also accounting for latent factors that may influence the observed counts. The model also accounts for the effect of covariates on the probability of belonging to the different components, thus enabling the integration of additional information that may influence the clustering structure. Furthermore, it is worth noticing that in the absence of meaningful row or column clusters, the model selection procedure may appropriately favor the simplest configuration, typically selecting $G = 1$ and/or $D = 1$, indicating no latent clustering structure.

The performance of the proposed model is demonstrated through a simulation study. Results of this study show that the model can be effectively used for dimensionality reduction, as well as for uncovering latent patterns in overdispersed count data. In particular, as the number of units and variables increases, the model is able to accurately recover the true parameter values, while maintaining good clustering performance for both units and variables. The simulation study further shows that the proposed method maintains strong performance even under model misspecification, highlighting its robustness and flexibility as an analytical tool of analysis.

While the current paper focuses on simulations, the proposed model can be readily applied to real-world overdispersed count data, such as genomic expression profiles, social network activity counts, or epidemiological incidence data. Future work could explore extensions to handle larger datasets, or adapt the framework to longitudinal or spatially structured count data.

References

1. Alfò, M., Marino, M.F., Martella, F.: Biclustering multivariate discrete longitudinal data. Stat. Comput. **34**, 42 (2024). https://doi.org/10.1007/s11222-023-10292-6
2. Aubert, J., Schbath, S., Robin, S.: Model-based biclustering for overdispersed count data with application in microbial ecology. Methods Ecol. Evol. **12**(6), 1050–1061 (2021). https://doi.org/10.1111/2041-210X.13582
3. Aubert, J.: cobiclust: Biclustering via Latent Block Model Adapted to Overdispersed Count Data. R package version 0.1.2 (2024). https://cran.r-project.org/web/packages/cobiclust/index.html
4. Bhatia, P. S., Iovleff, S., Govaert, G.: blockcluster: an R package for model-based co-clustering. J. Stat. Softw. **76**(9), 1–24 (2017). https://doi.org/10.18637/jss.v076.i09
5. Bock, H.: Simultaneous clustering of objects and variables. In: Tomassone, R. (ed.) Analyse des données et informatique, pp. 187–204. INRIA, Le Chesnay (1979)
6. Dempster, A.P., Laird, N.M., Rubin, D.B.: Maximum likelihood from incomplete data via the EM algorithm. J. R. Stat. Soc. Ser. B Stat. Methodol. **39**(1), 1–38 (1977)
7. Failli, D., Marino, M.F., Martella, F.: Finite mixtures of latent trait analyzers with concomitant variables for bipartite networks: an analysis of COVID-19 data. Multivar. Behav. Res. **59**(4), 801–817 (2024). https://doi.org/10.1080/00273171.2024.2335391

8. Failli, D., Marino, M.F., Martella, F.: A novel approach for biclustering bipartite networks: an extension of finite mixtures of latent trait analyzers. J. Classif. **42**, 492–516 (2025). https://doi.org/10.1007/s00357-025-09500-x
9. Gollini, I., Murphy, T.B.: Mixture of latent trait analyzers for model-based clustering of categorical data. Stat. Comput. **24**, 569–588 (2014). https://doi.org/10.1007/s11222-013-9389-1
10. Govaert, G., Nadif, M.: Latent block model for contingency table. Commun. Stat. Theory Methods **39**(3), 416–425 (2010). https://doi.org/10.1080/03610920903140197
11. Hubert, L., Arabie, P.: Comparing partitions. J. Classif. **2**, 193–218 (1985)
12. Martella, F., Alfò, M.: A finite mixture approach to joint clustering of individuals and multivariate discrete outcomes. J. Stat. Comput. Simul. **87**(11), 2186–2206 (2017). https://doi.org/10.1080/00949655.2017.1322593
13. Pinheiro, J.C., Bates, D.M.: Approximations to the log-likelihood function in the nonlinear mixed-effects model. J. Comput. Graph. Stat. **4**(1), 12–35 (1995)
14. Schwarz, G.: Estimating the dimension of a model. Ann. Stat. **6**(2), 461–464 (1978)

SAFE Models in Finance

Paolo Giudici[1], Alla Petukhina[2], and Alessandro Piergallini[1](✉)

[1] Department of Economics and Management, University of Pavia, Pavia, Italy
alessandro.piergallini01@universitadipavia.it
[2] Hochschule für Technik und Wirtschaft Berlin (HTW Berlin), Berlin, Germany

Abstract. We propose SAFE-inspired evaluation metrics for financial time-series forecasting using R^2 as a unifying scale. The framework is model-agnostic and evaluates forecasting models along three complementary dimensions: predictive accuracy, explainability, and robustness. Accuracy is measured through the out-of-sample coefficient of determination, explainability through the contribution of exogenous predictors, and robustness through the sensitivity of forecasts to controlled input perturbations. We further summarize these dimensions under progressive stress conditions by means of area-under-rank-graduation measures. The approach is illustrated in a Bitcoin forecasting application based on an ARIMAX model with macro-financial regressors, showing how the proposed metrics provide an interpretable and coherent assessment of model performance.

Keywords: Financial time series models · Sustainability · Accuracy · Explainability

1 Introduction

Artificial intelligence methods have become increasingly relevant in financial time-series forecasting, with machine learning and deep learning models widely adopted for both academic and applied purposes [4,12]. Their use in finance has expanded rapidly due to their strong predictive performance in complex, nonlinear environments [9,10]. However, alongside these gains, concerns related to transparency, reliability, and regulatory compliance have emerged, particularly in high-stakes applications governed by the European Artificial Intelligence Act [5,6].

Explainable AI (XAI) methods have been proposed to address these challenges, with several model-agnostic tools developed to interpret machine learning models in financial contexts. Nevertheless, their direct application to time-series data is problematic, as standard perturbation schemes may distort temporal dependencies and provide limited insight into model reliability [3,11,13].

Recently, the S.A.F.E. (Sustainability, Accuracy, Fairness, Explainability) framework has been proposed as a compliance-oriented approach for evaluating AI models, initially for regression [8] and classification tasks [7], and later extended to financial time series [8] and natural language processing [1]. While

P. Brito et al. (Eds.): IFCS 2026, STUDIES CLASS, pp. 178–183, 2026.
https://doi.org/10.1007/978-3-032-32009-4_22

SAFE principles provide a structured evaluation paradigm, their application to time-series forecasting models remains limited, particularly with respect to standard forecasting performance measures.

Motivated by these considerations, this paper proposes SAFE-inspired evaluation metrics for financial time-series forecasting. We focus on three key dimensions: predictive accuracy, explainability, and robustness. In particular, predictive accuracy is measured through the out-of-sample coefficient of determination (R^2_{acc}), which provides a familiar and interpretable performance metric widely used in the forecasting literature.

We illustrate the proposed framework in the context of Bitcoin price forecasting using an ARIMAX specification with a set of macro-financial regressors. This empirical application highlights how SAFE-inspired metrics can provide a unified and interpretable assessment of forecasting performance, variable importance, and model robustness in financial time-series settings.

2 SAFE R^2 Metrics

We aim to predict a time series $\mathbf{y} = (y_1, \ldots, y_T)^\top$ using a set of p regressors $F = \{F_1, \ldots, F_p\}$. For any subset $S \subseteq F$, denote the corresponding predictions by $\widehat{\mathbf{y}}^S = (\widehat{y}_1^S, \ldots, \widehat{y}_T^S)^\top$, and let $\overline{y} = \frac{1}{T}\sum_{t=1}^T y_t$. The reliability of the predictions will be assessed in terms of accuracy, robustness and explainability (see [2]).

Accuracy is measured through the out-of-sample coefficient of determination

$$R^2_{\text{acc}}(\mathbf{y}, \widehat{\mathbf{y}}) = 1 - \frac{\sum_{t=1}^T (y_t - \widehat{y}_t)^2}{\sum_{t=1}^T (y_t - \overline{y})^2}. \tag{1}$$

Robustness measures the stability of predictions under controlled perturbations. Given baseline predictions $\widehat{\mathbf{y}}$ and predictions obtained after perturbing the input variables $\widehat{\mathbf{y}}^*$, we define

$$R^2_{\text{rob}} = 1 - \frac{\sum_{t=1}^T (\widehat{y}_t - \widehat{y}_t^*)^2}{\sum_{t=1}^T (y_t - \overline{y})^2}. \tag{2}$$

Explainability evaluates the contribution of each regressor by comparing predictions from the full model with those obtained after removing that variable. For regressor $j \in F$ we define

$$R^2_{\text{expl}}(j) = \frac{\sum_{t=1}^T (\widehat{y}_t^F - \widehat{y}_t^{F\setminus\{j\}})^2}{\sum_{t=1}^T (y_t - \overline{y})^2}. \tag{3}$$

To evaluate the resilience of these three dimensions under progressive stress conditions, we compute sequences of Rank Graduation measures and summarize them through a common Area Under Rank Graduation metric

$$AURG^* = \frac{1}{P}\sum_{p=1}^{P} RG^*(p), \tag{4}$$

where $RG^*(p)$ denotes a generic Rank Graduation measure evaluated at step p. Depending on the dimension considered, $RG(p)$ corresponds to $RGA(p)$ for accuracy, $RGR(p)$ for robustness, and $RGE(p)$ for explainability, leading to the corresponding summary indices $AURGA$, $AURGR$, and $AURGE$.

The graduation parameter p represents different progressive stress mechanisms: for $AURGA$, observations are removed according to the predicted ranking; for $AURGR$, p corresponds to the normalized perturbation intensity applied to the inputs; for $AURGE$, regressors are progressively removed following their importance ranking.

3 Results

Before starting the analysis, we assess potential multicollinearity among the exogenous regressors by computing Variance Inflation Factors (VIFs). All variables exhibit VIF values close to one (all below 1.12), indicating negligible multicollinearity and supporting the joint inclusion of all regressors in the forecasting models.

Table 1 reports the estimated coefficients, statistical tests and diagnostics for the chosen ARIMAX model, ARIMAX(0,0,3).

Table 1. Coefficient estimates and diagnostics for ARIMAX(0,0,3).

Variable	Coef.	Std.Err.	z	p	Sig.	**Diagnostics**	
Gold	0.4114	0.0730	5.672	0.000	***	Observations	3168
Oil	0.0044	0.0140	0.327	0.744		LogLik	-8513.0
SP500	0.9056	0.0390	23.332	0.000	***	AIC	17044
USD_CNY	0.3539	0.2620	1.351	0.177		BIC	17099
USD_EUR	0.0789	0.1690	0.467	0.641		HQIC	17063.6
ma.L1	-0.0164	0.0120	-1.318	0.188		Ljung–Box Q(1)	0.01 (p=0.94)
ma.L2	0.0376	0.0140	2.690	0.007	**	Jarque–Bera	6544 (p=0.00)
ma.L3	-0.0083	0.0140	-0.584	0.559		Heterosk. H	0.40 (p=0.00)
σ^2	12.7217	0.1830	69.464	0.000	***	Skew/Kurt	-0.29 / 10.02

Table 1 indicates that, among the exogeneous variables, only Gold and SP500 are significant. Model diagnostics suggest no remaining serial correlation in the residuals, as indicated by the Ljung–Box test, supporting the adequacy of the ARIMAX structure. However, the Jarque–Bera test strongly rejects normality, with evidence of excess kurtosis, and the heteroskedasticity test indicates significant variance instability. We now calculate the one step ahead forecasts for the year 2025, using all previous data plus a rolling window, which adds one

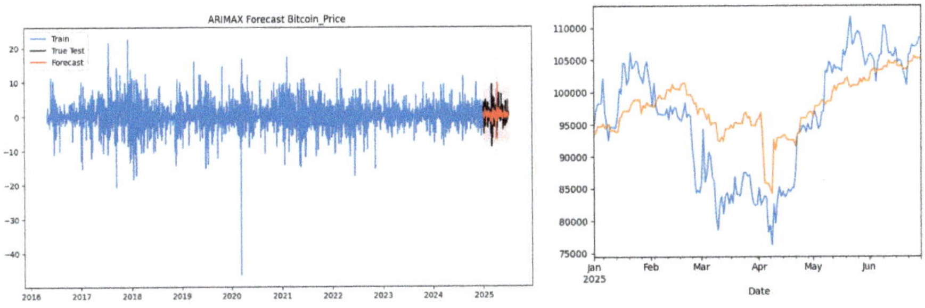

(a) ARIMAX return forecasts (b) ARIMAX price forecasts

Fig. 1. ARIMAX forecasts for returns (*log-diff* ×100) and reconstructed prices.

data point at a time, from the beginning of 2025 onwards. Figure 1a shows the results.

Figure 1a indicates that forecasts are not very good: they are less volatile than the actual values. Figures 1b show that the price forecast follow the actual trend quite closely.

Table 2 reports the corresponding SAFE predictive accuracy, explainability and robustness metrics for the ARIMAX model.

In terms of predictive accuracy, the ARIMAX specification exhibits limited explanatory power for returns ($R^2_{\text{acc}} = 0.0395$), reflecting the well-known difficulty of predicting high-frequency financial returns due to their large noise component. In contrast, predictive accuracy is substantially higher when forecasts are evaluated in terms of prices ($R^2_{\text{acc}} = 0.5039$), indicating that the model is more effective at capturing long-run price dynamics than short-term return fluctuations. Additionally, the model achieves a directional prediction accuracy (DPA) of approximately 54.7% on daily returns, slightly above the 50% benchmark implied by a random walk.

The explainability component highlights a highly concentrated importance structure. The overall explainability level is high, with R^2_{expl} values (summarized by $AURGE$) above 0.9 for both returns and prices, indicating that the SAFE decomposition captures most of the variation in predictive performance across coalitions. In terms of variable importance, the S&P500 index exhibits the largest R^2_{expl} contribution for returns, while Gold dominates the price formulation. This pattern is consistent with the estimated ARIMAX coefficients reported in Table 1, where only S&P500 and Gold appear statistically significant.

The robustness component evaluates the sensitivity of forecasts to perturbations in the exogenous regressors. Table 2 reports the Area Under the Robustness Curve (AURGR) for each predictor. For return forecasts, AURGR values are relatively homogeneous across regressors, indicating a similar degree of sensitivity of ARIMAX predictions to input perturbations. When forecasts are expressed in price levels, robustness increases for most variables due to the smoothing effect

Table 2. SAFE accuracy, explainability, robustness (AURGR) for the ARIMAX model.

Accuracy

	Return	Price
RMSE	2.4046	6354.3927
R^2_{acc}	0.0395	0.5039
$AURGA$	0.5412	0.8153

Explainability

Variable	Return	Price
SP500	**0.2095**	0.2791
Gold	0.0311	**0.4942**
USD_CNY	0.0009	0.0013
USD_EUR	0.0002	0.0033
Oil	1.2e-5	4.5e-5
AURGE	0.9317	0.9056

AURGR

Variable	Return	Price
SP500	0.3900	0.4197
Gold	0.3729	0.6594
USD_CNY	0.3857	0.5513
USD_EUR	0.3887	0.5282
Oil	0.3873	0.5551

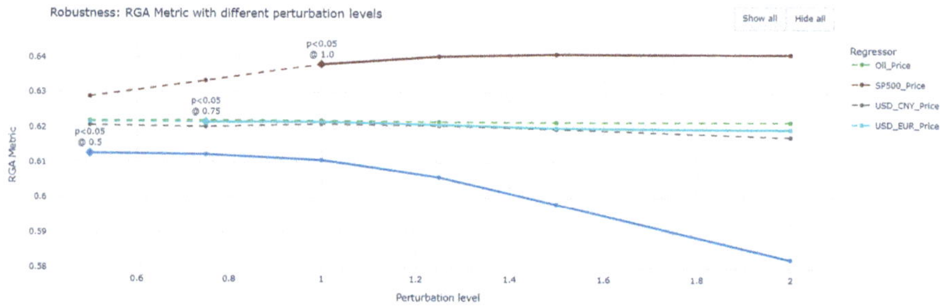

Fig. 2. RGA metric to regressor perturbations. Solid lines indicate the perturbation levels at which the perturbed forecast significantly differs from the not perturbed ARIMAX forecast (Diebold–Mariano test, $p < 0.05$).

induced by the cumulative reconstruction from returns. In this setting, Gold exhibits the highest robustness value, while the S&P500 appears to be the most sensitive predictor.

Figure 2 reports the evolution of the RGA metric under increasing perturbations. Most regressors (Oil, USD/CNY and USD/EUR) display very stable behaviour, indicating limited sensitivity of the forecasts to disturbances in these variables. In contrast, the model appears more sensitive to Gold and the S&P500. Gold shows a noticeable decline in RGA as perturbations increase, while the S&P500 exhibits a slight increase followed by stabilization of the metric. Vertical markers indicate the perturbation levels at which the perturbed forecasts become statistically different from the baseline ARIMAX forecast according to the Diebold–Mariano test ($p < 0.05$).

Acknowledgement. The paper is the result of a close collaboration between the three authors. A.P. worked on data curation, coding and paper drafting while P.G. supervised the work, and A.P. revised the draft paper.

References

1. Babaei, G., Giudice, O., Giudici, P., Maggi, A.: Safe natural language processing. In: Proceedings of the International Joint Conference on Neural Networks (IJCNN), pp. 1–9 (2025). https://doi.org/10.1109/IJCNN64981.2025.11228343
2. Babaei, G., Giudici, P., Raffinetti, E.: A rank graduation box for SAFE AI. Expert Syst. Appl. **259**, 125239 (2025). https://doi.org/10.1016/j.eswa.2024.125239
3. Cerneviciene, J., Kabasinskas, A.: Explainable artificial intelligence (XAI) in finance: a systematic literature review. Artif. Intell. Rev. **57**(8), 216 (2024). https://doi.org/10.1007/s10462-024-10854-8
4. Chen, Z., Ma, M., Li, T., Wang, H., Li, C.: Long sequence time-series forecasting with deep learning: a survey. Inf. Fusion **97**, 101819 (2023). https://doi.org/10.1016/j.inffus.2023.101819
5. European Commission: European artificial intelligence act comes into force (2024). https://ec.europa.eu/commission/presscorner/detail/en/ip_24_4123, press release, 1 August 2024
6. European Union: Regulation (EU) 2024/1689 on artificial intelligence (AI act) (2024). https://eur-lex.europa.eu/legal-content/EN/TXT/?uri=CELEX:32024R1689, official Journal of the European Union
7. Ghosh, I., Datta Chaudhuri, T., Babaei, G., Giudici, P., Raffinetti, E.: Predicting BRICS NIFTY50 returns using XAI and S.A.F.E AI lens. Front. Artif. Intell. **8**, 1668700 (2025). https://doi.org/10.3389/frai.2025.1668700
8. Giudici, P., Piergallini, A., Recchioni, M.C., Raffinetti, E.: Explainable artificial intelligence methods for financial time series. XXPhys. A **655**, 130176 (2024). https://doi.org/10.1016/j.physa.2024.130176
9. Huang, J., Chai, J., Cho, S.: Deep learning in finance and banking: a literature review and classification. Front. Bus. Res. China **14**(2), 121–144 (2020). https://doi.org/10.1186/s11782-020-00082-6
10. Kelly, B.T., Xiu, D.: Financial machine learning. Found. Trends Financ. **13**(3–4), 205–363 (2023). https://doi.org/10.1561/0500000064
11. Khan, F.S., Mazhar, S.S., Mazhar, K., AlSaleh, D.A., Mazhar, A.: Model-agnostic explainable artificial intelligence methods in finance: a systematic review. Artif. Intell. Rev. (2025). https://doi.org/10.1007/s10462-025-11215-9
12. Lim, B., Zohren, S.: Time-series forecasting with deep learning: a survey. Phil. Trans. R. Soc. A **379**(2194), 20200209 (2021). https://doi.org/10.1098/rsta.2020.0209
13. Minh, D., Wang, H.X., Li, Y.F., Nguyen, T.N.: Explainable artificial intelligence: a comprehensive review. Artif. Intell. Rev. **55**(5), 3503–3568 (2022). https://doi.org/10.1007/s10462-021-10088-y

Powered Mean Ensemble Learning

Paolo Giudici[1(✉)], Gloria Polinesi[2], Francesca Mariani[3], and Maria Cristina Recchioni[3]

[1] Department of Economics, University of Pavia, Pavia, Italy
gloria.polinesi@unimc.it
[2] Department of Economics and Law, University of Macerata, Macerata, Italy
f.mariani@staff.univpm.it
[3] Department of Economics and Social Sciences, Politechnic University of Marche, Ancona, Italy
m.c.recchioni@staff.univpm.it

Abstract. Ensemble learning is widely recognized as an effective strategy for improving predictive performance by combining multiple machine learning models. In this study, we consider ensembles obtained as weighted averages of different models: a linear regression, a random forest and a neural network, and propose to optimise their weights by maximising a convex combination between the predictive R^2 and the Rank Graduation accuracy, in a cross validation setting. We compare different types of averages: arithmetic, quadratic, geometric and harmonic. We illustrate our proposal by means of simulations and by the application to the Boston Housing Prices dataset. The optimised ensemble models achieve, under certain conditions, a lower mean squared error relative to the best single model.

Keywords: Ensemble weighting · Cramér-von Mises statistic · Rank Graduation Accuracy

1 Introduction

Ensemble learning, which combines multiple Machine Learning (ML) models, has proven to be a highly effective strategy for improving predictive performance in both classification and regression tasks. Since a single model is rarely capable of capturing the full underlying structure and complexity of real-world data, aggregating the predictions of several models often results in substantially more accurate and stable outcomes, reducing bias, variance, or both, as shown in the early works of [8] and [4];

From a methodological perspective, the construction of an ensemble model can be viewed as the integration of a diverse set of base learners to produce a final prediction for the target variable: a continuous value or a class label. The most widely used ensemble techniques are bagging [3], boosting [6], and stacking [23]; [21]. Bagging builds an ensemble model by training multiple independent models on bootstrapped samples of the original data and subsequently combining their

P. Brito et al. (Eds.): IFCS 2026, STUDIES CLASS, pp. 184–193, 2026.
https://doi.org/10.1007/978-3-032-32009-4_23

predictions through equal weight arithmetic averaging. This strategy is particularly effective in reducing variance and mitigating overfitting; however, the use of uniform weights often limits its predictive accuracy compared with more flexible ensemble approaches. Motivated by this fact, [15] use out-of-bag sample as an independent learning sample for the computation of linear discriminant functions. Boosting, by contrast, constructs ensemble models sequentially, assigning increased importance to observations that were previously mispredicted. In doing so, boosting directs learning toward more challenging cases and typically achieves improved predictive accuracy [10] although it may lead to a higher variance, and it is not easy to interpret. Stacking further enhances ensemble performance by learning how to optimally combine heterogeneous base learners through a meta-learner that assigns data-driven weights to individual model predictions. This adaptive weighting mechanism generally leads to superior predictive performance relative to the equal-weight averaging employed in bagging [24]. Against this advantage, the data driven determination of the weights may lead to higher variance, and is difficult to interpret.

Given the importance of model interpretability, it would be desirable to improve an averaging method, such as bagging, optimising the weights, rather than taking dependent models, as in boosting, or than building a meta model to learn the weights, as in stacking. The main challenge in doing so lies in determining how to accurately define the weights assigned to the individual base learners, when aggregating their predictions.

The simplest approach consists of averaging the outputs of all models, implicitly assuming that each learner contributes equally to the final prediction. Nevertheless, a substantial body of literature has shown that equal-weight averaging is often suboptimal and that carefully designed weighted ensembles can achieve significantly better predictive performance; see, for example, [2,9,18,19,22].

Motivated by these findings, [17] propose a novel ensemble learning framework for regression problems inspired by the classical portfolio optimization theory [16]. Their approach enables the estimation of ensemble weights by incorporating information on the prediction errors of the base learners, on the variability of those predictions, and on their covariance structure. This formulation is fully consistent with the bias-variance decomposition, providing a principled and theoretically grounded strategy for ensemble construction; see, for further details, [4,5].

In this paper, we propose a different optimization framework for estimating the optimal weights of an ensemble model. In our proposal, ensemble models are chosen to optimise not only predictive accuracy (the difference between the "true' and the predicted values), but also the predictive concordance (the difference between the "true" and the predicted ranks). For a binary target response, this implies optimising models with respect to both the Brier Score and the Area Under the Curve (AUC); for a continuous target response with respect to both the Mean Square Error (MSE) and the Rank Graduation Accuracy (RGA) measure of [13], which extends AUC to non binary variables.

While the MSE and the Brier Score evaluates predictive accuracy based on the magnitude of prediction errors, AUC and RGA assess performance in terms of rank concordance. This property makes them particularly robust to the presence of outliers and extreme observations.

Before optimising both predictive accuracy and concordance, it is appropriate to define them on a consistent scale. For MSE (and the Brier Score, a special case of the MSE), a natural normalisation is the predictive R^2:

$$\mathrm{R}^2(\mathrm{Y}, \hat{\mathrm{Y}}) = 1 - \frac{\sum_{i=1}^{n} (y_i - \hat{y}_i)^2}{nVar(Y)}. \tag{1}$$

Recently, [1] have shown that the RGA (of which AUC is a special case) can be expressed by means of a similar normalisation, in which the MSE at the numerator is replaced by the Cramér-von Mises distance [7] between the empirical cumulative distributions of true and predicted values, and the variance by the Gini index defined via the Lorenz curve $L(\cdot)$ [11]:

$$\mathrm{RGA}(Y, \hat{Y}) = 1 - \frac{CVM(F_Y, F_{\hat{Y}})}{G(Y)} \tag{2}$$

where

$$CVM(F_Y, F_{\hat{Y}}) = \int_{-\infty}^{\infty} \left| F_Y(u) - F_{\hat{Y}}(u) \right| dF_Y(u), \tag{3}$$

and

$$G(Y) = 1 - 2 \int_0^1 L(u)\, du.$$

By combining R^2 and RGA within a convex optimization framework, our approach enables a refined and flexible selection of ensemble weights that simultaneously controls two complementary dimensions of predictive quality: similarity in predicted values, captured by R^2 (or Brier) and similarity in predicted rankings, captured by RGA (or AUC):

As previously discussed, to maintain explainability of the results we will consider only ensemble models that are averages of individual models. We will consider not only the classic arithmetic average, employed in bagging, but other types of mean: the quadratic mean, the geometric mean and the harmonic mean. All special cases of powered means (see e.g. [14]):

$$M_p(x_1, \ldots, x_n) = \left(\frac{1}{n} \sum_{i=1}^{n} x_i^p \right)^{1/p}, \qquad p \neq 0.$$

For $p = 0$ the limit exists and defines the geometric mean:

$$M_0(x_1, \ldots, x_n) = \left(\prod_{i=1}^{n} x_i \right)^{1/n} = \lim_{p \to 0} M_p(x_1, \ldots, x_n).$$

Within each training iteration, the predictions from the base learners are aggregated using different mean-based schemes, including the harmonic, geometric, arithmetic, and quadratic means. The use of explicit and interpretable aggregation mechanisms further contributes to explainability by allowing a transparent assessment of how individual base learners influence the final ensemble prediction.

Weight optimisation seeks to achieve accuracy, and the use of a powered mean explainability. A safe machine learning model [12] should also be robust. To achieve robustness, the ensemble weights are obtained through a procedure that repeatedly trains the ensemble model on multiple randomly selected subsets of the data. This resampling-based strategy mitigates sensitivity to data perturbations and enhances the stability of the resulting predictions.

The resulting R^2 and RGA values are then averaged across all ensemble runs, ensuring that the final weighted combination achieves stable and robust predictive accuracy across diverse training samples, rather than overfitting to any single subset.

The overall predictive performance on the test sets is finally evaluated using the Mean Squared Error (MSE) criterion, if the response is continuous; or the Brier score, if the response is binary.

2 Methodology

Let

$$\underline{y}_j^c = [y_{1j}^c, y_{2j}^c, \dots, y_{nj}^c]^\top$$

denote the values of the target variable Y, and let

$$\underline{\hat{y}}_j^c = [\hat{y}_{1j}^c, \hat{y}_{2j}^c, \dots, \hat{y}_{nj}^c]^\top$$

be the corresponding predictions produced by the j-th machine-learning model on the c-th bootstrapped training set [3], for $j = 1, \dots, M$ and $c = 1, \dots, C$.

For each training set c, we construct ensemble predictions using different mean-aggregation schemes -harmonic, geometric, arithmetic, and quadratic-defined as follows:

$$\hat{y}_i^{c,s} = \left[\sum_{j=1}^{M} (y_i^c)^s w_j\right]^{\frac{1}{s}}, \qquad i = 1, \dots, n,\ s = -1, 1, 2. \tag{4}$$

When $s = 0$, Eq. 4 reduces to the geometric mean:

$$\underline{\hat{y}}^{c,0} = \prod_{j=1}^{M} (\underline{y}_j^c)^{w_j},$$

with $\sum_{j=1}^{M} w_j = 1$.

For each aggregation scheme $s = -1, 0, 1, 2$, we set $\hat{y}^c = \hat{y}^{c,s}$ and estimate ensemble weights by solving the optimization problem, whose objective function is a convex combination of two different quantities, each lying between 0 and 1, as follows:

$$\max_{\underline{w}\geq\underline{0}}(1-\alpha)\left[\frac{1}{C}\sum_{c=1}^{C}\left(1-\frac{\omega_{\underline{y}^c,\underline{\hat{y}}^c}}{G(\underline{y}^c)}\right)\right]+ \tag{5}$$
$$\alpha\left[\frac{1}{C}\sum_{c=1}^{C}\left(1-\frac{\mathrm{MSE}(\underline{y}^c,\underline{\hat{y}}^c)}{V(\underline{y}^c)}\right)\right]$$
$$\text{s.t. } \underline{w}^\top\underline{u}=1,$$

where $\underline{0}=[0,\ldots,0]$ and $\underline{u}=[1,\ldots,1]^\top$.

By maximizing Eq. (5) with respect to the base learners weight vector $\underline{w}$ for a given α, $0\leq\alpha\leq 1$, we aim to select ensemble weights ensuring predictive accuracy in terms of both ranking and values. Note that when $\alpha=0$ and $\alpha=1$, we consider only the first (ranking-based predicted accuracy) and second (value-based predicted accuracy) quantities, respectively.

Specifically, the term $\omega_{\underline{y}^c,\underline{\hat{y}}^c}$ measures the discrepancy between the target and the ensemble predictions on the c-th training set via the Cramér–von Mises (CvM) distance [7]:

$$\omega_{\underline{y}^c,\underline{\hat{y}}^c}=\frac{1}{|\psi|}\sum_{i=1}^{|\psi|}\left|F_{\underline{y}^c}(\psi_i)-F_{\underline{\hat{y}}^c}(\psi_i)\right|,$$

where $F_{\underline{y}^c}$ and $F_{\underline{\hat{y}}^c}$ denote the empirical distribution functions of $\underline{y}$ and $\underline{y}^c$, respectively, both evaluated at the same $|\psi|$ points of the grid $\psi=\{\psi_1,\ldots,\psi_{|\psi|}\}$, which is constructed as an equally spaced sequence ranging from $\min(\underline{y}^c)$ to $\max(\underline{y}^c)$ with step size 0.001.

The denominator $G(\underline{y}^c)$ represents the *Gini index* of the target distribution. Consequently,

$$1-\frac{\omega_{\underline{y}^c,\underline{\hat{y}}^c}}{G(\underline{y}^c)}$$

is the *Rank Graduation Accuracy (RGA)* (see [1], for details).

The quadratic loss term

$$\mathrm{MSE}(\underline{y}^c,\underline{\hat{y}}^c)=\frac{1}{n}\sum_{i=1}^{n}(y_i^c-\hat{y}_i^c)^2$$

is normalized by the variance $V(\underline{y}^c)$, so that

$$1-\frac{\mathrm{MSE}(\underline{y}^c,\underline{\hat{y}}^c)}{V(\underline{y}^c)}$$

yields the coefficient of determination R^2.

The objective function in problem (5) is therefore a convex combination of two predictive accuracy metrics: RGA and R^2. The structure of this loss function closely resembles the logic behind the elastic net [20]–a regularized regression method that linearly combines the L_1 and L_2 penalties to balance sparsity and stability. Analogously, our approach combines a rank-based distributional similarity RGA (i.e., L_1), with a variance-based accuracy measure R^2 (i.e., L_2), balanced through the parameter α.

The core idea behind problem (5) is that the ensemble weights are selected to simultaneously maximize predictive accuracy and stability across the distribution of the observed and predicted variables.

3 Empirical Findings

An experimental study is conducted to demonstrate the competitive performance of the methodological approach outlined in Sect. 2[1]. We utilize the available Boston data set from the MASS R package. This data set comprises information collected by the U.S. Census Service regarding $n = 506$ housing values in the suburbs of Boston, as shown in Table 1. The dependent variable is the median value of owner-occupied homes (MEDV), expressed in thousands of dollars.

Table 1. Description of target and features in the dataset.

Variable	Description
CRIM	Per capita crime rate by town
ZN	Proportion of residential land zoned for lots over 25,000 sq. ft.
INDUS	Proportion of non-retail business acres per town
CHAS	Charles River dummy variable (1 if tract bounds river; 0 otherwise)
NOX	Nitric oxides concentration (parts per 10 million)
RM	Average number of rooms per dwelling
AGE	Proportion of owner-occupied units built prior to 1940
DIS	Weighted distances to five Boston employment centers
RAD	Index of accessibility to radial highways
TAX	Full-value property-tax rate per $10,000
PTRATIO	Pupil-teacher ratio by town
B	$1000(B_k - 0.63)^2$ where B_k is the proportion of blacks by town
LSTAT	Percentage lower status of the population
MEDV	Median value of owner-occupied homes in $1000's (Target variable)

[1] Code is available upon request.

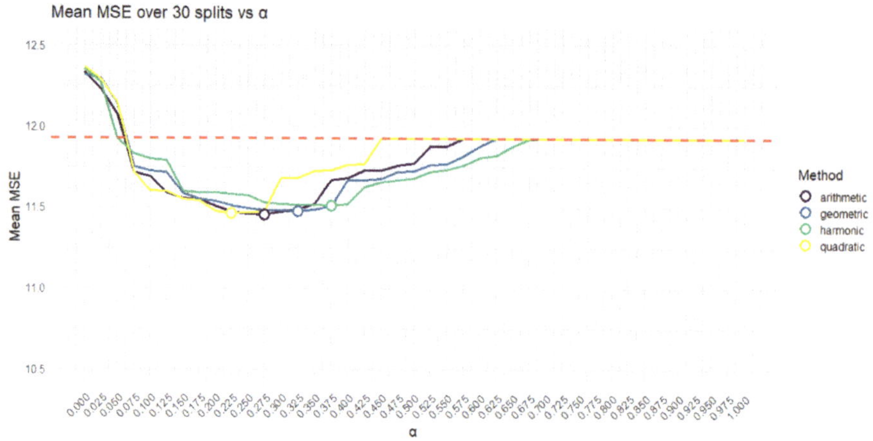

(a)

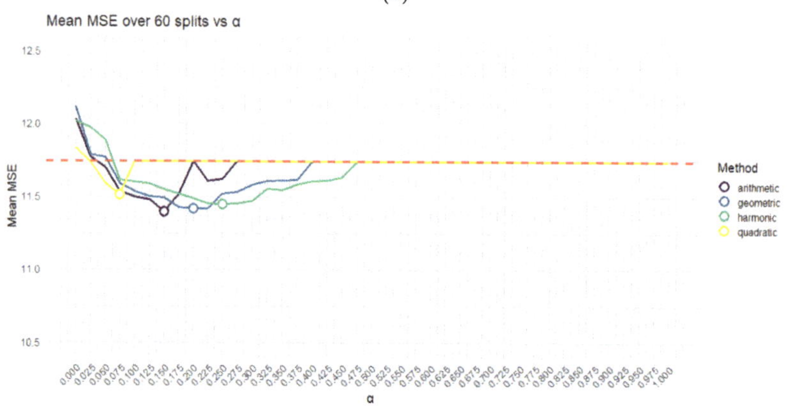

(b)

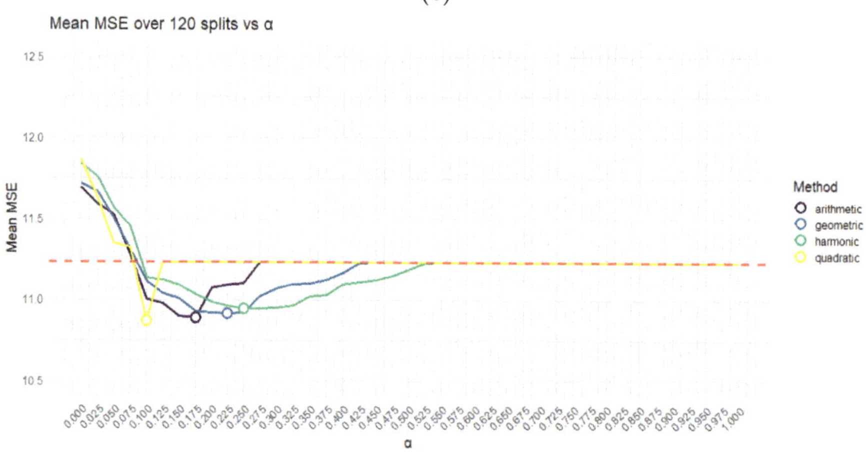

(c)

Fig. 1. Mean MSE on the test set computed using ensemble models based on different mean-aggregation schemes, shown for 30 (a), 60 (b), and 120 (c) splits for $0 \leq \alpha \leq 1$. The dashed horizontal line represents the performance of the rf model.

To generate the predicted values, we employ three different machine learning (ML) models: linear regression (lm), feedforward neural network (nnet), and random forest (rf).

While rf uses its default settings[2], the nnet model is configured with a maximum of 200 iterations, a weight decay of 0.01 to mitigate overfitting, and 5 nodes in the hidden layer. All three models are trained in multiple training scenarios to assess robustness and stability. Specifically, we consider $C = 30, 60$, and 90 distinct training sets, each constructed by randomly selecting 80% of the observations from the original data set. This repeated sampling strategy enables us to evaluate the consistency of model performance across different subsamples of the data and to mitigate the influence of any single random split.

We compute the optimal weights for the ensemble model according to Eq. (5) for $0 \leq \alpha \leq 1$ and $s = -1, 0, 1, 2$. To evaluate the empirical advantages of our proposed ensemble model, we average its mean squared error (MSE) across the $C = 30, 60, 90$ test sets (i.e., the out-of-bag samples withheld from the optimization in 5) and compare it with the MSE of the best-performing model, namely the rf model (Fig. 1).

Figure 1 shows the mean MSE on the test set as a function of α for different mean-aggregation schemes and different numbers of splits. The dashed horizontal line represents the performance of the best individual base learner, the rf model. The ensemble model outperforms the best model only for intermediate values of α, i.e., away from the extreme cases $\alpha = 0$ and $\alpha = 1$.

At extremes, the ensemble does not provide any performance gain in terms of MSE. As α approaches the value 0.5, the ensemble progressively collapses to the best model, regardless of the number of splits considered, with all the weight assigned to the rf learner.

4 Conclusions

This study introduced a weighted ensemble learning framework that integrates multiple mean-aggregation schemes and optimizes ensemble weights through a convex combination of two complementary accuracy criteria: traditional predictive performance, measured by R^2, and a rank based distance derived from the Cramér-von Mises statistic. By assessing these measures across numerous train-test splits, the proposed method explicitly emphasizes robustness and guards against overfitting to any single data partition.

Using the Boston Housing Prices dataset as a benchmark, we demonstrated that combining linear regression, a feedforward neural network, and a random forest within the framework yields ensemble models with more stable and, in several cases, improved predictive accuracy relative to individual base learners. In particular, the resulting ensembles achieved lower mean squared error under specific configurations determined by the parameter α, while avoiding undue reliance on a single model or on value-based fit alone.

[2] Number of tree to grow equal to 500 and number of variables randomly sampled as candidates at each split equal to 4.

Overall, the findings highlight the value of incorporating both predictive accuracy and distributional alignment when determining ensemble weights. This approach provides a principled way to construct ensembles that balance performance and robustness, making them better suited for applications where stability across data splits and resilience to distributional variability are essential. Future work may extend this framework to several datasets, alternative distributional metrics, and more complex model classes to further explore its versatility and practical impact.

References

1. Auricchio, G., Bernardelli, A.E., Giudici, P., Toscani, G.: On rank graduation metrics for high dimensional ordinal data. arXiv preprint arXiv:2511.23100 (2025)
2. Bhasuran, B., Murugesan, G., Abdulkadhar, S., Natarajan, J.: Stacked ensemble combined with fuzzy matching for biomedical named entity recognition of diseases. J. Biomed. Inf. **64**, 1–9 (2016)
3. Breiman, L.: Bagging predictors. Mach. Learn. **24**(2), 123–140 (1996)
4. Brown, G., Wyatt, J., Harris, R., Yao, X.: Diversity creation methods: a survey and categorisation. Inf. Fusion **6**(1), 5–20 (2005)
5. Brown, G., Wyatt, J.L., Tino, P., Bengio, Y.: Managing diversity in regression ensembles. J. Mach. Learn. Res. **6**(9), 1621–1630 (2005)
6. Chen, C., Xiong, Z., Tian, X., Wu, F.: Deep boosting for image denoising. In: Proceedings of the European Conference on Computer Vision (ECCV), pp. 3–18 (2018)
7. Cramér, H.: On the composition of elementary errors: first paper: mathematical deductions. Scand. Actuar. J. **1928**(1), 13–74 (1928)
8. Dietterich, T.G.: Ensemble methods in machine learning. In: Kittler, J., Roli, F. (eds.) MCS 2000. LNCS, vol. 1857, pp. 1–15. Springer, Heidelberg (2000). https://doi.org/10.1007/3-540-45014-9_1
9. Ekbal, A., Saha, S.: Stacked ensemble coupled with feature selection for biomedical entity extraction. Knowl.-Based Syst. **46**, 22–32 (2013)
10. Ganaie, M.A., Hu, M., Malik, A.K., Tanveer, M., Suganthan, P.N.: Ensemble deep learning: a review. Eng. Appl. Artif. Intell. **115**, 105151 (2022)
11. Gini, C.: Measurement of inequality of incomes. Econ. J. **31**(121), 124–125 (1921)
12. Giudici, P., Raffinetti, E.: Safe artificial intelligence in finance. Finan. Res. Lett. **56**, 104088 (2023)
13. Giudici, P., Raffinetti, E.: RGA: a unified measure of predictive accuracy: P. giudici, e. raffinetti. Adv. Data Anal. Classificat. **19**(1), 67–93 (2025)
14. Hardy, G.H., Littlewood, J.E., Pólya, G.: Inequalities, 2nd edn. Cambridge University Press, Cambridge (1952)
15. Hothorn, T., Lausen, B.: Bagging combined classifiers. In: Classification, Clustering, and Data Analysis: Recent Advances and Applications, pp. 177–184. Springer, Heidelberg (2002)
16. Markowitz, H.: Modern portfolio theory. J. Finan. **7**(11), 77–91 (1952)
17. Pérez-Rodríguez, J., Fernández-Navarro, F., Ashley, T.: Estimating ensemble weights for bagging regressors based on the mean-variance portfolio framework. Expert Syst. Appl. **229**, 120462 (2023)

18. Peykani, P., Mohammadi, E., Emrouznejad, A., Pishvaee, M.S., Rostamy-Malkhalifeh, M.: Fuzzy data envelopment analysis: an adjustable approach. Expert Syst. Appl. **136**, 439–452 (2019)
19. Shahhosseini, M., Hu, G., Pham, H.: Optimizing ensemble weights and hyperparameters of machine learning models for regression problems. Mach. Learn. Appl. **7**, 100251 (2022)
20. Shen, W., Wang, J., Ma, S.: Doubly regularized portfolio with risk minimization. In: Proceedings of the AAAI Conference on Artificial Intelligence, vol. 28, no. 1 (2014)
21. Wang, Z., Liang, Z., Zeng, R., Yuan, H., Srinivasan, R.S.: Identifying the optimal heterogeneous ensemble learning model for building energy prediction using the exhaustive search method. Energy Build. **281**, 112763 (2023)
22. Winham, S.J., Freimuth, R.R., Biernacka, J.M.: A weighted random forests approach to improve predictive performance. Stat. Anal. Data Min. ASA Data Sci. J. **6**(6), 496–505 (2013)
23. Wolpert, D.H.: Stacked generalization. Neural Netw. **5**(2), 241–259 (1992)
24. Yao, J., Zhang, X., Luo, W., Liu, C., Ren, L.: Applications of stacking/blending ensemble learning approaches for evaluating flash flood susceptibility. Int. J. Appl. Earth Obs. Geoinf. **112**, 102932 (2022)

New Integrative Approach to Lung Cancer Risk Prediction Using Fuzzy Logic, Bayesian Networks, and Fuzzy c-Means Clustering

Lida Hooshyar[1], Ryan Godin[1], Alireza Khastan[2], Rui Borges[3], and Nadia Tahiri[1](✉)

[1] Département d'informatique, Université de Sherbrooke, Sherbrooke, QC, Canada
{Lida.Hooshyar,Ryan.Godin,Nadia.Tahiri}@USherbrooke.ca
[2] Mathematics Department, Institute for Advanced Studies in Basic Sciences (IASBS), Zanjan, Iran
khastan@iasbs.ac.ir
[3] School of Mathematics and Statistics, University of St Andrews, St Andrews, UK
rb251@st-andrews.ac.uk

Abstract. Most diseases, including lung cancer, result from interactions between genetic and environmental factors. Traditional Bayesian networks often use binary or discretized variables, limiting sensitivity to intermediate states and individual variability. By integrating fuzzy logic, variables are assigned continuous membership degrees, allowing gradual uncertainty propagation and smoother probability estimates. Fuzzy clustering of gene expression and methylation profiles identifies gene groups linked to cancer risk, connecting molecular variability to disease probability. This approach enables a more nuanced assessment of genetic and environmental contributions to disease, supporting individualized risk evaluation.

Availability and Implementation: All Python scripts, as well as simulated and real data and illustrated results, are freely available at the following GitHub repository: Fuzzy-Bayesian-Network.

Keywords: Fuzzy logic · Bayesian networks · Gene-Environment interactions · Soft clustering · Lung cancer risk prediction

1 Introduction

Disease causation is complex, involving genetic, environmental, and lifestyle factors. Understanding these interactions is crucial for guiding research, improving diagnosis, and developing treatments [2]. Many diseases, including lung cancer, result from multiple genes and environmental exposures interacting in nonlinear ways. Capturing these interactions is challenging due to high-dimensional genomic data, individual heterogeneity, and biological uncertainty.

P. Brito et al. (Eds.): IFCS 2026, STUDIES CLASS, pp. 194–201, 2026.
https://doi.org/10.1007/978-3-032-32009-4_24

Machine learning, particularly Bayesian networks (BNs), provides a framework for modeling these relationships [3]. BNs encode conditional dependencies among variables, enabling modeling of gene-gene and gene-environment interactions. They have been applied in bioinformatics for gene expression analysis, protein signaling networks, protein-protein interactions, and pedigree analysis, and in medicine, forensics, and ecology for causal inference and prediction. Fault tree analysis similarly maps logical connections between potential failures and outcomes, including disease prediction.

Traditional BNs and fault trees rely on binary logic, assigning variables to fixed categories. This can obscure continuous and overlapping factors like age, lifestyle, or gene expression. For instance, "young" varies among individuals, and genes may act differently depending on context; oncogenes promote proliferation, tumor suppressors constrain it, and some genes have dual roles [1]. Binary classification may therefore overlook heterogeneity, limiting personalized predictions.

Fuzzy logic addresses these limitations by representing uncertainty and graded membership. Integrating fuzzy membership functions and fuzzy c-means clustering into BNs allows continuous modeling, captures overlapping gene behaviors, and propagates uncertainty effectively [6]. Fuzzy Bayesian Networks (FBNs) extend BNs to handle subjective evidence, fuzzy inputs, and outputs, enabling nuanced inference in complex biological systems—particularly relevant in bioinformatics, where molecular data are inherently uncertain and overlapping.

The remainder of the paper is organized as follows: Sect. 2 presents the FBN methodology, Sect. 3 applies it to a Gene Expression Omnibus (GEO) dataset, and Sect. 4 concludes with potential directions for future research.

2 Method

The study integrates three fuzzy components using two data types. Personal and clinical variables are modeled with fuzzy membership functions, while genetic features (oncogenes and tumor suppressors) are processed using fuzzy c-means clustering. The resulting membership degrees describe positive or negative associations with the inferred outcome and are combined with conditional probabilities through fuzzy Bayesian computation. The network outputs a risk value for each sample.

2.1 Fuzzy Membership Functions

Membership functions are first defined for the variables used in the analysis, including personal, clinical, or environmental attributes depending on the dataset. Gaussian and sigmoidal functions are used because they provide smooth transitions and represent gradual variation and uncertainty more adequately than piecewise linear functions.

Each continuous variable x_i is represented by a fuzzy set $\mathcal{X}$ with membership function $\mu_F^{(s)}(x_i) \in [0,1]$. For a feature set $F = \{f_1, \ldots, f_m\}$ with l states, the Gaussian membership function is

$$\mu_F^{(s)}(x_i) = e^{-\frac{(x_i - \bar{x}_F)^2}{2\sigma_F^2}}, \; \forall i \in [1, \ldots, n], \; \forall s \in [1, \ldots, l], \tag{1}$$

where $\bar{x}_F$ and σ_F denote the mean and standard deviation of feature F across all samples. Increasing and decreasing sigmoidal functions are

$$\mu_F^{(s)}(x_i) = \frac{1}{1 + e^{-\alpha_F(x_i - \beta_F)}}, \tag{2}$$

$$\mu_F^{(s)}(x_i) = 1 - \frac{1}{1 + e^{-\alpha_F(x_i - \beta_F)}}, \tag{3}$$

where α_F controls the slope and β_F the inflection point, both estimated from the data distribution.

Each value x_i is mapped to a vector of membership degrees

$$\mu_F^{(s)}(x_i) : x_i \to \mathcal{A} = \{\mu_F^{(1)}(x_i), \ldots, \mu_F^{(l)}(x_i)\}, \tag{4}$$

which forms the fuzzy states of an input node in the FBN.

2.2 Fuzzy *c*-Means Clustering

Fuzzy c-means clustering is applied to a second feature set $F' = \{f_1, \ldots, f_{m'}\}$ for the same samples. Features are clustered separately according to positive and negative correlation trends. Along two attribute dimensions, clusters typically form lowlow, lowhigh, highlow, and highhigh patterns, although the number of cluster types depends on the number of attributes.

The algorithm outputs cluster membership degrees

$$\mu_{F'}^{(k)}(x_i) : x_i \to \mathcal{B} = \{\mu_{F'}^{(1)}(x_i), \ldots, \mu_{F'}^{(c)}(x_i)\}, \tag{5}$$

which define fuzzy node states for the FBN, where c is the number of clusters.

2.3 Fuzzy Bayesian Network and Inference

The fuzzy sets $\mathcal{A}$ and $\mathcal{B}$ serve as node inputs in the FBN. Conditional probability between fuzzy sets is computed from membership degrees as

$$P(\mathcal{A} \mid \mathcal{B}) = \frac{\sum_{i=1}^{n} \sum_{s=1}^{l} \sum_{k=1}^{c} \mu_{\mathcal{A}}^{(s)}(x_i)\, \mu_{\mathcal{B}}^{(k)}(x_i)\, P(\mu_{\mathcal{B}}^{(k)}(x_i) \mid \mu_{\mathcal{A}}^{(s)}(x_i))\, P(\mu_{\mathcal{A}}^{(s)}(x_i))}{P(\mathcal{B})}, \tag{6}$$

where s and k index membership states and all membership degrees and probabilities lie in $[0,1]$.

The marginal fuzzy probability is

$$P(X) = \sum_{i=1}^{n} \sum_{s=1}^{l} \mu_X^{(s)}(x_i)\, P(\mu_X^{(s)}(x_i)). \quad (7)$$

Membership degrees quantify partial belonging of each state, and associated classical probabilities weight their contribution in inference.

3 Data

The analysis uses the public GEO dataset GSE32867 [4], containing paired DNA methylation and gene expression profiles from lung adenocarcinoma and adjacent normal tissues. From 289 original samples, 61 incomplete cases were removed, leaving 114 paired samples ($n = 228$) with complete molecular and personal variables, including age, smoking status, and physical activity.

Genetic data are organized as methylation and expression matrices ($p \times n$) covering oncogenes (positively correlated with cancer) and tumor suppressor genes (negatively correlated). Four submatrices were derived: oncogene methylation and expression ($p' \times n$) and tumor suppressor methylation and expression ($p'' \times n$), with $p = p' + p''$.

Expression and methylation measurements correspond to platforms GPL6884 and GPL8490. Genes are identified by ILMN probe codes and cg CpG identifiers. Only lung cancerrelevant genes commonly reported in genomics studies were retained to support reproducibility [5].

4 Results

This section applies the proposed method to the dataset described in the *Data* section and reports the results for tumor and normal samples. Model performance is evaluated by comparison with a binary counterpart. The dataset contains annotated tumor and normal samples and supports assessment of discrimination ability. Using the fuzzyBayesian approach, higher estimated risk probabilities are expected for tumor samples.

Membership degrees of variables are computed using Gaussian, sigmoid, and inverse sigmoid functions. This representation assigns each sample a degree of belonging within fuzzy categories and defines an individual profile. The variables include age, physical activity, and smoking status, commonly used in lung cancer studies. The method remains adaptable to alternative variable sets.

Sigmoid, inverse sigmoid, and Gaussian membership functions describe lower, upper, and central regions of each variable domain. Their combined use provides a continuous representation of uncertainty and gradual transitions between fuzzy categories.

Fuzzy c-means clustering is then applied separately to genes positively correlated with tumor progression (oncogenes) and negatively correlated genes (tumor suppressors), using methylation and expression levels. Genes are grouped into hypomethylation or hypermethylation and low or high expression, forming four clusters: low expressionhypomethylation, low expressionhypermethylation, high expressionhypomethylation, and high expressionhypermethylation.

Figure 1 shows clustering results for oncogenes. The x-axis represents samples and the y-axis membership degrees across four clusters. Normal samples are shown in blue and tumor samples in red.

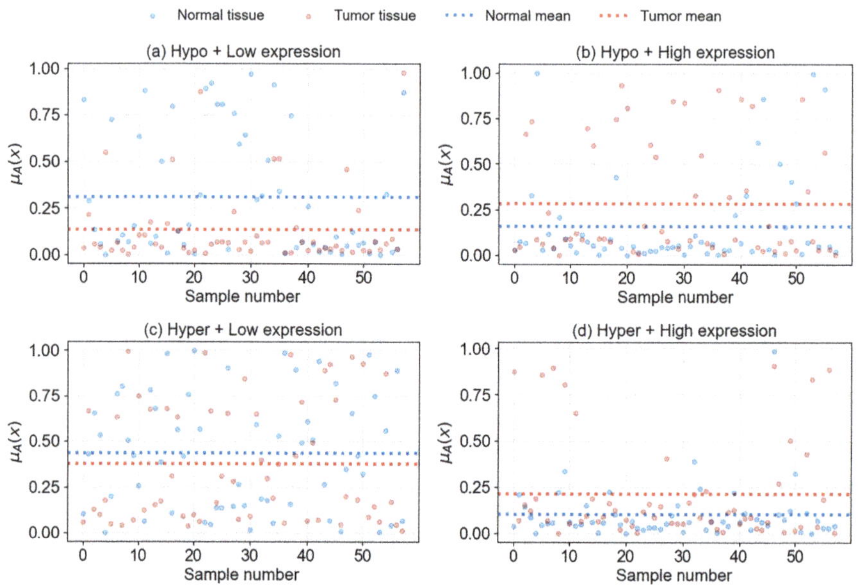

Fig. 1. Fuzzy membership degrees of individual samples based on oncogene profiles across four clusters. The y-axis denotes membership degree and the x-axis the sample number. Blue and red dots indicate normal and tumor samples. Panels (a)(d) correspond to the four methylationexpression combinations.

Cluster 1 shows higher mean membership for normal samples than tumor samples, consistent with oncogenes of low expression and limited tumor contribution. Cluster 2 shows higher membership for tumor samples and corresponds to highly expressed oncogenes associated with tumor progression. Cluster 3 presents similar medium membership values for both sample types, consistent with hypermethylated and weakly expressed oncogenes. Cluster 4 shows higher tumor membership and corresponds to methylated but transcriptionally active oncogenes.

Figure 2 reports analogous clustering for tumor suppressor genes using the same methylation and expression criteria.

Membership degrees and clustered gene features are integrated into a fuzzy Bayesian network (FBN) to compute risk probabilities. The network models dependencies among age (A), physical activity (Ph), smoking (S), gene profile (G), and lung cancer (L). Unlike a conventional BN, the FBN uses membership degrees instead of crisp states (Fig. 3).

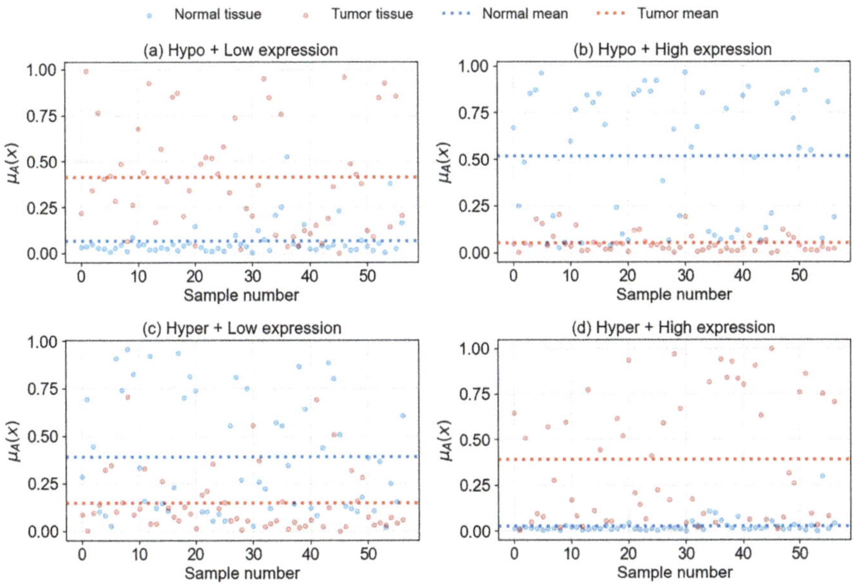

Fig. 2. Fuzzy membership degrees of samples based on tumor-suppressor gene profiles across four clusters. Blue and red dots denote normal and tumor samples.

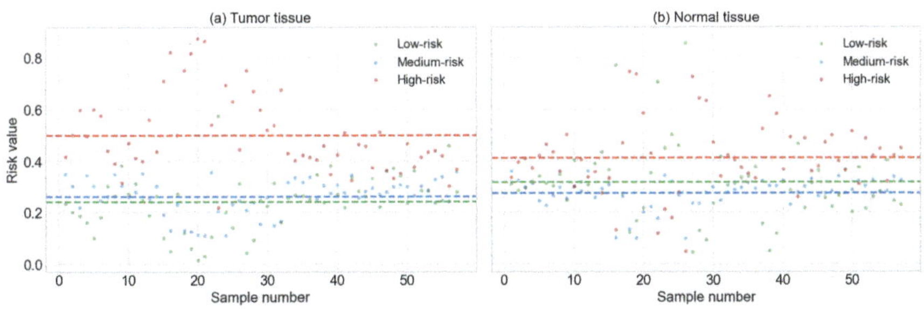

Fig. 3. Risk probabilities obtained from the FBN for tumor and normal samples. Colors denote low, medium, and high risk.

Equation 8 gives the fuzzy joint probability:

$$P(\tilde{L} \mid \tilde{A}, \tilde{Ph}, \tilde{S}, \tilde{G}) = \frac{\sum_{i=1}^{n} \prod_{x_i \in X} \mu_{\tilde{X}}(x_i) \cdot P(\tilde{L} \mid x_i) \cdot P(x_i)}{P(\tilde{A}, \tilde{Ph}, \tilde{S}, \tilde{G})}, \tag{8}$$

where $x_i = (a_i, ph_i, s_i, g_i)$ denotes the ith sample values and $\mu_{\tilde{X}}(x_i)$ the corresponding membership degrees. Conditional probability tables are defined according to lung cancer risk studies.

4.1 Evaluation of Binary and Fuzzy Bayesian Networks

Binary BN and FBN were compared on the same dataset (n = 114 with two replicates per sample). The FBN uses fuzzy c-means clustering and produces continuous risk vectors, while the BN relies on k-means and hard assignments.

Predictive performance was measured using Accuracy, Precision, Recall, and F1 score (see Table 1).

Table 1. Comparison of binary and FBN models using standard classification metrics.

Models	TP	Accuracy	Precision	Recall	F1 score
BN	66	57.8%	1.00	0.57	0.73
FBN	95	83.4%	1.00	0.83	0.91

The FBN improves accuracy from 57.8% to 83.4% and increases correctly classified instances. Precision equals 1.00 for both models, while recall rises from 0.57 to 0.83, leading to an F1 increase from 0.73 to 0.91. These results are consistent with fuzzy modeling, where soft boundaries reduce information loss near cluster limits. Although computational cost is higher, predictive reliability is improved in the presence of overlapping and gradual patterns.

5 Conclusion and Future Works

This study presents an approach integrating fuzzy logic with BNs to address uncertainties in biological data for lung cancer prediction. Fuzzy membership functions are defined for variables such as age, physical activity, and smoking status, while gene profiles are incorporated through clustering. The analysis was performed on a dataset obtained from the GEO portal. The resulting network represents interactions between genetic and environmental factors, extending beyond binary representations to capture the variability inherent in biological data and enabling individualized assessment. The analysis shows that the FBN achieves higher predictive performance than the binary Bayesian network (BN), with accuracy increasing from 57.8% to 83.4% and F1 score from 0.73 to 0.91. Recall improved from 0.57 to 0.83, indicating more effective detection of true positive cases. As future work, we plan to extend the FBN by incorporating additional omics layers, including proteomics and metabolomics, to improve prediction accuracy. We will also refine the membership functions and explore potential applications in real-time clinical decision-making.

Code and Data Availability. All code, processed data matrices, and analysis scripts used in this study are publicly accessible. The complete repository, including instructions for reproducing the analyses, is available at https://github.com/lidahooshyar/Fuzzy-Bayesian-Network.

References

1. Datta, N., Chakraborty, S., Basu, M., Ghosh, M.K.: Tumor suppressors having oncogenic functions: the double agents. Cells **10**(1), 46 (2020)
2. Kraft, P., Yen, Y.C., Stram, D.O., Morrison, J., Gauderman, W.J.: Exploiting gene-environment interaction to detect genetic associations. Hum. Hered. **63**(2), 111–119 (2007)
3. Rodin, A.S., Boerwinkle, E.: Mining genetic epidemiology data with bayesian networks i: Bayesian networks and example application (plasma apoe levels). Bioinformatics **21**(15), 3273–3278 (2005)
4. Selamat, S.A., et al.: Genome-scale analysis of DNA methylation in lung adenocarcinoma and integration with MRNA expression. Genome Res. **22**(7), 1197–1211 (2012)
5. Søes, S., et al.: Hypomethylation and increased expression of the putative oncogene elmo3 are associated with lung cancer development and metastases formation. Oncoscience **1**(5), 367 (2014)
6. Zadeh, L.A.: Fuzzy sets. Inf. Control **8**(3), 338–353 (1965)

A URV-Based Biclustering for Uncovering Glioblastoma Molecular Signatures

Francesca Martella(✉) and Monia Ranalli

Dipartimento di Scienze Statistiche, Sapienza Università di Roma, Rome, Italy
francesca.martella@uniroma1.it

Abstract. Microarray technology often generates ordinal detection calls (Absent, Marginal, Present) that are frequently treated as continuous or nominal, losing their inherent ranked structure. We propose a model-based biclustering framework based on the Underlying Response Variable (URV) approach. By modeling these calls as discretizations of latent Gaussian mixtures, we identify localized molecular signatures. Our model employs a flexible factorial covariance structure where gene partitions can vary across patient clusters, offering a more nuanced view than rigid Latent Block Models (LBM). Using a pairwise composite likelihood estimation, we analyze 22777 genes from the TCGA-GBM project, identifying distinct clusters that perfectly align with disease status.

1 Introduction

High-throughput genomic technologies, such as microarrays, have revolutionized our understanding of complex biological systems by allowing the simultaneous measurement of thousands of genes. Traditionally, gene expression analysis has focused on continuous intensity values. While continuous intensities remain the dominant representation in transcriptomic analysis, detection calls have long been used as a complementary analytical approach. In particular, they provide a reliability-driven representation of gene expression, focusing on whether a signal can be confidently distinguished from background noise rather than on its exact magnitude. This perspective is especially relevant in microarray settings characterized by high technical variability, where distinguishing true expression from background signal may be more robust than modeling weak quantitative differences.

Building on this framework, several normalization algorithms have been developed to translate raw intensity measurements into discrete "detection calls". Among the most widely used is MAS 5.0, which generates calls that categorize gene expression into three ordinal states: Absent (A), Marginal (M), and Present (P). These calls reflect the statistical confidence that a gene's signal is significantly different from background noise. While these calls are inherently ranked, they lack fixed intervals, posing a significant challenge for standard clustering algorithms designed for continuous data. Importantly, the use of detection calls is not intended to replace continuous modeling approaches. Rather, it

P. Brito et al. (Eds.): IFCS 2026, STUDIES CLASS, pp. 202–209, 2026.
https://doi.org/10.1007/978-3-032-32009-4_25

reflects a different focus: identifying biologically meaningful activation patterns that are robust to measurement noise. In this sense, ordinal modeling complements continuous-expression analysis by emphasizing signal reliability and structural patterns over minor intensity differences.

Several studies have explicitly used detection calls either for filtering, reliability assessment, or direct modeling purposes. For instance, [9] discuss the use of detection calls to improve reliability in Affymetrix data analysis; [1] highlight how presence/absence calls can be used to assess transcript reliability particularly for low-abundance genes; [5] introduce a microarray CGH method for presence/absence calling in non-model genomes.

The evolution of microarray analysis has moved from simple one-mode clustering toward biclustering (or co-clustering), which aims to identify localized patterns where a subset of genes behaves consistently across a subset of samples. Early distance-based biclustering methods, such as double k-means [13], were computationally efficient but lacked the probabilistic flexibility required to handle the discrete and ordinal nature of genomic detection calls. Consequently, model-based approaches have gained prominence for their ability to provide a theoretically grounded alternative for complex data structures. These proposals have been developed for various data types, including continuous data [8] and binary or count data [11] and references within it.

Within the framework of latent variable modeling, two primary paradigms exist for handling ordinal data: Item Response Theory (IRT) and the Underlying Response Variable (URV) approach. IRT models assume that the probability of observing a specific category is an analytic function of latent variables, typically assuming a nominal latent structure like in Latent Class Analysis (LCA) [4]. Conversely, the URV approach—adopted in this study—treats ordinal variables as discretizations of latent continuous Gaussian variables [6,10]. This allows for the preservation of the ordinal ranking while utilizing the flexibility of Gaussian mixture models to uncover latent partitions [3,7,12]. In our context, the URV formulation provides a principled bridge between the latent continuous biological signal and its observed ordinal categorization. Rather than artificially imposing continuity, the model acknowledges that the observed data arise from a thresholding mechanism applied to an underlying signal, which is consistent with the preprocessing strategy used to generate detection calls.

Despite recent advances, such as the Latent Block Model (LBM) for ordinal data [2], many existing methods enforce a rigid grid structure where the row and column partitions are assumed to be independent. This rigidity often fails to capture the biological reality of genomic data, where gene dependencies may vary significantly between different tissue types or disease states. Furthermore, the analysis of high-dimensional matrices (e.g., the 22777 genes in the TCGA-GBM project) often becomes computationally infeasible due to the need to calculate high-dimensional integrals.

In this paper, we adopt a model-based biclustering framework that addresses these limitations through a flexible URV approach. Our model introduces a factorial representation of the data [8], allowing the variable partition to vary across unit clusters, which provides a more parsimonious and biologically interpretable representation of genomic signatures. To overcome the "curse of dimensionality", we employ a pairwise composite likelihood estimation [12], which significantly enhances computational feasibility by focusing on bivariate marginals rather than the full joint density.

The remainder of this paper is organized as follows. Section 2 details the methodological framework, including the thresholding model and the factorial covariance structure. Section 3 describes the estimation procedure via the pairwise EM algorithm and discusses model selection. Section 4 presents the results of the application to the TCGA-GBM microarray dataset. Finally, Sect. 5 offers some concluding remarks.

2 Methodological Framework

The proposed framework treats observed ordinal variables as manifestations of latent continuous variables.

Let x_j be an ordinal variable (gene call) for $j = 1, \ldots, J$. We assume x_j is a discretization of a latent continuous variable y_j through a thresholding model:

$$x_j = c_j \Leftrightarrow \gamma^{(j)}_{c_j-1} \leq y_j < \gamma^{(j)}_{c_j} \tag{1}$$

where $-\infty = \gamma^{(j)}_0 < \gamma^{(j)}_1 < \cdots < \gamma^{(j)}_{C_j} = +\infty$ are non-observable thresholds. Here, C_j denotes the number of categories for gene call j. For the TCGA-GBM dataset, $C_j = 3$ for all j, corresponding to the detection calls (A, M, P). The upper boundary $\gamma^{(j)}_{C_j}$ is fixed to $+\infty$ to identify the model and cover the latent variable support.

The latent vector $\mathbf{y}$ is assumed to follow a finite Gaussian mixture model:

$$f(\mathbf{y}; \boldsymbol{\Theta}) = \sum_{g=1}^{G} p_g \phi(\mathbf{y}; \boldsymbol{\eta}_g, \boldsymbol{\Sigma}_g) \tag{2}$$

where p_g is the cluster weight and $\boldsymbol{\eta}_g$ and $\boldsymbol{\Sigma}_g$ are the component-specific mean and covariance.

To achieve biclustering, we specify a factorial structure for $\boldsymbol{\Sigma}_g$:

$$\boldsymbol{\Sigma}_g = \mathbf{A}_g \mathbf{A}'_g + \mathbf{D}_g \tag{3}$$

where $\mathbf{A}_g = \{a_{jgq}\}$ is a binary row-stochastic matrix that assigns variable j to segment q in cluster g ($g = 1, \ldots G$; $q = 1, \ldots Q$, with $Q = \max_{g=1,\ldots,G} Q_g$). Specifically, the generic element a_{gjq} is equal to 1 if and only if the j−th variable belongs to the q−th segment in the g-th cluster, and 0 otherwise. Crucially, the partition $\mathbf{A}_g$ is cluster-specific, allowing for a flexible representation where different gene groups characterize different patient clusters (segments). $\mathbf{D}_g$ represents the cluster-specific diagonal matrix of unique variances. More precisely, $\mathbf{D}_g = \text{diag}(d_{1g}, \ldots, d_{Jg})$ where each $d_{jg} > 0$. Crucially, the partition

$\mathbf{A}_g$ is cluster-specific, allowing for a flexible representation where different gene groups characterize different patient clusters. Our methodology builds upon the URV framework proposed by [10] and [6], extending it to a biclustering context through component-specific factorial structures. The use of pairwise likelihood for ordinal mixtures follows the approach of [12], while the parsimonious representation of the covariance matrix is inspired by the factor analyzer mixtures for gene expression data introduced by [8].

3 Estimation and Model Selection

Full maximum likelihood estimation is infeasible as the number of ordinal variables increases due to high-dimensional integrals. We maximize the pairwise log-likelihood based on bivariate marginals:

$$p\ell(\boldsymbol{\psi};\mathbf{x}) = \sum_{j<h} \sum_{c_j,c_h} n_{c_j c_h}^{(jh)} \log \left[\sum_{g=1}^{G} p_g \pi_{c_j c_h}^{(jh)}(\boldsymbol{\eta}_g^{(jh)}, \boldsymbol{\Sigma}_g^{(jh)}, \boldsymbol{\Gamma}^{(jh)}) \right], \tag{4}$$

where $\boldsymbol{\psi} = \{p_1, \ldots, p_{G-1}, \boldsymbol{\eta}_1, \ldots, \boldsymbol{\eta}_G, \mathbf{A}_1, \ldots, \mathbf{A}_G, \mathbf{D}_1, \ldots, \mathbf{D}_G, \boldsymbol{\Gamma}\}$ is the set of model parameters, $\boldsymbol{\Gamma}$ is the set of threshold vectors $\gamma^{(j)}$ $(j = 1, \ldots, J)$, and $\pi_r(\boldsymbol{\mu}_g, \boldsymbol{\Sigma}_g, \boldsymbol{\Gamma})$ is the probability of response pattern $\mathbf{x}_r$ in cluster g. This approach is computationally scalable as it only requires evaluating integrals of bivariate normal distributions. Parameters are updated via a pairwise EM algorithm.

The pairwise likelihood approach offers a pragmatic solution to the 'curse of dimensionality' typical of genomic data. By substituting the full joint density with the product of bivariate marginals, we reduce the complexity from J-dimensional to 2-dimensional integrals. While this results in a loss of statistical efficiency compared to the full MLE – as higher-order dependencies are not explicitly modeled – the resulting estimators are consistent and asymptotically normal under mild regularity conditions [12]. In our application, this trade-off is essential, as the full likelihood would be numerically intractable.

We use the composite Integrated Completed Likelihood criterion (cICL) to select G and Q:

$$\text{cICL} = -2p\ell(\hat{\boldsymbol{\psi}}; \boldsymbol{Z}, \boldsymbol{X}) + 2\text{Entropy} + \delta \log(N) \tag{5}$$

where N is the sample size, $\boldsymbol{Z}$ is the cluster membership matrix, $\delta = 3JG - 2J + (C_J - 1)J + G - 1$ represents the free parameters, and Entropy term represents the penalty for classification uncertainty.

A further important point of the proposed model that is worth to be discussed is parameter identifiability.

Here, the threshold parameters $\boldsymbol{\Gamma}$ do not change over the components and thus this allows us to apply the constraint on the means and the variances of the latent variables fixed to 0 and 1 to only one component (the reference component) rather than to the whole mixture to be identified.

This is sufficient to estimates both thresholds and component parameters if all the observed variables have three categories at least and when groups are known. However, the accuracy of threshold parameter estimates depends on the presence of all categories in the reference group.

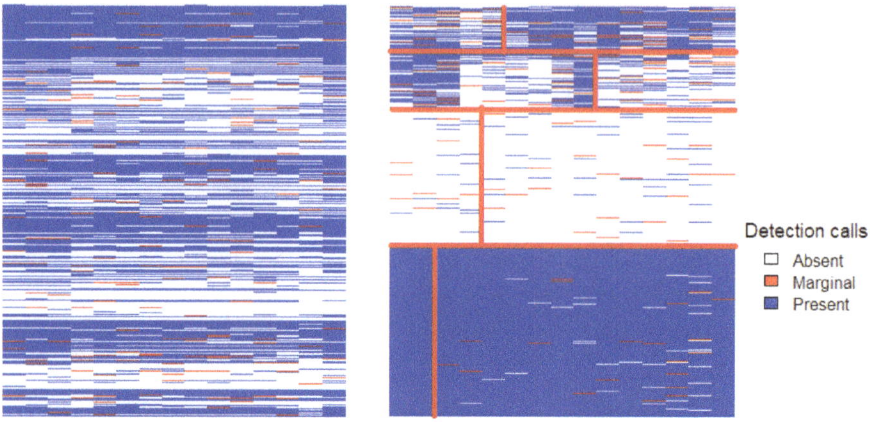

Fig. 1. Original data (left) and biclustering result into 4×2 blocks (right)

Adopting a pairwise likelihood estimation approach, the number of essential parameters equals the number of parameters of a log linear model with only two factor interaction terms. Thus, it means that we can estimate a lower number of parameters compared to a full maximum likelihood approach. In detail, given a $C_1 \times C_2 \times \ldots \times C_J$ contingency table, the necessary condition for the identifiability is that the number of model parameters has to be at most equal to

$$\sum_{j=1}^{J}(C_j - 1) + \sum_{j=1}^{J-1} \sum_{h=j+1}^{J} (C_j - 1)(C_h - 1).$$

Finally, the factorial reparameterization of component-specific covariance of the latent variable involving $\mathbf{A}_g$ is unique since the matrix $\mathbf{\Lambda}_g$ does not suffer from the identifiability issues, apart from clusters labels permutations, differently from the factor loadings matrix in the factor analysis model.

4 Application: TCGA-GBM Microarray Data

We analyzed 22777 genes across 15 individuals (10 glioblastoma, 5 normal) from the TCGA-GBM project.

The proposed algorithm was applied for $G = 1, 2, \ldots, 6$ and $Q = 1, 2, \ldots, 4$. The optimal solution identified by c-ICL criterion is given by $G = 4$ clusters and $Q = 2$ segments (see Fig. 1). Figure 2 highlights the main characteristics of each cluster that are briefly described in the following.

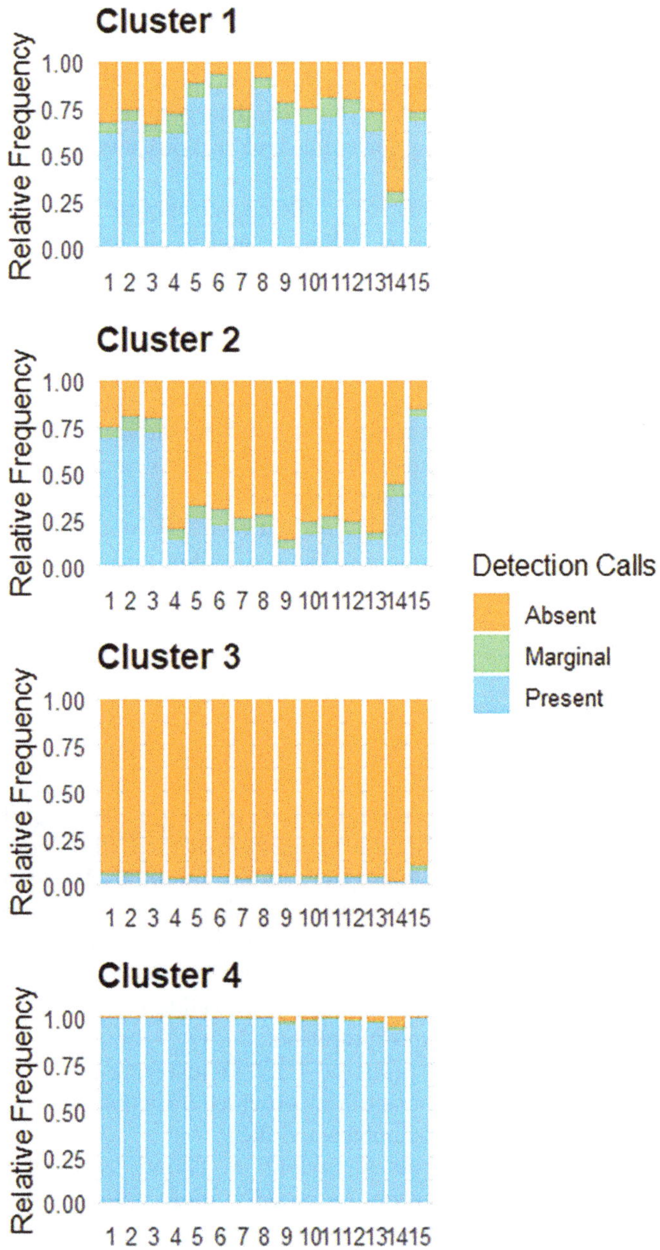

Fig. 2. Relative frequencies of individual detection calls by Cluster

- **Cluster 1 (Disease Signature):** Smallest cluster ($n = 2785$ genes) showing a clear separation between normal and tumor samples, identifying genes associated with glioblastoma mechanisms.

- **Cluster 2 (Heterogeneous Regulation)** is similar to Cluster 1 in terms of size ($n_2 = 2898$). It captures genes that are generally active in normal tissue but show heterogeneous expression in tumor samples.
- **Cluster 3 (Silenced Genes)** is the second largest ($n = 7379$) dominated by "Absent" calls, identifying genes inactive in both healthy and cancerous brain tissue.
- **Cluster 4 (Global Expression)** is the second-largest cluster ($n = 9215$) where genes are predominantly "Present" across all samples, representing baseline metabolic activity.

5 Conclusion

The proposed URV biclustering model provides a scalable and flexible framework for the analysis of high-dimensional genomic data, made computationally feasible through the use of pairwise composite likelihood.

By explicitly allowing for dependencies between row and column partitions, the model is able to identify coherent biclusters that correspond to localized molecular structures in the data. In particular, this approach reveals how specific groups of genes exhibit consistent expression patterns within well-defined subsets of individuals, such as tumor versus normal samples or clinically homogeneous patient subgroups.

By focusing on these localized patterns rather than global clustering structures, the proposed method facilitates the discovery of biologically meaningful molecular signatures and captures patient-specific heterogeneity that would likely remain undetected using conventional clustering techniques. This added level of resolution is especially valuable in genomic studies, where complex interactions and heterogeneous disease mechanisms are common. It should be noted that the empirical application is based on a limited sample of subjects, which may affect the stability of the identified clusters and the generalizability of the results. Therefore, these findings should be interpreted as exploratory, and further studies with larger sample sizes are needed to validate the observed patterns and strengthen their biological interpretation.

Overall, we believe that the proposed URV biclustering model represents a valuable methodological contribution and provides a flexible and scalable tool for the analysis of complex high-dimensional data.

While its relevance is particularly evident in genomic and other biological applications, the proposed framework is broadly applicable to a wide range of fields characterized by high-dimensional, and possibly heterogeneous units and variables. Potential applications include psychometrics and social sciences, where ordinal responses are common, as well as marketing and consumer research.

The empirical application is based on a limited sample of subjects ($N = 15$), which may affect the stability of the identified clusters and the generalizability of the results. Therefore, the findings should be interpreted as exploratory. Further studies with larger sample sizes are needed to validate these patterns and strengthen their biological interpretation.

References

1. Archer, K.J., Reese, S.E.: Detection call algorithms for high-throughput gene expression microarray data. Brief. Bioinform. **11**(2), 244–252 (2009)
2. Corneli, M., Bouveyron, C., Latouche, P.: Co-clustering of ordinal data via latent continuous random variables and not missing at random entries. J. Comput. Graph. Stat. **29**(4), 771–785 (2020)
3. Everitt, B.: A finite mixture model for the clustering of mixed-mode data. Stat. Probab. Lett. **6**(5), 305–309 (1988)
4. Goodman, L.A.: Exploratory latent structure analysis using both identifiable and unidentifiable models. Biometrika **61**(2), 215–231 (1974)
5. Jonker, M.J., et al.: Absence/presence calling in microarray-based CGH experiments with non-model organisms. Nucleic Acids Res. **42**(11), e94 (2014)
6. Jöreskog, K.G.: New developments in LISREL: analysis of ordinal variables using polychoric correlations and weighted least squares. Qual. Quant. **24**(4), 387–404 (1990)
7. Lubke, G., Neale, M.: Distinguishing between latent classes and continuous factors with categorical outcomes: class invariance of parameters of factor mixture models. Multivar. Behav. Res. **43**(4), 592–620 (2008)
8. Martella, F., Alfò, M., Vichi, M.: Biclustering of gene expression data by an extension of mixtures of factor analyzers. Int. J. Biostat. **4**(1) (2008)
9. McClintick, J.N., Edenberg, H.J.: Effects of filtering by present call on analysis of microarray experiments. BMC Bioinform. **7**, 49 (2006)
10. Muthén, B.: A general structural equation model with dichotomous, ordered categorical, and continuous latent variable indicators. Psychometrika **49**(1), 115–132 (1984)
11. Pledger, S., Arnold, R.: Multivariate methods using mixtures: correspondence analysis, scaling and pattern-detection. Comput. Stat. Data Anal. **71**, 241–261 (2014)
12. Ranalli, M., Rocci, R.: Mixture models for ordinal data: a pairwise likelihood approach. Stat. Comput. 1–19 (2016)
13. Vichi, M., Kiers, H.A.: Factorial k-means analysis for two-way data. Comput. Stat. Data Anal. **37**(1), 49–64 (2001)

Matching Between Partition Incidence Matrices and Decision Tree Classifier Criteria

Boris Mirkin[1,2(✉)] and Azizbek Bekniyazov[1]

[1] School of Data Analysis and Artificial Intelligence, National Research University "Higher School of Economics", Moscow, Russian Federation
bmirkin@hse.ru

[2] School of Mathematical and Computer Sciences, Birkbeck University of London, London, UK

Abstract. This paper proposes a family of scoring criteria for building classification trees. Our family is generated over rescaling weights of target categories according to a model matching the input and target partitions. We prove that popular Gini and Chi-squared criteria belong to the family. We experimentally show, using both real-world and synthetic datasets, that use of match scoring criteria can lead to superior results over those obtained using Entropy and other popular criteria.

Keywords: Classifier · classification tree · scoring criteria · partition match model

1 Introduction

Classifier decision tree model is popular because of its logical and visual clarity. An important part of the feature-splitting process for tree building is the criterion for choosing both a feature to split and a cut-off over the feature. Three popular criteria: Entropy, Gini index and Pearson's Chi-squared, constitute the core of most popular decision tree producing programs, C4.5 [10], CART [2] and CHAID [4], respectively. For the time being, several significant improvements to them have been proposed including methodologies for building oblique and multivariate cut-off criteria [7]; investigation of statistical nature of the criteria with probability modeling [5]; models and methods for building optimal cut-off trees rather than just single splits [3]. Yet, however, nothing has been done with respect to investigation of theoretical fundament for the splitting criteria. Indeed, the criteria have been picked up from fragments of different frameworks, and no general framework to generate criteria – the more so to compare them – has been developed so far.

This paper makes a step in the direction of overcoming the gap. We present a simple linear model relating the target partition with the predictor partition which is being built. This model involves a set of parameters for scaling frequencies of target categories. We derive a formulaic expression for the contribution of the predictor partition to the summary variance of the target categories, which depends on the scaling parameters and, in this way, generates a family of association indices which can be used as classification

P. Brito et al. (Eds.): IFCS 2026, STUDIES CLASS, pp. 210–217, 2026.
https://doi.org/10.1007/978-3-032-32009-4_26

decision tree criteria. We prove that the Gini and Chi-square indices belong to this family. We pick up several other criteria from our family and experimentally verify that some of them can be superior at some real-world and synthetic datasets.

2 A Model for Matching Partition Incidence Matrices

Consider a finite set I of N objects $i \in I$ and two partitions, $R = \{R_1, R_2, \dots, R_K\}$ and $S = \{S_1, S_2, \dots, S_V\}$, over I. Assign 0/1 incidence matrices, $Z = (z_{ik})$ of dimension $N \times K$ and $X = (x_{iv})$ of dimension $N \times V$, to these partitions, R and S, respectively. Elements of these matrices are defined so that $z_{ik} = 1$ if $i \in R_k$, and $z_{ik} = 0$ if $i \notin R_k$; and, similarly, $x_{iv} = 1$ if $i \in S_v$, and $x_{iv} = 0$ if $i \notin S_v$. In what follows these two play different roles. We assume that partition S, as well as matrix X, correspond to a target categorical feature over I, while partition R, as well as its matrix Z, correspond to a cluster partition over I which is to be devised to predict, or approximate, feature X from some other data whose composition is not important at these definitions.

Then the matrix X can be considered a quantified form of the target feature obtained by enveloping its categories into "dummies", the binary columns of matrix X. In analogy with (bi)linear cluster models described in [8, 9], one can relate these partitions with a model like

$$y_{iv} = \sum_{k=1}^{K} z_{ik} c_{kv} + e_{iv} \tag{1}$$

where y_{iv} are elements of matrix X standardized according to equations

$$y_{iv} = \frac{x_{iv} - a_v}{b_v}$$

in which a_v is a shift parameter, and b_v a rescaling parameter for the 0/1 – dummy representing the v-th part of partition S, and e_{iv} are residuals to be minimized over unknown c_{kv} values. Assume that the scale shift a_v is the average value in v-th column of matrix X, which is, obviously, the proportion of part R_v in I, further denoted as p_v. One may refer to model (1) as R-S-match model. By minimizing the sum of squares of residuals e_{iv} in (1), we arrive at expression $F(S, c) = \sum_{k=1}^{K} |S_k| < c_k, c_k >$ for that part of the data scatter T, which is explained by the association between R and S according to model in (1), to be maximized. This can be further reformulated as

$$F(R, S) = N \sum_{k=1}^{K} \sum_{v=1}^{V} \frac{(p_{kv} - p_k p_v)^2}{p_k b_v^2} \tag{2}$$

where p_k, p_v and p_{kv} are, respectively, the proportion of objects in S_k, in R_v, and $S_k \cap R_v$. Further on we consider its relative part, $f(R, S) = \frac{F(R,S)}{N}$, which is a family generated by various values of the rescaling factor b_v.

Obviously, at $b_v = p_v^{1/2}$, the value (2) is equal to the Chi-squared Pearson association index used in the popular CHAID decision tree package [4]. Moreover, it is not difficult to prove that another popular association measure, the Gini index used in CART [2], is also a member of the family, at $b_v = 1$.

We consider a few more specifications for the b_v parameter in (2); four of them are presented in Table 1 to supplement its rows 1 and 2 corresponding to already mentioned Pearson and Gini indices.

Table 1. Values of the scaling b_v parameter in the match association criterion in (2).

No	b_v
1	1
2	$p_v^{0.5}$
3	$[p_v(1-p_v)]^{0.5}$
4	$-\log(p_v)$
5	$p_v^{0.5}\log(p_v)$
6	$-p_v\log(p_v)$

3 Setting of Experiments

The goal of experimental computations is to check whether any match association index from the list in Table 1 can be better at building classifier decision trees than any of the three popular criteria mentioned in Sect. 1: the Entropy, the Gini index (number 1 in Table 1), and the Chi-squared (number 2 in Table 1). Besides the set of six classification criteria in Table 1 plus Entropy, the experiment comprises:

1. Datasets at which decision trees are to be built.
2. An algorithm for building classification trees.
3. Criteria for assessment of the results.

Let us consider them in turn.

3.1 Datasets

We use both real-world data and synthetic data. Real-world datasets are obtained from the UCI Irvine repository [1]. The list of them, together with their characteristics, is presented in Table 2. Due to space limitations, only a subset of the experimental results is reported in this paper; complete results for all the datasets are available at GitHub site https://github.com/buxazich/Datagenerator.

We generate synthetic data sets by using data generator proposed in [6], see also [11]. It produces an $N \times V$ data matrix X of Gaussian clusters in the following steps:

A. We specify the number of objects, N, the number of features, V, and the number of clusters, K. We also define the minimum cluster size, n, which may be needed to operate with stochastic data models. The product $K * n$ is to be less than N: there remain $N - K * n$ objects to randomly distribute among the K clusters to be generated. In this way, the number of elements N_k is specified at each of the clusters to be generated.

B. For each cluster$k$, $k = 1, 2, \ldots, K$, its center c_k is generated as a V-dimensional vector of components generated from a uniform distribution over interval $(\alpha - 1,\ 1 - \alpha)$ where α is a user-defined real,$0 < \alpha < 1$, to characterize the intermix of clusters generated. The larger the squeeze parameterα, the nearer to each other are the centers and the more interspersed are elements of different clusters.

Table 2. Datasets from UCI Irvine repository used in our experiments

Dataset		N	V	K
1	Breast Cancer (Prognostic)	194	32	2
2	Balance Scale	625	4	3
3	Car Evaluation	1728	6	4
4	Connectionist Bench Sonar	208	60	2
5	Contraceptive Method Choice	1473	9	3
6	Hayes Roth	132	4	3
7	Hepatitis	80	19	2
8	Ionosphere	351	34	2
9	Glass	214	9	6
10	Congressional Voting Records	232	16	2
11	Spambase	4601	57	2
12	Mushroom	8214	22	7

A. To generate k-th cluster, we generate a vector of standard deviations $g_k = (g_{kv})$ randomly from interval [0.05, 0.10], $k = 1, 2, \ldots, K$. Then we generate N_k V-dimensional vectors whose v-th components are taken from a Gaussian distribution $N(0, g_{kv})$ with 0 expectation and standard deviation g_{kv}, and add each of them to the center c_k. In this way, we obtain a matrix X_k of dimension $N_k \times V$ of elements of cluster k.
B. We combine the matrices X_k together into a final $N \times V$ data matrix X, so that the hidden partition of row indices is known to the user.

In the follow-up, all the synthetic datasets have $N = 2000$ objects; the minimum cluster size is defined as 50. There are two feature space dimensions $V = 6$ and 15, three cluster numbers $K = 4, 8$, and 15, and two squeezing parameter values $\alpha = 0.5$ (clusters well separated) and $\alpha = 0.85$ (clusters are well interspersed). Therefore, there are 12 = 2*3*2 parameter settings.

3.2 Computation Algorithm

For each parameter combination, we generate 50 datasets. At each of the datasets, we run a binary tree classifier algorithm at the number of splits not more than $t = 3$ or $t = 4$, with the size of a cluster to be split not less than 10. The cluster to be split is chosen according to its contribution to the data scatter, computed according to formula $w_k = |S_k|\langle c_k, c_k \rangle$. The split is performed on feature v and its value r_v to maximize the scoring function, either entropy or one of functions 1–6 in Table 1. After this is done over all 50 data tables generated, we report the average value and standard deviation for each of the scoring functions. At each of the real-world datasets, we split it randomly into two subsets, the training one, 75%, of the objects, and the testing one, 25% of the objects. We build the decision tree over the training set, and then we compare that with the target partition on the testing set. Then we average them over 50 independent runs.

3.3 Assessment of Results

To quantitatively assess the performance of classification and clustering algorithms, a set of evaluation metrics is commonly employed, each characterizing different aspects of model behavior and predictive reliability.

We follow in this to the approaches developed in machine learning. When the classification task involves just two classes, positive and negative, the outcomes of binary classification are conventionally counted as true positives or true negatives if they are correct, and false negatives or false positives, if they are not. By denoting the respective numbers by TP, TN, FN, and FP, the popular evaluation metrics are Accuracy, Recall, Precision and the harmonic average of the latter two, F1. Target partitions usually have several classes (categories). In such a case, the metrics above apply at each class taken as the positive while all the rest are considered as negative, after which the results are averaged. We also apply a genuine partition-to-partition similarity metric, the popular Adjusted Rand Index (ARI) [12].

4 Experimental Results

Overall, the goal of our computations was to see whether our match association indexes are relevant at all. In this section, we present evidence for the positive answer to the question, Moreover, we will see that the offhand indexes 3 through 6 in Table 1 can be superior to over the popular ones at some specific situations.

4.1 Results at Real-World Data

In what follows, we report average scoring results over 50 runs of the algorithm; standard deviations are rather small, reaching not more than 10–15% of the averages, and, thus, are omitted from the tables. First, we notice that all b_v variants achieve identical performance and consistently outperform the Entropy baseline across all five assessment scores (Table 3).

An interesting case emerges at the real-world dataset Car Evaluation (Table 4). At decision trees of depth 3, there is always one winner overall. However, when we move to a better association by increasing the tree depth to 4, the picture changes considerably (see Table 4). There is no single winner anymore. Version 6. $b_v = -p_v\log(p_v)$ wins.

Table 3. Average assessment scores for the Spambase dataset at tree depth = 3.

Spambase							
Metric	Entropy	b_v					
		1	2	3	4	5	6
Accuracy	0.869	**0.882**	**0.882**	**0.882**	**0.882**	**0.882**	**0.882**
Precision	0.876	**0.883**	**0.883**	**0.883**	**0.883**	**0.883**	**0.883**
Recall	0.869	**0.882**	**0.882**	**0.882**	**0.882**	**0.882**	**0.882**
F1	0.867	**0.881**	**0.881**	**0.881**	**0.881**	**0.881**	**0.881**
ARI	0.543	**0.582**	**0.582**	**0.582**	**0.582**	**0.582**	**0.582**

over Accuracy and Recall. Entropy wins over Precision, while versions 3. $b_v = \left[p_v(1-p_v)\right]^{0.5}$, 4. $b_v = \log(p_v)$, and 5. $b_v = p_v^{0.5}\log(p_v)$, win over both F1 and ARI.

The reported results show that indeed, any of our offhand scoring functions can be superior to popular measures at appropriate datasets.

Table 4. Average assessment scores for the Car Evaluation dataset at tree depth = 4.

Car Evaluation							
Metric	Entropy	b_v					
		1	2	3	4	5	6
Accuracy	0.834	**0.844**	0.842	**0.844**	**0.844**	**0.844**	**0.844**
Precision	**0.852**	0.845	0.843	0.846	0.846	0.846	0.845
Recall	0.834	**0.844**	0.842	**0.844**	**0.844**	**0.844**	**0.844**
F1	0.831	0.837	0.834	**0.838**	**0.838**	**0.838**	0.837
ARI	0.651	0.682	0.667	**0.684**	**0.684**	**0.684**	0.679

4.2 Results at Synthetic Datasets

The following Tables 5–6 present patterns of match association scores at relatively modest number of features, 6, when moving from 4-part target partitions (Table 5) to 15-part target partition (Table 6).

Table 5 corresponds to the case at which clusters are well separated, and the number of features is greater than the number of target clusters, so that the decision tree approximates the target partition very closely, so that all the scoring functions reach 0.98 or 0.99, that are very close to the maximum value 1. The winning match association functions here are versions 3. $b_v = \left[p_v(1-p_v)\right]^{0.5}$ and 5. $b_v = p_v^{0.5}\log(p_v)$ from Table 1, already mentioned with respect to Table 4.

Table 5. Average assessment scores for the synthetic dataset at 6 features, 4 well separated clusters, and tree depth = 3.

V, K, α, depth = (6, 4, 0.5, 3)							
Metric	Entropy	b_v					
		1	2	3	4	5	6
Accuracy	0.992	0.993	0.993	**0.994**	**0.994**	**0.994**	0.993
Precision	0.992	0.993	0.993	**0.994**	**0.994**	**0.994**	**0.994**
Recall	0.992	0.993	0.993	**0.994**	**0.994**	**0.994**	0.993
F1	0.992	0.993	0.993	**0.994**	**0.994**	**0.994**	0.993
ARI	0.979	0.982	0.982	**0.983**	**0.983**	**0.983**	0.982

Table 6 relates to increasing the number of target clusters to a large value, $K = 15$, while leaving the number of features the same, 6. This leads to dramatic fall in precision, to 0.3 values, thus decreasing F1 value as well. There is no single winner here:

Entropy is the best over Accuracy, Recall and ARI whereas scoring criterion 4. $b_v = \log(p_v)$ is the best over Precision and F1.

Table 6. Average assessment scores for the synthetic dataset at 6 features, 15 well separated clusters, and tree depth = 3.

V, K, α, depth = (6, 15, 0.5, 3)							
Metric	Entropy	b_v					
		1	2	3	4	5	6
Accuracy	**0.512**	0.457	0.470	0.486	0.508	0.460	0.469
Precision	0.308	0.290	0.327	0.302	**0.349**	0.291	0.325
Recall	**0.512**	0.457	0.470	0.486	0.508	0.460	0.469
F1	0.371	0.332	0.358	0.355	**0.391**	0.333	0.358
ARI	**0.567**	0.432	0.408	0.490	0.490	0.443	0.404

These examples show that there should be some regularities over synthetic data caused by an interplay of three main forces: the dataset dimension V, the number of target clusters K, and the squeezing parameter α.

5 Conclusion

This paper proposes a family of scoring criteria to supplement the currently popular criteria of Entropy, Gini index and Chi-squared for building classification trees. Our family is generated by a least-squares model to match the input and target partitions represented by their incidence matrices. We prove that both Gini and Chi-squared criteria belong to the match association family. We propose a few offhand match scoring criteria belonging to our family and we experimentally show that some of them can be superior to those popular ones in some situations such as, for example, when the number of target classes is larger than the number of features.

However, the results of the computational experiments depend on the characteristics of the datasets used. Therefore, the generalizability of the observed improvements may be limited and further validation on a broader range of experiments is needed.

Among directions for further development, we can indicate the issue of finding conditions at which this or that representative of our match association criteria family is superior to other criteria for classification tree building. Another direction would be developing algorithms for enhancing classification tree techniques by using representatives of our match association family. This can involve such directions as random forest classifiers or boosting.

The authors are grateful to the anonymous reviewers for helpful comments. Support from the Basic Research Program of HSE University is gratefully acknowledged.

References

1. Asuncion, A., Newman, D.:UCI Machine Learning Repository (2007)
2. Breiman, L., Friedman, J.H., Olshen, V., Stone, C.J.: Classification and Regression Trees. Wadsworth, Belmont (1984)
3. Demirović, E., et al.: Murtree: optimal decision trees via dynamic programming and search. J. Mach. Learn. Res. **23**(26), 1–47 (2022)
4. Kass, G.V.: An exploratory technique for investigating large quantities of categorical data. Appl. Stat. **29**, 119–127 (1980)
5. Klusowski, J.M., Tian, P.M.: Large scale prediction with decision trees. J. Am. Stat. Assoc. **119**(545), 525–537 (2024)
6. Kovaleva, E.V., Mirkin, B.G.: Bisecting K-means and 1D projection divisive clustering: a unified framework and experimental comparison. J. Classif. **32**(3), 414–442 (2015). https://doi.org/10.1007/s00357-015-9186-y
7. Loh, W.Y.: Fifty years of classification and regression trees. Int. Stat. Rev. **82**(3), 329–348 (2014). https://doi.org/10.1111/insr.12016
8. Mirkin, B.G.: A sequential fitting procedure for linear data analysis models. J. Classif. **7**(2), 167–195 (1990). https://doi.org/10.1007/bf01908715
9. Mirkin, B.: Core Data Analysis. Springer, Cham (2019). https://doi.org/10.1007/978-3-030-00271-8
10. Quinlan, J.R.: C4.5: Programs for Machine Learning. Elsevier, Amsterdam (2014)
11. Rykov, A., De Amorim, R.C., Makarenkov, V., Mirkin, B.: Inertia-based indices to determine the number of clusters in k-means: an experimental evaluation. IEEE Access **12**, 11761–11773 (2024)
12. Steinley, D.: Properties of the Hubert-Arable adjusted Rand index. Psychol. Methods **9**(3), 386 (2004). https://doi.org/10.1037/1082-989X.9.3.386
13. Sut, N., Simsek, O.: Comparison of regression tree data mining methods for prediction of mortality in head injury. Expert Syst. Appl. **38**(12), 15534–15539 (2011). https://doi.org/10.1016/j.eswa.2011.06.006

One-Sample Hypothesis Test of Mean Interval for Interval-Valued Data

Fernando Montes and Anuradha Roy(✉)

Department of Statistics and Data Science, The University of Texas at San Antonio, San Antonio, TX, USA
Fernando.Montes2@my.utsa.edu,
Anuradha.Roy@utsa.edu
https://caicc.utsa.edu/faculty/profiles/roy-anuradha.html

Abstract. A parametric hypothesis test of the mean interval for interval-valued data in one population is studied under the assumption that both $p-$dimensional upper and $p-$dimensional lower bounds have the same variance-covariance matrix. An orthogonal transformation is employed to obtain an equivalent hypothesis test of $p-$dimensional mean interval in terms of a normal $p-$dimensional vector of mid-points and a log-normal $p-$dimensional vector of ranges. The normal mean vector is tested using Hotelling's T^2, while the log-normal vector of ranges is tested via the construction of a generalized pivotal quantity in a Monte Carlo simulation. Independence of the two tests allows the use of the Benjamini-Hochberg multiple testing procedure to control the false discovery rate. Simulations are conducted to compute the empirical Type I errors of the test statistic at nominal significance levels 0.01 and 0.05. The test is investigated with a real-world dataset.

Keywords: Generalized pivotal quantity · Hypothesis test · Interval-valued data · Multivariate log-normal · Orthogonal transformation · Type I errors

1 Matrix of Intervals

Let $I[\boldsymbol{Y}]$ represent $(n \times p)-$dimensional interval-valued data matrix, where n denotes the number of sampling units, p denotes the number of variables

$$I\left[\boldsymbol{Y}\right] = \begin{bmatrix} \boldsymbol{y}_1' \\ \vdots \\ \boldsymbol{y}_n' \end{bmatrix} = \begin{bmatrix} I[y]_{1,1}, I[y]_{1,2} & \cdots & I[y]_{1,p} \\ \vdots & \ddots & \vdots \\ I[y]_{n,1}, I[y]_{n,2} & \cdots & I[y]_{n,p} \end{bmatrix},$$

and each element is an interval. A generic interval $I[y]_{i,j} \equiv [y_{i,1\_j}, y_{i,2\_j}]$ $\forall$ $i = 1, \ldots, n$, $j = 1, \ldots, p$ and we assume $y_{i,1\_j} \leq y_{i,2\_j}$ with probability 1. For each element, the first subscript represents the observation unit/sample. The second subscript represents the variable. Each sample is characterized by p (interval-valued) variables.

P. Brito et al. (Eds.): IFCS 2026, STUDIES CLASS, pp. 218–226, 2026.
https://doi.org/10.1007/978-3-032-32009-4_27

After rearranging the ith sample by grouping together the first p upper bounds of the intervals and then the remaining p lower bounds of the intervals, a typical sample in $2p$−dimensional vector form looks like

$$\boldsymbol{y}_i = (y_{i,2\_1}, \ldots, y_{i,2\_p}, y_{i,1\_1}, \ldots, y_{i,1\_p})'; \quad i = 1, \ldots, n.$$

Therefore, $\overline{\boldsymbol{y}} = (1/n \sum_{i=1}^{n} \boldsymbol{y}_i)$. Now, the $(2p \times 1)$−dimensional random samples $\boldsymbol{y}_1, \boldsymbol{y}_2, \ldots, \boldsymbol{y}_n$ (all arranged as the first p upper bounds and then the next p lower bounds) are independent and identically distributed with $(2p \times 1)$−dimensional mean vector $\boldsymbol{\mu}_y = \begin{bmatrix} \boldsymbol{\mu}_y^+ \\ \boldsymbol{\mu}_y^- \end{bmatrix}$ and $2p \times 2p$ dimensional variance-covariance matrix $\boldsymbol{\Sigma}_y = \begin{bmatrix} \boldsymbol{U}_0 & \boldsymbol{U}_1 \\ \boldsymbol{U}_1 & \boldsymbol{U}_0 \end{bmatrix}$. The notation $\boldsymbol{\mu}_y^+$ represents the $p \times 1$ dimensional mean vector of the upper bounds, and $\boldsymbol{\mu}_y^-$ represents the $p \times 1$ dimensional mean vector of the lower bounds. The matrix $\boldsymbol{U}_0$ is a $p \times p$ positive definite (PD) symmetric matrix, and $\boldsymbol{U}_1$ is a $p \times p$ symmetric matrix subject to the constraints that $\boldsymbol{U}_0 + \boldsymbol{U}_1$ and $\boldsymbol{U}_0 - \boldsymbol{U}_1$ are PD matrices, so that $\boldsymbol{\Sigma}_y$ is also PD (for a proof, see Lemma 2.1 in [9]). The parameter matrices $\boldsymbol{U}_0$ and $\boldsymbol{U}_1$ are unstructured and the matrix $\boldsymbol{U}_1$ is symmetric by construction; see [6] for details.

The two $p \times p$−dimensional blocks $\boldsymbol{U}_0$ in $\boldsymbol{\Sigma}_y$ represent the variance-covariance matrix of the p variables at the upper bounds as well as at the lower bounds of the intervals, whereas the two $p \times p$−dimensional off-diagonal blocks $\boldsymbol{U}_1$ in $\boldsymbol{\Sigma}_y$ represent the covariance matrix of the p variables between the lower and upper bounds of the intervals.

1.1 Orthogonal Transformation of the Covariance Matrix $\boldsymbol{\Sigma}_y$

Let us consider the orthogonal matrix $\boldsymbol{\Gamma}_0 = (\boldsymbol{H}_2' \otimes \boldsymbol{I}_p)$, where $\boldsymbol{H}_2' = [\boldsymbol{h}_1, \boldsymbol{h}_2] = \begin{bmatrix} \frac{1}{\sqrt{2}} & \frac{1}{\sqrt{2}} \\ \frac{1}{\sqrt{2}} & -\frac{1}{\sqrt{2}} \end{bmatrix}$ is an orthogonal matrix and $\boldsymbol{I}_p$ is the $p \times p$ identity matrix. The $(2p \times 2p)$−dimensional orthogonal matrix $\boldsymbol{\Gamma}_0$ diagonalizes $\boldsymbol{\Sigma}_y$ as $\boldsymbol{\Gamma}_0 \boldsymbol{\Sigma}_y \boldsymbol{\Gamma}_0' =$ Diag $[\boldsymbol{\Delta}_2; \boldsymbol{\Delta}_1]$, where $\boldsymbol{\Delta}_2 = \boldsymbol{U}_0 + \boldsymbol{U}_1$ and $\boldsymbol{\Delta}_1 = \boldsymbol{U}_0 - \boldsymbol{U}_1$. Let $\boldsymbol{\Gamma}_0' = [\boldsymbol{\Gamma}_{0,1} : \boldsymbol{\Gamma}_{0,2}]$, where both the component matrices are $(2p \times p)$−dimensional. Let

$$\boldsymbol{\Gamma}_0 \boldsymbol{y} = \begin{bmatrix} \boldsymbol{\Gamma}_{0,1}' \\ \boldsymbol{\Gamma}_{0,2}' \end{bmatrix} \boldsymbol{y} = \begin{bmatrix} \boldsymbol{\Gamma}_{0,1}' \boldsymbol{y} \\ \boldsymbol{\Gamma}_{0,2}' \boldsymbol{y} \end{bmatrix} := \begin{bmatrix} \boldsymbol{y}_{21} \\ \boldsymbol{y}_{22} \end{bmatrix},$$

where $\boldsymbol{y}_{21}$ and $\boldsymbol{y}_{22}$ represent the midpoints and the midranges between the lower bounds and the corresponding upper bounds of the intervals (see [7]). Note that we say represent because $\boldsymbol{y}_{21}$ and $\boldsymbol{y}_{22}$ are not exactly the midpoint and midrange of the intervals. Nevertheless, they are proportional up to a constant and so are representative of the midpoint and midrange.

Now, $\mathrm{Var}(\boldsymbol{\Gamma}_0 \boldsymbol{y}) = \boldsymbol{\Gamma}_0 \mathrm{Var}(\boldsymbol{y}) \boldsymbol{\Gamma}_0' =$ Diag $[\boldsymbol{\Delta}_2; \boldsymbol{\Delta}_1]$. So, the two $(p \times 1)$ vectors $\boldsymbol{y}_{21}$ and $\boldsymbol{y}_{22}$ are uncorrelated and $\mathrm{Var}(\boldsymbol{y}_{21}) = \boldsymbol{\Delta}_2$ and $\mathrm{Var}(\boldsymbol{y}_{22}) = \boldsymbol{\Delta}_1$ (see [7] and [8]). So, we see that the variance-covariance matrices of $\boldsymbol{y}_{21}$ and $\boldsymbol{y}_{22}$ are $\boldsymbol{\Delta}_2$ and $\boldsymbol{\Delta}_1$, and they are not calculated from the mid-points and mid-ranges.

Therefore, $\boldsymbol{y}_{21} \sim ((\boldsymbol{h}_1' \otimes \boldsymbol{I}_p)\boldsymbol{\mu}_y, \boldsymbol{\Delta}_2)$ and $\boldsymbol{y}_{22} \sim ((\boldsymbol{h}_2' \otimes \boldsymbol{I}_p)\boldsymbol{\mu}_y, \boldsymbol{\Delta}_1)$. We now assume that $\boldsymbol{y}_{21}$ follows a $p-$variate normal distribution and $\boldsymbol{y}_{22}$ follows a $p-$variate log-normal distribution to circumvent the positive domain of the entries of $\boldsymbol{y}_{22}$. That is, $\ln(\boldsymbol{y}_{22})$ follows a normal distribution. Due to the normality assumption, $\boldsymbol{y}_{21}$ and $\ln(\boldsymbol{y}_{22})$ are independent. Let $\boldsymbol{x} = \ln(\boldsymbol{y}_{22})$. Therefore,

$$\boldsymbol{y}_{21} \sim N_p\big((\boldsymbol{h}_1' \otimes \boldsymbol{I}_p)\boldsymbol{\mu}_y, \boldsymbol{\Delta}_2\big) \quad \text{and let} \quad \ln(\boldsymbol{y}_{22}) = \boldsymbol{x} \sim N_p\big(\boldsymbol{\mu}_x, \boldsymbol{\Sigma}_{x,\boldsymbol{\Delta}_1}\big).$$

1.2 Multivariate Log-Normal Distribution

From the previous section $\ln(\boldsymbol{y}_{22}) = \boldsymbol{x} \sim N_p\big(\boldsymbol{\mu}_x, \boldsymbol{\Sigma}_{x,\boldsymbol{\Delta}_1}\big)$, with

$$\boldsymbol{\mu}_x = \begin{bmatrix} \mu_{x_1} \\ \vdots \\ \mu_{x_p} \end{bmatrix}, \quad \text{and} \quad \boldsymbol{\Sigma}_{x,\boldsymbol{\Delta}_1} = \begin{bmatrix} \sigma_{11}, \sigma_{12} \dots \sigma_{1p} \\ \vdots \quad \ddots \quad \vdots \\ \sigma_{p1}, \sigma_{p2} \dots \sigma_{pp} \end{bmatrix}.$$

$$\text{Let} \quad E(\boldsymbol{y}_{22}) = (\boldsymbol{h}_2' \otimes \boldsymbol{I}_p)\boldsymbol{\mu}_y = \boldsymbol{\mu}_{y_{22}}^- = \begin{bmatrix} \mu_{y_{22,1}}^- \\ \vdots \\ \mu_{y_{22,p}}^- \end{bmatrix} = \begin{bmatrix} \exp(\mu_{x_1} + 0.5\sigma_{11}) \\ \vdots \\ \exp(\mu_{x_p} + 0.5\sigma_{pp}) \end{bmatrix}$$

$$\text{and} \quad \mathrm{Var}(\boldsymbol{y}_{22}) = \boldsymbol{\Delta}_1 = \begin{bmatrix} \delta_{11}, \delta_{12} \dots \delta_{1p} \\ \vdots \quad \ddots \quad \vdots \\ \delta_{p1}, \delta_{p2} \dots \delta_{pp} \end{bmatrix},$$

where $\delta_{st} = \exp[\mu_{x_s} + \mu_{x_t} + 0.5(\sigma_{ss} + \sigma_{tt})][\exp(\sigma_{st}) - 1]$, s,t = 1, ..., p.

The mean vector $\boldsymbol{\mu}_x$ and the variance-covariance matrix $\boldsymbol{\Sigma}_{x,\boldsymbol{\Delta}_1}$ of the log-normal distribution can be calculated from the known parameters $\boldsymbol{\mu}_{y_{22}}^-$ and $\boldsymbol{\Delta}_1$.

2 Hypothesis Test

The authors in [3] and [5] developed two independent tests of equality of fuzzy means based on a fuzzy metric to model the interval-valued data. The parametric approach to test the mean interval for $p-$dimensional interval-valued data was first proposed by [2]. An exact test of the mean interval for interval-valued data with only one interval-valued variable was developed by [8]. A new parametric test for $p-$dimensional interval-valued data which comprises of an exact test and a generalized pivotal test was proposed by [7]. This article is a continuing study of our previous work [7]. In this article, we present some simulation studies to see the effect of the sample size n, the dimensions p, and the nominal significance levels (NSLs) α. However, we only report the results of the simulation study for $p = 3$ for the sake of brevity.

We test the following hypothesis

$$H_0 : \underset{2p\times 1}{\boldsymbol{\mu}_y} = \begin{bmatrix} \boldsymbol{\mu}_y^+ \\ \boldsymbol{\mu}_y^- \end{bmatrix} = \underset{2p\times 1}{\boldsymbol{\mu}_0} = \begin{bmatrix} \boldsymbol{\mu}_0^+ \\ \boldsymbol{\mu}_0^- \end{bmatrix} \quad \text{vs.} \quad H_1 : \underset{2p\times 1}{\boldsymbol{\mu}_y} \neq \underset{2p\times 1}{\boldsymbol{\mu}_0}, \tag{1}$$

where $\underset{2p\times 1}{\boldsymbol{\mu}_0}$ is the population mean vector. After multiplying both sides of the Hypothesis (1) by $\boldsymbol{\Gamma}_0$, we see from [7] that Hypothesis (1) is equivalent to testing independently the following two hypotheses:

$$H_{01} : \underset{p\times 1}{(\boldsymbol{h}_1' \otimes \boldsymbol{I}_p)\boldsymbol{\mu}_y} = \underset{p\times 1}{(\boldsymbol{h}_1' \otimes \boldsymbol{I}_p)\boldsymbol{\mu}_0} \text{ vs. } H_{11} : \underset{p\times 1}{(\boldsymbol{h}_1' \otimes \boldsymbol{I}_p)\boldsymbol{\mu}_y} \neq \underset{p\times 1}{(\boldsymbol{h}_1' \otimes \boldsymbol{I}_p)\boldsymbol{\mu}_0}, \quad (2a)$$

$$H_{02}^{LN} : \underset{p\times 1}{\boldsymbol{\mu}_x} = \ln\big(\underset{p\times 1}{(\boldsymbol{h}_2' \otimes \boldsymbol{I}_p)\boldsymbol{\mu}_0}\big) \text{ vs. } H_{12}^{LN} : \underset{p\times 1}{\boldsymbol{\mu}_x} \neq \ln\big(\underset{p\times 1}{(\boldsymbol{h}_2' \otimes \boldsymbol{I}_p)\boldsymbol{\mu}_0}\big). \quad (2b)$$

Hypothesis (1) is rejected whenever one of the Hypotheses (2a) and (2b) is rejected. Because the tests are independent and performed simultaneously, the overall Type I error rate can be corrected using Benjamini-Hochberg (BH) correction [1]. To test the Hypothesis (2a), Hotelling's T^2 test could be employed and the test statistic is as follows

$$T^2 = (\overline{\boldsymbol{y}} - \boldsymbol{\mu}_0)'(\boldsymbol{h}_1 \otimes \boldsymbol{I}_p)\Big(\frac{1}{n}\widehat{\boldsymbol{\Delta}}_2\Big)^{-1}(\boldsymbol{h}_1' \otimes \boldsymbol{I}_p)(\overline{\boldsymbol{y}} - \boldsymbol{\mu}_0) \sim T^2_{p,n-1}. \quad (3)$$

To test the Hypothesis (2b), the algorithm proposed by Lin [4] can be employed. Lin [4] provided a generalized pivotal quantity that allows us to test this mean, given by

$$\boldsymbol{T} = \overline{\boldsymbol{x}} - \Big(\frac{\boldsymbol{W}}{n}\Big)^{1/2}\Big(\frac{\boldsymbol{\Sigma}_x}{n}\Big)^{-1/2}(\overline{\boldsymbol{X}} - \boldsymbol{\mu}_x) + 0.5\,\mathrm{Diag}(\boldsymbol{W}) = \overline{\boldsymbol{x}} - \Big(\frac{\boldsymbol{W}}{\mathrm{n}}\Big)^{1/2}\boldsymbol{Z} + 0.5\,\mathrm{Diag}(\boldsymbol{W}).$$

where $\boldsymbol{W}$ is a simulated $p-$variate random variable whose observed value is the sample covariance matrix. We construct $\boldsymbol{W}$ via $\boldsymbol{W} = \boldsymbol{a}^{1/2}\boldsymbol{R}^{-1}\boldsymbol{a}^{1/2}$, where $\boldsymbol{R} \sim W_p(n-1, \boldsymbol{I}_p)$ is a Wishart random variable. The matrix $\boldsymbol{a}$ is the observed covariance matrix for the log-transformed data, and $\boldsymbol{Z} \sim N_p(\boldsymbol{0}, \boldsymbol{I}_p)$.

Monte Carlo simulations are run to generate pivotal quantities, which we then compare to a critical value cutoff $\boldsymbol{\eta}$, and the proportion of values above this cutoff represents a generalized $p-$value. See [7] for the algorithm.

2.1 Simulation Studies

We carry out Monte Carlo simulations to calculate empirical significance levels (ESLs) for NSLs $\alpha \in \{0.01, 0.05\}$ and for $p = 3$. We simulate a distribution with mean $\boldsymbol{\mu} = 0 \times \mathbf{1}_3$ and parameter matrices $\boldsymbol{U}_0$ and $\boldsymbol{U}_1$:

$$\boldsymbol{U}_0 = \begin{pmatrix} 2\ 1\ 2 \\ 1\ 4\ 3 \\ 2\ 3\ 5 \end{pmatrix} \quad \text{and} \quad \boldsymbol{U}_1 = \begin{pmatrix} 0.40\ 0.11\ 0.40 \\ 0.11\ 0.60\ 0.15 \\ 0.40\ 0.15\ 0.60 \end{pmatrix}.$$

For each sample size $n \in \{20, 50, 100, 200, 500, 1000\}$, we simulate 50,000 independent datasets from the specified multivariate normal/log-normal model with parameter matrices $\boldsymbol{U}_0$ and $\boldsymbol{U}_1$. For every simulated dataset, we compute two test statistics and their associated p-values: (i) Hotelling's T^2 test and (ii) Lin's simulation-based test using 10,000 internal simulations per dataset. The p-values

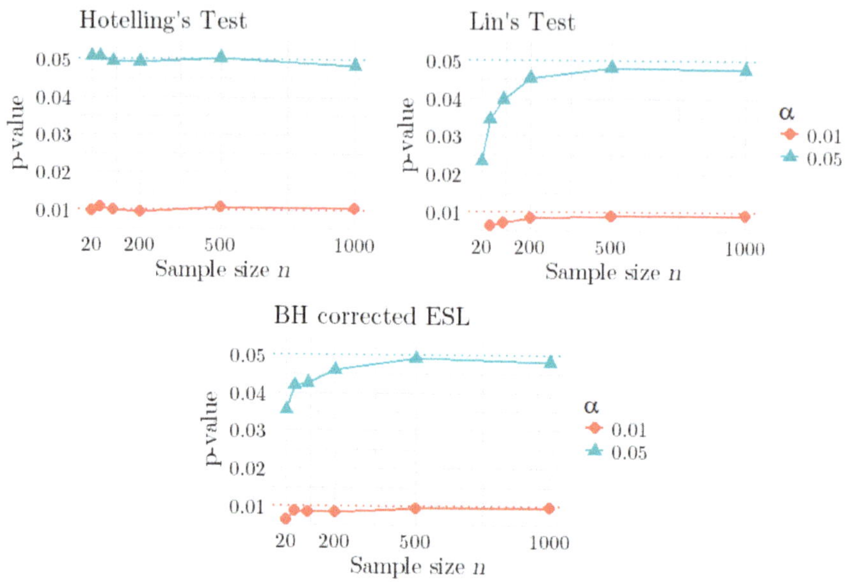

Fig. 1. Convergence of ESL to NSL α as n increases

rejection proportion at each significance level across all 50,000 replicates provides an ESL estimate $\hat{\alpha}$ for each test.

Hotelling's T^2 test has ESLs that fluctuate tightly around the NSL α across all sample sizes and dimensions, showing its small-sample reliability. Meanwhile, Lin's simulation-based log-normal test has a slight downward bias at small sample sizes, with ESLs asymptotically approaching the NSL as n increases. The results demonstrate that the Hotelling's test has a robust Type I error, while Lin's method is asymptotically correct but requires moderate to large sample sizes to attain the nominal α-level. See Fig. 1.

Table 1 presents the ESLs for NSLs 0.01 and 0.05 for different test statistics, Lin's ESL, Hotelling's T^2 ESL, and BH's corrected ESL and the 95% BH corrected confidence intervals (CIs) for NSL α for various sample sizes when $p = 3$. As expected, for fixed α, the BH's corrected ESL converges to NSL α as the sample size n increases.

Benjamini-Hochberg's corrected ESL always underestimates the corresponding NSL except for large sample sizes. Also, the 95% BH corrected CIs do not contain the NSL α, rather the CIs are marginally below the α value. This warns about the accuracy of the BH's corrected ESL values for small to medium sample sizes. The 95% CIs contain the corresponding NSLs for large sample sizess, i.e., for $n = 500$ and 1000.

Table 1. ESLs for different test statistics and 95% BH corrected CIs for α for various sample sizes when $p = 3$ and NSLs are equal to 0.01 and 0.05

	NSL	ESL			95% CI for α
n	(α)	Lin	Hotelling	BH	(BH corrected)
20	0.01	0.0031	0.0098	0.0064	(0.0057, 0.0071)
	0.05	0.0234	0.0507	0.0354	(0.0338, 0.0370)
50	0.01	0.0064	0.0108	0.0087	(0.0078, 0.0095)
	0.05	0.0344	0.0507	0.0419	(0.0402, 0.0437)
100	0.01	0.0072	0.0099	0.0084	(0.0076, 0.0092)
	0.05	0.0397	0.0493	0.0425	(0.0407, 0.0443)
200	0.01	0.0084	0.0094	0.0084	(0.0076, 0.0092)
	0.05	0.0454	0.0493	0.0460	(0.0442, 0.0479)
500	0.01	0.0088	0.0107	0.0094	(0.0086, 0.0103)
	0.05	0.0482	0.0503	0.0492	(0.0473, 0.0511)
1000	0.01	0.0093	0.0106	0.0097	(0.0088, 0.0106)
	0.05	0.0477	0.0485	0.0482	(0.0464, 0.0501)

2.2 An Example

We analyze a real dataset, a snapshot of which is given in Table 2. These data present measurements (in cm) of three features ($p = 3$), namely the width of the Pileus (cap), the length and the thickness of the Stipe (stalk) of a variety of edible mushroom species. The values of these features were extracted from "The Fungi of California" species index (https://www.mykoweb.com/CAF/species_index.html). A sample of 105 edible mushrooms was selected for our population of interest.

Table 2. Mushroom data

Species	Pileus (Cap) Width	Stipe (Stem) Length	Stipe (Stem) Thickness
Agaricus fissuratus	[6.0, 21.0]	[4.0, 14.0]	[1.00, 3.50]
Agaricus benesi	[4.0, 8.0]	[5.0, 11.0]	[1.00, 2.00]
Agaricus bernardii	[7.0, 16.0]	[4.0, 7.0]	[3.00, 4.50]
Agaricus bitorquis	[5.0, 15.0]	[4.0, 10.0]	[2.00, 4.00]

We expect samples drawn from this population of 105 edible mushrooms to generate large p-values from the testing procedure, indicating failure to reject the null hypothesis. Suppose we draw many random samples from this population and test the Hypothesis (1) for each random sample. If the test performs well, we would expect to see a roughly uniform distribution of p-values for small values of n. We would also expect larger p-values more frequently for larger values of n, as those samples begin to approach the population size. We will illustrate this by showing the distribution of the p-values for permuted samples at the end of this section.

We explain the testing procedure for a single test run. First, we rearrange the data so that the first three columns contain the maximum values of each interval and the last three columns contain the minimum values of each interval. Next, we apply the orthogonal matrix $\boldsymbol{\Gamma}_0$ to the hypothesized value in the null hypothesis (1), $\boldsymbol{\mu}_0 = (11.0190, 9.7857, 2.4248, 3.8971, 4.000, 1.1352)'$, so that the result of the transformation represents the midpoint and range of each interval as described in Sect. 1. To test the effectiveness of our method, we draw four random samples (intervals) of size $n \in \{11, 26, 53, 79\}$ from our population. These sample sizes n represent 10%, 25%, 50%, and 75% of the population size. The parameter matrices $\boldsymbol{U}_0$ and $\boldsymbol{U}_1$ are estimated using the formulas given in Sect. 3 of [6]. Given the mean vector and variance-covariance matrix for the log-normal data, we can estimate the mean vector and variance-covariance matrix for the associated normal data using the transformations depicted in Sect. 1.2, and vice versa. We give the estimated values $\widehat{\boldsymbol{\Delta}}_2$, $\widehat{\boldsymbol{\Delta}}_1$ and $\widehat{\boldsymbol{\Sigma}}_{x,\boldsymbol{\Delta}_1}$ for the $n = 53$ sample:

$$\widehat{\boldsymbol{\Delta}}_2 = \begin{bmatrix} 26.07 & 14.00 & 7.15 \\ 14.00 & 21.73 & 3.85 \\ 7.15 & 3.85 & 2.92 \end{bmatrix}, \quad \widehat{\boldsymbol{\Delta}}_1 = \begin{bmatrix} 8.40 & 3.15 & 1.68 \\ 3.15 & 4.93 & 0.82 \\ 1.68 & 0.82 & 0.56 \end{bmatrix}, \quad \widehat{\boldsymbol{\Sigma}}_{x,\boldsymbol{\Delta}_1} = \begin{bmatrix} 0.31 & 0.15 & 0.31 \\ 0.15 & 0.25 & 0.19 \\ 0.31 & 0.19 & 0.48 \end{bmatrix}.$$

That is, the variance-covariance matrix $\widehat{\boldsymbol{\Sigma}}_{x,\boldsymbol{\Delta}_1}$ is calculated from $\widehat{\boldsymbol{\mu}}_0$ and $\widehat{\boldsymbol{\Delta}}_1$ using the conversion formulas given in Sect. 1.2.

Testing for the log-normal portion of the data in Hypothesis (2b), the Hypothesis H_{02}^{LN} becomes as follows:

$$H_{02}^{LN} : \ln\left((\boldsymbol{h}_2' \otimes \boldsymbol{I}_p)\boldsymbol{\mu}_y\right) = \begin{bmatrix} 1.6166 \\ 1.4088 \\ -0.0923 \end{bmatrix} \quad \text{vs.} \quad H_{12}^{LN} : \ln\left((\boldsymbol{h}_2' \otimes \boldsymbol{I}_p)\boldsymbol{\mu}_y\right) \neq \begin{bmatrix} 1.6166 \\ 1.4088 \\ -0.0923 \end{bmatrix}.$$

The right-hand side of the above hypothesis is calculated using the following

$$\underset{p\times 1}{\boldsymbol{\mu}_x} = \ln\left((\boldsymbol{h}_2' \otimes \boldsymbol{I}_p)\underset{p\times 1}{\boldsymbol{\mu}_0})\right) = (1.6166, 1.4088, -0.0923)'.$$

We use Hotelling's T^2 and Lin's algorithm (with 100,000 iterations) to test the Hypothesis (1) for the sample intervals. Table 3 displays the results of each test on the random samples of size $n = \{11, 26, 53, 79\}$ of the Mushroom data. The quantities P_N and P_{LN} are the p-values from the Hotelling's T^2 test and Lin's test respectively. The mean vectors are for the upper and lower bounds of the sample data. Hotelling's T^2 p-value is calculated using $\widehat{\boldsymbol{\Delta}}_2$ and Lin's test p-value is calculated using $\widehat{\boldsymbol{\Sigma}}_{x,\boldsymbol{\Delta}_1}$.

Notice that despite the large sample size (relative to the population size) Lin's test can produce low p-values (see the $n = 53$ sample test results). This is

Table 3. .

	Mean Vector of Upper/Lower Bounds	p-values	
		P_N	P_{LN}
Population $\boldsymbol{\mu}_0$ →	$(11.0190, 9.7857, 2.4248, 3.8971, 4.0000, 1.1352)'$	–	–
Sample size ↓	$\widehat{\boldsymbol{\mu}}_0$ ↓		
11	$(10.0909, 8.3636, 1.7364, 3.7273, 3.4091, 0.8545)'$	0.1398	0.8229
26	$(11.9231, 10.0000, 2.7192, 4.1923, 4.0000, 1.1885)'$	0.9225	0.5430
53	$(10.8868, 10.1415, 2.5585, 4.0943, 4.2075, 1.2075)'$	0.6914	0.2415
79	$(11.1519, 9.8101, 2.4076, 3.9051, 3.9304, 1.1316)'$	0.9825	0.9319

a feature of sampling variability and is not unexpected. To illustrate this fact, we performed the following experiment: We permuted 1000 random samples for each sample size $n = \{11, 26, 53, 79\}$ and tested them using Lin's procedure (1000 iterations). We then collected and displayed the p-values in Fig. 2. As our sample size approaches the population size, the test fails to reject far more often. However, this does not exclude low p-values from occurring during the testing procedure.

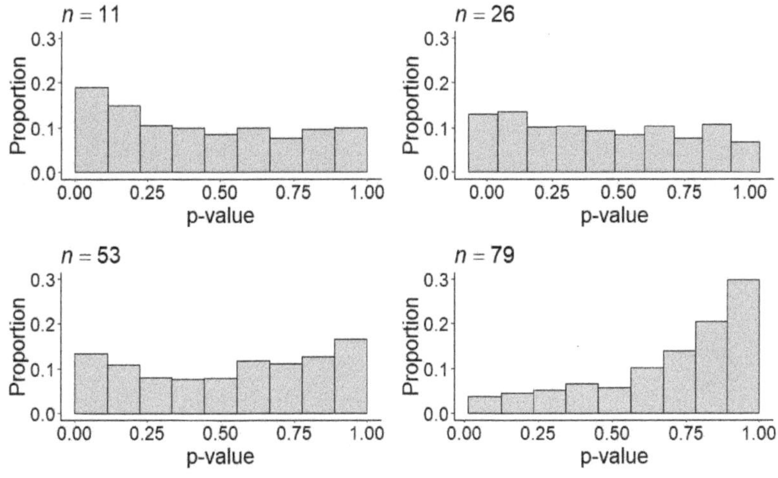

Fig. 2. 1000 simulated p-values for sample sizes $n = 11, 26, 53, 79$

References

1. Benjamini, Y., Hochberg, Y.: Controlling the false discovery rate: a practical and powerful approach to multiple testing. J. R. Stat. Soc. Ser. B Stat Methodol. **57**, 289–300 (1995). https://doi.org/10.1111/j.2517-6161.1995.tb02031.x

2. Brito, P., Duarte Silva, A.P.: Modelling interval data with normal and skew-normal distributions. J. Appl. Stat. **39**(1), 3–20 (2012). https://doi.org/10.1080/02664763.2011.575125
3. González Rodríguez, G., Colubi, A., Gil, M.A., D'Urso, P.: An asymptotic two dependent samples test of equality of means of fuzzy random variables. In: Proceedings of the 17th Conference of IASC-ERS, COMPSTAT 2006, Rome, Italy, pp. 689-695 (2006), https://doi.org/10.1007/978-3-7908-1709-6
4. Lin, S.H.: Comparing the mean vectors of two independent multivariate log-normal distributions. J. Appl. Stat. **41**(2), 259–274 (2014). https://doi.org/10.1080/02664763.2013.838669
5. Montenegro, M., Casal, M.R., Lubiano, M.A., Gil, M.A.: Two-sample hypothesis tests of means of a fuzzy random variable. Inform. Sci. **133**(1–2), 89–100 (2001). https://doi.org/10.1016/S0020-0255(01)00078-0
6. Roy, A.: Two-stage principal component analysis on interval-valued data using patterned covariance structures. Adv. Data Anal. Classif. (2026). https://doi.org/10.1007/s11634-025-00650-9
7. Roy, A., Montes, F.: Hypothesis testing of mean interval for p-dimensional interval-valued data. In: Trejos, J., Chadjipadelis, T., Grané, A., Villalobos, M. (eds) Data Science, Classification, and Artificial Intelligence for Modeling Decision Making. IFCS 2024. Studies in Classification, Data Analysis, and Knowledge Organization, pp. 161–169. Springer, Cham (2025). https://doi.org/10.1007/978-3-031-85870-3_18
8. Roy, A., Klein, D.: Testing of mean interval for interval-valued data. Commun. Stat.-Theor. M. **49**(20), 5028–5044 (2020). https://doi.org/10.1080/03610926.2019.1612915
9. Roy, A., Leiva, R.: Estimating and testing a structured covariance matrix for three-level multivariate data. Commun. Stat.-Theor. M. **40**(11), 1945–1963 (2011). https://doi.org/10.1080/03610921003672212

Adaptive Spectral Clustering in Shared-Frailty Survival Models for Risk-Based Patients Stratification

Alessandra Ragni[1](✉), Lara Cavinato[1], and Francesca Ieva[1,2]

[1] MOX, Department of Mathematics, Politecnico di Milano, Piazza Leonardo da Vinci 32, 20133 Milan, Italy
{alessandra.ragni,lara.cavinato,francesca.ieva}@polimi.it
[2] Health Data Science Centre, Human Technopole, Viale Rita Levi-Montalcini 1, 20157 Milan, Italy

Abstract. Stratifying patients according to their risk of experiencing an event of interest, while accounting for institutional effects induced by hierarchical data structures, contributes to improving clinical decision-making in real-world studies involving time-to-event outcomes. In this work, we analyze administrative healthcare data from the Enhance-Heart study that include heart failure patients hospitalized for COVID-19 in the Lombardy Region (Italy) over a period of approximately one and a half years, to identify patient subgroups with similar event risk while accounting for hospital-level dependence. To do so, we propose a likelihood-based survival modeling framework that integrates a parametric shared frailty model with a spectral clustering approach, enabling joint estimation of risk-driven patient clusters and survival parameters within a unified penalized optimization framework. Our results uncover clinically interpretable patient subgroups and not negligible between-hospital variability in mortality risk, demonstrating the value of jointly modeling patient stratification and hierarchical dependence in time-to-event data.

Keywords: shared frailty models · hierarchical data · spectral clustering · penalization · survival analysis

1 Introduction

Real-world studies in clinical settings that involve time-to-event data often benefit from the identification of patient subgroups with similar risk categories, thereby supporting clinical decision-making (Collins and Varmus, 2015). However, the frequent presence of dependence induced by shared group membership (e.g., hospital-level effects) calls for modeling frameworks that jointly accommodate patient stratification and hierarchical dependence. Motivated by the *Enhance-Heart* study of heart failure patients hospitalized for COVID-19 in Lombardy Region in Northern Italy, where both patient risk profiles and hospital

P. Brito et al. (Eds.): IFCS 2026, STUDIES CLASS, pp. 227–234, 2026.
https://doi.org/10.1007/978-3-032-32009-4_28

resources influenced mortality (Janke et al.2021), we address the need for survival models that can simultaneously stratify patients and account for between-hospital variability, based on their underlying risk (hazard) of experiencing the event.

Survival models typically account for such dependence through shared frailty terms (Duchateau and Janssen 2008), but they do not identify latent patient subgroups. Conversely, recent survival-informed clustering approaches uncover heterogeneous risk profiles (Liu et al. 2020; Chapfuwa et al. 2020), but often neglect hierarchical structure. This leaves a gap for methods that integrate patient stratification and group-level dependence within a unified survival framework.

In this work, we propose a novel extension of parametric shared frailty models that jointly estimates patient clusters, defined by similar hazards of experiencing the event, and survival parameters within a unified framework. Patient stratification is achieved through a risk-driven spectral clustering approach embedded directly in a penalized full-likelihood optimization, allowing similarity structures to be learned adaptively from the estimated hazard. The resulting clusters are interpretable in terms of event risk that explicitly accounts for group-level dependence, providing a flexible tool for survival analysis in clustered data.

The remainder of the paper is organized as follows. Section 2 introduces the proposed model, Sect. 3 describes the dataset and preprocessing steps, and Sect. 4 presents the main findings. Finally, Sect. 5 concludes with a discussion and directions for future research.

2 The Model

2.1 Shared-Frailty Survival Model and Notation

We consider right-censored[1] survival data with a grouped structure, i.e., units $i = 1, \ldots, n_g$ nested within groups $g = 1, \ldots, M$, with total number of units $N := \sum_{g=1}^{M} n_g$. Let $\mathcal{Z} = \{(y_{gi}, \delta_{gi}, \boldsymbol{x}_{gi}) \mid i = 1, \ldots, n_g,\ g = 1, \ldots, M\}$ be the observed data, where y_{gi} is the observed follow-up time, δ_{gi} indicates whether the event of interest is observed, and $\boldsymbol{x}_{gi} \in \mathbb{R}^{d \times 1}$ is the covariates vector. To model the hazard rate of unit i in group g while accounting for group-level heterogeneity, we use the *shared-frailty model* (Duchateau and Janssen 2008)

$$h_{gi}(t \mid u_g) = h_0(t; \boldsymbol{\psi}) u_g \exp(\boldsymbol{x}'_{gi} \boldsymbol{\beta}) \tag{1}$$

where $\boldsymbol{\beta} \in \mathbb{R}^{d \times 1}$ is the vector of fixed-effect coefficients, $u_g \sim f_U(\cdot \mid \theta)$ is the frailty for group g, being θ the frailty variance, and $h_0(t; \boldsymbol{\psi})$ is a parametric baseline hazard. The marginal loglikelihood, obtained by integrating out the frailty distribution over its support, can be expressed using the Laplace transform

[1] We assume independent and noninformative censoring.

$\mathcal{L}(s)$ $(s \geq 0)$ as

$$\ell(\boldsymbol{\psi}, \boldsymbol{\beta}, \theta \mid \mathcal{Z}) = \sum_{g=1}^{M} \Bigg[\Bigg[\sum_{i=1}^{n_g} \delta_{gi} \left(\log(h_0(y_{gi}; \boldsymbol{\psi})) + \boldsymbol{x}'_{gi}\boldsymbol{\beta} \right) \Bigg] + \log \Bigg[(-1)^{d_g} \mathcal{L}^{(d_g)} \Bigg(\sum_{i=1}^{n_g} H_0(y_{gi}; \boldsymbol{\psi}) \exp(\boldsymbol{x}'_{gi}\boldsymbol{\beta}) \Bigg) \Bigg] \Bigg] \tag{2}$$

with $H_0(t; \boldsymbol{\psi}) = \int_0^t h_0(s; \boldsymbol{\psi}) ds$, $d_g := \sum_{i=1}^{n_g} \delta_{gi}$ is the number of events in group g and $(-1)^k \mathcal{L}^{(k)}(s) := (-1)^k \frac{d^k}{ds^k} \mathcal{L}(s)$ (Munda et al. 2012).

2.2 Joint Clustering and Parameter Estimation

We propose a unified framework that performs clustering of the units into C clusters where survival parameters and clustering structure are learned jointly, integrating a parametric shared frailty survival model with an adaptive spectral clustering penalty. The clustering structure is not fixed a priori but is inferred from the data through a learned similarity graph that depends on the survival model parameters. Specifically, similarity between two units i and j, respectively, in groups g and h, namely (g, i) and (h, j), is quantified through the frailty-adjusted distance

$$d_{gi,hj}(\boldsymbol{\beta}, \theta, \mathcal{Z}) = \left\| (\boldsymbol{x}'_{gi}\boldsymbol{\beta} + \log u_g) - (\boldsymbol{x}'_{hj}\boldsymbol{\beta} + \log u_h) \right\|,$$

which measures squared differences in log-hazard contributions, where $\log h_0(t; \boldsymbol{\psi})$ cancels out in pairwise comparisons. Here, the frailty terms $\log u_g$ and $\log u_h$ are computed by their conditional expectations given the observed data.

Joint estimation of survival parameters and clustering structure is achieved by solving the following optimization problem, that combines the negative marginal log-likelihood of the shared frailty survival model with a spectral clustering penalty:

$$\begin{aligned} \min_{\boldsymbol{\beta}, \boldsymbol{\psi}, \theta, \mathbf{S}, \mathbf{F}} \quad & -\ell(\boldsymbol{\psi}, \boldsymbol{\beta}, \theta \mid \mathcal{Z}) \\ & + \gamma \Big\{ \sum_{g,h=1}^{M} \sum_{i=1}^{n_g} \sum_{j=1}^{n_h} \left(d^2_{gi,hj}(\boldsymbol{\beta}, \theta, \mathcal{Z})\, s_{gi,hj} + \mu\, s^2_{gi,hj} \right) + 2\lambda Tr(\mathbf{F}'\mathbf{L}_{\mathbf{S}}\mathbf{F}) \Big\} \\ \text{s.t.} \quad & \sum_{h=1}^{M} \sum_{j=1}^{n_h} s_{gi,hj} = 1, \quad \mathbf{s}_{gi} \geq 0, \quad \forall i = 1, \ldots, n_g,\ g = 1, \ldots, M, \\ & \mathbf{F}'\mathbf{F} = \mathbf{I},\ \mathbf{F} \in \mathbb{R}^{N \times C} \end{aligned} \tag{3}$$

The similarity matrix $\mathbf{S} \in \mathbb{R}^{N \times N}$ encodes a probabilistic neighborhood structure, where each entry $s_{gi,hj}$ represents the affinity between two units and is constrained to lie on the probability simplex. The quadratic term $\mu\, s_{gi,hj}^2$ regularizes the affinity matrix and prevents degenerate neighborhood assignments as described in Nie et al. (2014), while the tuning parameter γ controls the relative contribution of clustering versus survival estimation, and is inspired by the framework of Liu et al. (2020). The matrix $\mathbf{F} \in \mathbb{R}^{N\times C}$ plays the role of the cluster indicator matrix, being the scalar C the number of clusters. Finally, to induce a partition into C clusters, the affinity matrix is interpreted as the adjacency matrix of a weighted graph with Laplacian $\mathbf{L_S}$. The trace term $\mathrm{Tr}(\mathbf{F}'\mathbf{L_S}\mathbf{F})$, weighted by λ, corresponds to a spectral relaxation that promotes a graph structure with C well-separated connected components.

Estimation Procedure. Given the large number of unknowns, we solve the optimization in Eq. (3) via block coordinate descent, iterating until convergence through three alternating updates: (i) update $\mathbf{F}$ while keeping $\mathbf{S}$, $\boldsymbol{\beta}, \boldsymbol{\psi}, \theta$ fixed; (ii) update $\boldsymbol{\beta}, \boldsymbol{\psi}, \theta$ with $\mathbf{F}, \mathbf{S}$ fixed; (iii) update $\mathbf{S}$ while holding $\mathbf{F}, \boldsymbol{\beta}, \boldsymbol{\psi}, \theta$. Step (i) reduces to a spectral problem, whose solution is given by the C eigenvectors of the Laplacian $\mathbf{L_S}$ corresponding with its smallest eigenvalues. Step (ii) performs penalized frailty survival estimation, following the framework of Munda et al. (2012), and adding the similarity-based penalty. Step (iii) updates $\mathbf{S}$ by solving a constrained quadratic program that decomposes across units and admits a closed-form solution via the KKT conditions for each row $\mathbf{s}_{gi}$: $s_{gi,hj} = \max\{0, \alpha_{gi} - w_{gi,hj}/(2\mu_{gi})\}$ where sparsity is enforced by retaining only the k nearest neighbors, i.e., exactly k entries per row are nonzero, with self-similarity excluded.

Thanks to this constraint, the scalars α_{gi} and μ_{gi} admit closed-form expressions: $\alpha_{gi} = 1/k + \sum_{p=1}^{k} \hat{w}_{gi,p}/(2k\mu_{gi})$ and $\mu_{gi} = k\hat{w}_{gi,k+1}/2 - \sum_{p=1}^{k} \hat{w}_{gi,p}/\ 2$, where $\hat{\mathbf{w}}_{gi}$ denotes the sorted distances (excluding the self-distance). Indeed, this result follows from enforcing the simplex constraint $\mathbf{s}_{gi}'\mathbf{1} = 1$ over the k nonzero entries, which yields a linear system in (α_{gi}, μ_{gi}) and thus fully determines the solution.

Furthermore, the penalty λ is initialized to a large value and then adaptively adjusted at each iteration as proposed in (Nie et al. 2017), so as to drive the graph toward exactly C components. The parameter k controls the sparsity of $\mathbf{S}$ and thereby the density of the graph: a too small k may fragment the graph into more than C components, while a too large k tends to over-connect it. The adaptive tuning of λ compensates for this, but k should nonetheless be chosen to ensure basic connectivity without sacrificing sparsity. For additional details, see Ragni et al. (2026).

Sensitivity to the Choice of $\boldsymbol{k}$ and $\boldsymbol{\gamma}$. To assess sensitivity to the choice of k and γ to cluster membership and parameters estimation, we refer to the simulation study in Ragni et al. (2026), where clustered survival data are generated inducing the presence of 10 groups and using a shared Gamma frailty. Covariate-driven linear predictors induce three well-separated latent clusters, from which survival times are generated under a Weibull baseline hazard, $H_0(t) = \xi^\rho t^\rho$

($\rho, \xi > 0$) with parameters $\psi = [\rho, \xi]$, and are subject to administrative censoring, across varying penalization and model settings.

Our model identifies the three induced clusters with 100% frequency for $\gamma \leq 0.1$. For larger values of γ, the identification rate decreases slightly, reaching a minimum of 95% at $\gamma = 0.4$.

The boxplots in Fig. 1 show that bias generally increases with γ, with the frailty variance θ slightly underestimated even in the unpenalized case (consistent with prior work), while most parameters exhibit stable variability across γ, except for the Weibull parameter ξ, whose spread increases noticeably as γ increases.

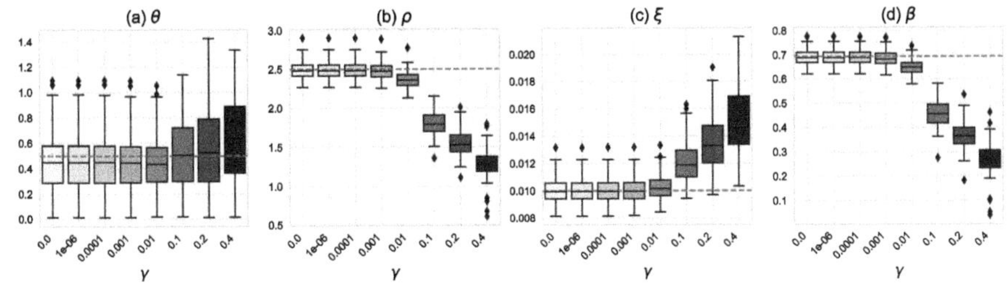

Fig. 1. Boxplots of parameter estimates $(\theta, \rho, \xi, \beta)$ across penalization levels γ (x-axis) under *administrative* censoring for $k = 20$ and $C = 3$. The true parameter values are shown as a piecewise-constant horizontal line.

3 The Dataset

We apply the proposed methodology to the *Enhance-Heart* dataset, an administrative healthcare database from the Lombardy Region (Italy) linked to the regional COVID-19 registry[2]. The cohort comprises 3,086 patients diagnosed with heart failure and hospitalized for COVID-19 between *31 January 2020*[3] and *18 June 2021*[4] across 32 hospitals. Death within 90 days of hospitalization is the event of interest, with patients surviving beyond this period or dying later treated as censored. This setting provides a realistic and challenging framework to assess the proposed approach in the presence of both patient heterogeneity and hierarchical structure given by hospitals.

Patient-level covariates include age, gender, comorbidity burden, and respiratory conditions, while hospital-level dependence is explicitly modeled through a parametric frailty model. Comorbidity burden is measured using an adaptation of Corrao et al. (2017) that excludes respiratory diseases relevant to COVID-19. Respiratory conditions are captured as a binary indicator equal to 1 if a

[2] https://www.dati.lombardia.it/.
[3] Start of the national COVID-19 state of emergency in Italy.
[4] Implementation of the Green Pass regulations governing reopening.

patient has more than one condition among chronic obstructive pulmonary disease, bronchitis, pneumonia, or respiratory failure.

4 Results

We evaluated multiple model configurations, varying frailty distributions (Gamma, Inverse Gaussian), baseline hazards (Weibull, Exponential), penalization parameters, and the number of clusters, selecting the optimal configuration via the highest mean silhouette score. The chosen model uses an Inverse Gaussian frailty, Weibull baseline, $\gamma = 10^{-1}$, $k = 150$, and $C = 3$ clusters, achieving a silhouette of 0.609.

Regression results indicate all covariates – age, gender, comorbidity burden, and respiratory conditions – have statistically significant effects. The estimated frailty variance ($\theta = 0.573$) confirms substantial hospital-level variability.

The three identified patient clusters differ in age, comorbidity, and respiratory burden: Cluster 1 includes the oldest, most comorbid patients; Cluster 2 has intermediate comorbidity with the highest respiratory burden; Cluster 3 contains the youngest, least comorbid patients. Cluster membership varies across hospitals, with some dominated by a single cluster and others showing mixed composition, highlighting notable within-hospital heterogeneity that our approach captures.

5 Discussion and Future Work

In this work, we introduce a likelihood-based framework that jointly performs units clustering and survival modeling in hierarchical time-to-event data, incorporating shared frailty to account for within-group dependence. Applied to the Enhance-Heart cohort of heart-failure patients hospitalized for COVID-19, the method identifies three clinically interpretable clusters. The non-negligible frailty variance further suggests that unobserved institutional factors affect survival outcomes beyond observed covariates, highlighting the value of jointly modeling clustering and hierarchical dependence.

Limitations include sensitivity to the neighborhood size k and to the prespecified number of clusters C, as well as parametric assumptions on the frailty distribution and baseline hazard. In particular, the use of a common baseline hazard across clusters may introduce bias if risk groups follow genuinely distinct hazard trajectories. Extending the framework to cluster-specific baselines is however non-trivial within our joint estimation scheme: cluster membership would need to be known when fitting the survival model in Step (ii), yet the optimal cluster configuration is only determined at convergence, and cluster-specific baselines would additionally make the frailty estimates $\log u_g$ – entering the pairwise distances – cluster-dependent.

Regarding the selection of C, the mean silhouette score is naturally evaluated post-hoc from the learned cluster assignments. Stability-based alternatives such as the GAP statistic (Tibshirani et al. 2001) or consensus clustering (Monti et al. 2003) could offer complementary evidence; however, their adaptation to our setting is non-trivial. The GAP statistic requires specifying an appropriate reference distribution, which may be not obvious for our setting, while consensus clustering involves repeated subsampling that may disrupt the grouped structure and affect frailty estimation. In addition, both approaches would substantially increase the computational burden. For our specific case, the choice $C = 3$ is also supported by findings reported in Caldera et al. (2025), in addition to the clinical interpretability of the resulting clusters, which differ meaningfully in age, comorbidity burden, and respiratory conditions.

Future work will investigate more flexible, potentially nonparametric, formulations – such as cluster-specific baseline hazards and alternative similarity measures – aimed at better characterizing risk profiles across patient subgroups, as well as lighter stability proxies for the selection of C.

Acknowledgments. The authors acknowledge the financial support from the Department of Mathematics of Politecnico di Milano through the grant 'Dipartimento di Eccellenza 2023-2027' by MUR, Italy. A.R. and L.C. are funded by the National Plan for NRRP Complementary Investments 'Advanced Technologies for Human-centred Medicine' (PNC0000003). This work is part of the *Enhance-Heart* project: Efficacy evaluatioN of the therapeutic-care patHways, of the heAlthcare providers effects aNd of the risk stratifiCation in patiEnts suffering from HEART failure.

References

Collins, F.S., Varmus, H.: A new initiative on precision medicine. N. Engl. J. Med. **372**(9), 793–795 (2015). https://doi.org/10.1056/NEJMp1500523

Janke, A.T., Mei, H., Rothenberg, C., Becher, R.D., Lin, Z., Venkatesh, A.K.: Analysis of hospital resource availability and Covid-19 mortality across the united states. J. Hosp. Med. **16**(4), 211–214 (2021). https://doi.org/10.12788/jhm.3539

Duchateau, L., Janssen, P.: The Frailty Model. Springer, Cham (2008)

Liu, C., Cao, W., Wu, S., Shen, W., Jiang, D., Yu, Z., Wong, H.S.: Supervised graph clustering for cancer subtyping based on survival analysis and integration of multiomic tumor data. IEEE/ACM Trans. Comput. Biol. Bioinf. **19**(2), 1193–1202 (2020). https://doi.org/10.1109/TCBB.2020.3010509

Chapfuwa, P., Li, C., Mehta, N., Carin, L., Henao, R.: Survival cluster analysis. In: Proceedings of the ACM Conference on Health, Inference, and Learning, pp. 60–68 (2020). https://doi.org/10.1145/3368555.3384465

Munda, M., Rotolo, F., Legrand, C.: parfm: Parametric frailty models in R. J. Stat. Softw. **51**, 1–20 (2012). https://doi.org/10.18637/jss.v051.i11

Nie, F., Wang, X., Huang, H.: Clustering and projected clustering with adaptive neighbors. In: Proceedings of the 20th ACM SIGKDD International Conference on Knowledge Discovery and Data Mining, pp. 977–986 (2014). https://doi.org/10.1145/2623330.2623726

Nie, F., Cai, G., Li, X.: (2017) Multi-view clustering and semi-supervised classification with adaptive neighbours. In: Proceedings of the AAAI conference on artificial intelligence, vol. 31 (2017). https://doi.org/10.1609/aaai.v31i1.10909

Ragni, A., Cavinato, L., Ieva, F.: Penalized likelihood optimization for adaptive neighborhood clustering in time-to-event data with group-level heterogeneity. arXiv preprint arXiv:2601.07446 (2026)
Corrao, G., et al.: Developing and validating a novel multisource comorbidity score from administrative data: a large population-based cohort study from Italy. BMJ Open **7**(12), e019,503 (2017). https://doi.org/10.1136/bmjopen-2017-019503
Tibshirani, R., Walther, G., Hastie, T.: Estimating the number of clusters in a data set via the gap statistic. J. Roy. Stat. Soc.: Ser. b (Stat. Methodol.) **63**(2), 411–423 (2001). https://doi.org/10.1111/1467-9868.00293
Monti, S., Tamayo, P., Mesirov, J., Golub, T.: Consensus clustering: a resampling-based method for class discovery and visualization of gene expression microarray data. Mach. Learn. **52**(1), 91–118 (2003). https://doi.org/10.1023/A:1023949509487
Caldera, L., et al.: Cluster-weighted modeling of lifetime hierarchical data for profiling Covid-19 heart failure patients. arXiv preprint arXiv:2507.12230 (2025)

Archetypal Networks for Clustering Classroom Social Structures

Giancarlo Ragozini[1(✉)], Roberto Rondinelli[1], and Maria Prosperina Vitale[2]

[1] Department of Political Science, University of Naples Federico II, Naples, Italy
{giragoz,roberto.rondinelli}@unina.it

[2] Department of Political and Social Studies, University of Salerno, Salerno, Italy
mvitale@unisa.it

Abstract. Schools and classrooms are complex social systems in which individuals and contexts continuously interact. Peer relationships are one of the main components of this system that shape students' behaviors, beliefs and choices. Classifying classrooms based on their relational dynamics across different social layers can inform more effective educational interventions. The present contribution proposes the use of Archetypal Analysis to cluster school-based network data. This method enables the synthesis of multiple classroom and school networks into a small number of "pure" and interpretable network configurations, highlighting prototypical relational structures associated with various post-diploma choices, particularly university enrollment. Using primary data collected in a sample of high school students in Southern Italy, the approach clusters friendship and different kinds of support networks among classmates by considering structural and topological network indices, along with contextual variables, such as family background and school type.

Keywords: Archetypal Analysis · School Networks · Multiplex Networks · Peer Relationships · Educational Inequalities

1 Introduction

Schools and classrooms are complex social systems in which individuals and context continuously interact [1]. From childhood through high school, students spend the majority of their time in these contexts, making relationships among classmates a central component of their daily lives. These relationships are relevant both quantitatively and qualitatively, persisting throughout the school years and often extending beyond them. During adolescence and emerging adulthood, peer relationships become increasingly important as sources of social support and as influential factors in shaping students' behaviors, beliefs, and choices, particularly with regard to future-oriented decisions [2].

Educational choices after high school represent a crucial turning point in students' life courses and are deeply embedded in social, institutional, and relational

P. Brito et al. (Eds.): IFCS 2026, STUDIES CLASS, pp. 235–242, 2026.
https://doi.org/10.1007/978-3-032-32009-4_29

contexts. A large body of research has shown that post-diploma trajectories, particularly university enrollment, are strongly stratified by family background and socioeconomic resources [3,4]. Students from families with high cultural and socioeconomic backgrounds are likely to pursue higher education, while disadvantaged students are disproportionately absorbed in vocational pathways and early labor market entry. Educational aspirations and attainments are not only shaped by family socioeconomic status but are also strongly affected by peer effects operating within school contexts [5–7]. Recent network-based studies have emphasized how friendship and support ties in school classes are structured by homophily and inequality mechanisms, contributing to differentiated educational outcomes [8]. Classrooms and schools operate as bounded social systems, where network structure, group composition, and peer selection processes jointly shape students' perceived opportunities and aspirations [9]. Traditional methods for grouping network nodes according to their relational patterns are generally based on clustering and blockmodeling [10].

In this contribution, a novel perspective is adopted to group a set of networks at the school level. To this end, Archetypal Analysis (AA) [11], and its extension to network data [12], is employed to capture the heterogeneity of social interactions among classmates. Archetypal Networks (ANs) are designed to synthesize a set of comparable networks into a small set of representative and "pure" configurations. By focusing on prototypical forms, the approach offers a new way to group and compare classes and schools, to link relational structures with family background and institutional tracking, and to explore how different social and educational contexts contribute to unequal student trajectories.

2 Comparing Networks Through Archetypal Analysis

The ANs, extending AA to a collection of networks, provide a statistical framework for the comparative analysis and synthesis of multiple networks within a common scenario [12]. The main objective is to extract a small set of representative network structures - referred to as archetypal networks - that capture the most distinctive and "pure" patterns observed across a collection of comparable networks. The method builds on AA, a multivariate technique originally developed to identify "pure types" on the boundary of a data cloud. Because archetypes are located on the convex hull of the data, they can be considered the extremes of the data scatter. This notion of extremeness defines the archetypes in terms of purity [13].

Formally, let us consider a set of N networks $\{\mathcal{G}_i\}_{i=1,\ldots,N}$, $\mathcal{G}_i = (\mathcal{V}_i, \mathcal{A}_i)$, with $\mathcal{V}_i$ the set of nodes and $\mathcal{A}_i \subseteq \mathcal{V}_i \times \mathcal{V}_i$ the set of directed arcs for each network i of the collection. The proposed procedure consists of three steps. In the first step, each network $\mathcal{G}_i$ is mapped into the Euclidean space E^p by computing a vector $\mathbf{x}_i$ of p features that characterize the whole network structure and topology. Given the feature matrix $\mathbf{X}$, in the second step the usual AA is applied to find a set of m archetypes $\mathbf{a}_j$. Finally, in the third step, the ANs are identified and profiled by using the tools proper of AA.

In the first step, in order to map the classroom networks in E^p, it is necessary to select p network indices and features describing and characterizing in a complete and substantive way the observed networks. The mapping and the relative solution of the ANs should be easy to interpret and strongly related to the field of application. Scholars [14,15] show a similar mapping approach within a cluster framework. The choice of the appropriate structural and topological indicators (e.g., centrality, cohesion, fragmentation, distance-based measures) should be strictly guided by the application context.

In the second step, the analysis seeks to identify the archetypes. Given the matrix $\mathbf{X} \in \Re^{n \times p}$ containing the set of the p-vectors obtained in the mapping step, AA looks for m vectors $\mathbf{a}_j$ that are convex combinations of the $\mathbf{x}_i$'s and such that each data point $\mathbf{x}_i$ is a convex combination of the vectors $\mathbf{a}_j$'s that satisfy the following conditions:

$$\begin{aligned} \mathbf{a}'_j = \beta'_j \mathbf{X}, &\quad j = 1, \ldots, m, \\ \beta_{ji} \geq 0 \;\; \forall j, i, &\quad \beta'_j \mathbf{1} = 1 \;\; \forall j; \end{aligned} \tag{1}$$

$$\begin{aligned} \mathbf{x}'_i = \gamma'_i \mathbf{A}, &\quad i = 1, \ldots, n, \\ \gamma_{ij} \geq 0 \;\; \forall i, j, &\quad \gamma'_i \mathbf{1} = 1 \;\; \forall i. \end{aligned} \tag{2}$$

with $\mathbf{A} = (\mathbf{a}'_1, \ldots, \mathbf{a}'_m)$, $\mathbf{A} \in \Re^{m \times p}$, the archetype matrix, and $\beta'_j = (\beta_{j1}, \ldots, \beta_{jn})$ and $\gamma'_i = (\gamma_{i1}, \ldots, \gamma_{im})$, the convex combination coefficients. Since an exact solution of Eqs. (1) and (2) exists only when the number of archetypes m equals the number V of vertices of the data convex hull [13], the archetypes are determined by solving the following minimization problem:

$$\min_{\mathbf{\Gamma}, \mathbf{B}} RSS(m) = \min_{\mathbf{\Gamma}, \mathbf{B}} \|\mathbf{X} - \mathbf{\Gamma} \mathbf{B}' \mathbf{X}\|_F \tag{3}$$

holding all the other conditions on the coefficients, with $\mathbf{\Gamma}$ and $\mathbf{B}$ the matrices containing the convex combination coefficients β_j and γ_i defined in Eqs. (1) and (2), and $\|\mathbf{Y}\|_F = \sqrt{tr(\mathbf{Y}\mathbf{Y}')}$ the Frobenius norm for a generic matrix $\mathbf{Y}$. Solutions to this minimization problem depend on m and are not nested across different values of m. The $RSS(m)$ in Eq. (3) illustrates the synthesizing power of the archetypes. It is a decreasing function of m, attaining its maximum at $m = 1$ and tending toward zero as m approaches the total number of vertices on the convex hull. As a result, archetypal networks capture extreme and well-separated structural configurations, positioned on the boundary of the empirical distribution of network features, rather than representing average or "typical" cases.

In the third step, each archetype $\mathbf{a}_j$ is associated with the corresponding AN $\mathcal{G}(\mathbf{a}_j)$. By examining the γ_i coefficients, it is possible to establish a correspondence between a network $\mathcal{G}_i$ and an archetype $\mathbf{a}_j$. Specifically, the AN $\mathcal{G}(\mathbf{a}_j)$ is defined as the network $\mathcal{G}_i$, whose feature vector $\mathbf{x}_i$ is the closest to the archetypes

$\mathbf{a}_j$ (i.e., the network for which $\gamma_{ij} \approx 1$). A substantial body of literature proposes methods to graphically and analytically use the archetypes to profile data [16–19].

Compared to clustering-based approaches of network data, ANs emphasize structural differentiation. Whereas cluster centroids summarize central tendencies, archetypes capture contrasting network forms and provide meaningful benchmarks for comparisons. The coefficients linking each observed network to the archetypes offer a graded and interpretable representation of similarity, enabling fuzzy classification, visualization in simplex spaces, and the identification of intermediate or hybrid network structures.

3 Applying Archetypal Analysis to School Network Data

The proposed approach is particularly well suited for the analysis of network data collected in educational contexts, where networks are naturally organized into hierarchical groupings, such as classrooms, schools, or school districts. In these contexts, multiple relational layers (friendship ties, support networks, advice-seeking relations) can be observed within bounded social settings. These networks may differ substantially across classrooms in terms of cohesion, centralization, inclusiveness, and group formation. The ANs approach provides a way to compare these structures and to identify prototypical forms of social interactions among students. A key advantage of this perspective lies in its flexibility regarding the choice of network indicators. Networks can be classified not only in structural terms but also in relation to broader social contexts. Indeed, in addition to classical structural measures, the mapping step can incorporate network-level variables derived from students' family and social backgrounds, such as parents' educational level, migration background, or neighborhood characteristics. Institutional characteristics (type of high school and students' university choices) can also be included in the analysis. In this way, ANs can be interpreted as configurations linking peer relations, social context, and educational trajectories.

The empirical application focuses on a sample of 28 high schools with 91 classrooms in Southern Italy,[1] where classes and schools represent bounded social contexts shaping students' aspirations and choices. For each classroom, four types of relationships were collected, reflecting friendship and social support ties among classmates [20,21]. The dataset also contains individual and contextual characteristics, including parents' educational level, type of school (lyceum, technical, vocational), and post-diploma choices or intentions (university enrollment versus entry into the labor market). From this perspective, ANs can help to identify: *i)*highly cohesive and inclusive classroom networks associated with strong peer support and favorable educational outcomes; *ii)* centralized or segmented network structures potentially linked to social inequality or unequal access to

[1] The data come from a survey financially supported by the PRIN 2022 project "*From high school to university: Assessing peers' influence in educational inequalities and performances.*" For details see: https://prin.unica.it/peer-up/.

social support; and *iii)*fragmented networks that may signal risks of social exclusion or weak integration within classrooms. By serving as benchmarks [13], ANs also facilitate comparisons across schools or educational tracks, supporting both descriptive analysis and policy-oriented evaluation. More broadly, the approach offers a powerful exploratory tool for understanding how relational structures, social context, and institutional settings jointly shape students' educational trajectories. Peer networks can act as meso-level mechanisms reinforcing norms, expectations, and information flows related to future choices after high school.

3.1 First Findings

Given the aim of clustering discussed above, the mapping step incorporates network indicators. It should be clarified that centrality measures are computed at the actor level, whereas AN focuses on the network at the aggregate level. Hence, the distribution of the actor-level measures is synthesized using the first two moments (i.e., the mean value and the variance). Specifically, the mapping includes: *i)* average centrality measures (in-degree, out-degree, and betweenness) along with their variability; *ii)* indicators of cohesion and fragmentation (number of isolated nodes, vulnerability, and modularity); and *iii)* measures of reciprocity, transitivity, and triadic closure. Homophily measures based on gender, family background, and university choices are excluded in order to focus on network structure and topology.

The scree plot in Fig. 1a, reporting the RSS for a number of archetypes ranging from 2 to 10, suggests that three archetypes provide a reasonable choice. A secondary jump occurs at six archetypes. In the following, the three-archetype solution is analyzed through the simplex plot of the γ_{ij} coefficients (Fig. 1b) and the corresponding profile plot (Fig. 2). This latter displays, through bar charts, the percentile profiles of each archetype with respect to the network features used to map graphs in E^p. The simplex plot, corresponding to a ternary diagram in the case of three archetypes, is a barycentric representation used to visualize compositional data, such as the γ_{ij} coefficients, which satisfy $\sum_j \gamma_{ij} = 1$. In this framework, component proportions are mapped to positions within an equilateral simplex (triangle), where the distance from each side corresponds to the value of each variable. Each point in the simplex plot represents a network, located according to its barycentric coordinates γ_{ij}. Points are colored according to the type of relationship: appraisal support (yellow dots), emotional support (green triangles), scholastic support (blue diamonds), and friendship (red squares). More precisely, looking at the percentile profile plot in Fig. 2, the first archetype (AN1) and the third archetype (AN3) exhibit contrasting profiles, with opposite percentiles across all variables, whereas the second archetype (AN2) displays a more mixed configuration. Specifically, the three archetypes are characterized as follows:

- AN1 is very close to two observed data points—one corresponding to the appraisal support (yellow dot) and the other to the emotional support (green triangle)—which are nearly overlapping in the archetypal space. These networks represent relationships among classmates in two vocational institutes,

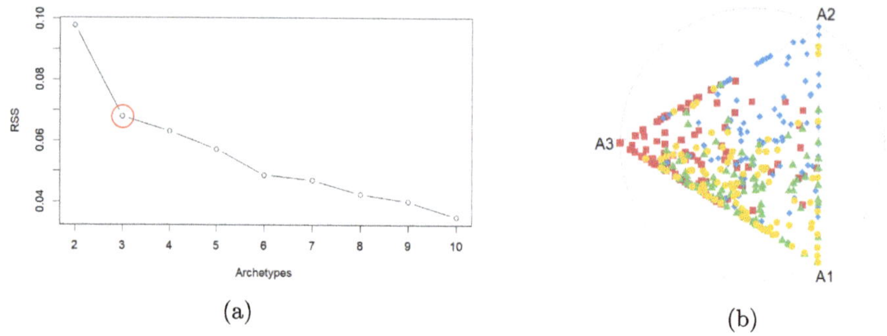

Fig. 1. (a) Scree plot of Archetypal Analysis; (b) Simplex plot

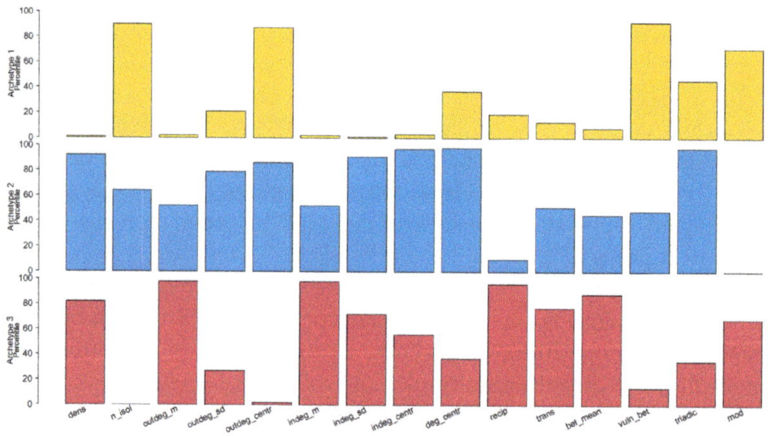

Fig. 2. Percentile plot of archetypal profiles.

one located in Naples and the other in Salerno. The profile plot suggests that both networks are characterized by low density, a high number of isolates, low average degree, and very limited triadic closure. With regard to gender composition, the emotional support network is composed entirely of male students. Overall, these structural characteristics point to a very limited capacity to provide emotional support. Similarly, the appraisal network exhibits weak supportive properties, indicating that both types of relationships are poorly developed in these contexts.

- AN2 refers to scholastic support networks among almost all male classmates attending a technical institute in Naples. This archetype is characterized by a markedly cohesive structural configuration. Specifically, the profile plot shows that networks of this type exhibit high density, the absence of isolates, and high levels of both modularity and transitivity. Such a combination of structural properties indicates a well-integrated and highly interactive network, suggesting broad access to scholastic support resources. High density and

transitivity facilitate efficient information exchange and collaborative learning processes. High modularity, in turn, suggests the presence of distinct subgroups, possibly reflecting study groups, classroom divisions, or friendship circles.
- AN3 captures friendship relationships and is characterized by a highly cohesive and structurally robust configuration. In particular, the network exhibits very high density, high average in-degree and out-degree, high levels of reciprocity, and low vulnerability. Such structural features point to a highly interconnected social environment, with balanced and stable interpersonal relationships. These networks are observed in high schools (lyceum) characterized either by a balanced gender composition or by an all-female student body. Moreover, these schools are associated with a high propensity for university enrollment and are typically attended by students from relatively advantaged family backgrounds. Such conditions are likely to promote not only higher levels of social integration but also shared norms, expectations, and aspirations, further reinforcing the density and reciprocity of friendship ties.

These preliminary results obtained from clustering classrooms based on their relational functioning across different network layers may inform the design of more effective educational interventions [22], such as university guidance programs and learning activities aimed at enhancing student soft skills.

References

1. Hardman, M.: Is complexity theory useful in describing classroom learning? In: Hudson, B., Meyer, M.A., (eds.) Beyond Fragmentation: Didactics, Learning and Teaching in Europe, pp. 355–366. Barbara Budrich Publishers, Opladen & Farmington Hills, MI (2011)
2. Sica, L.S., Crocetti, E., Ragozini, G., Aleni Sestito, L., Serafini, T.: Future-oriented or present-focused? The role of social support and identity styles on "futuring" in Italian late adolescents and emerging adults. J. Youth Stud. **19**(2), 183–203 (2016). https://doi.org/10.1080/13676261.2015.1059925
3. Crosnoe, R., Muller, C.: Family socioeconomic status, peers, and the path to college. Soc. Probl. **61**(4), 602–624 (2014). https://doi.org/10.1525/sp.2014.12255
4. Usala, C., Sulis, I., Porcu, M.: Exploring the impact of high schools, socioeconomic factors, and degree programs on higher education success in Italy. Big Data Res. (2025). https://doi.org/10.1016/j.bdr.2025.100539
5. Alexander, C.N., Campbell, E.Q.: Peer influences on adolescent educational aspirations and attainments. Am. Sociol. Rev. **29**(4), 568–575 (1964). https://doi.org/10.2307/2091205
6. Smith, E.: Peer preferences and educational decisions: heterogeneous associations across student socioeconomic status. Br. J. Sociol. Educ. **44**(2), 374–393 (2023). https://doi.org/10.1080/01425692.2022.2154641
7. Vitale, M.P., et al.: Networks, schools, and higher education: linking social capital and student pathways beyond secondary school. Ital. J. Appl. Stat. (2025). https://doi.org/10.36253/ijas-16656

8. An, W.: Friendship network formation in Chinese middle schools: patterns of inequality and homophily. Soc. Netw. **68**, 218–228 (2022). https://doi.org/10.1016/j.socnet.2021.07.003
9. Lubbers, M.J.: Group composition and network structure in school classes: a multilevel application of the p* model. Soc. Netw. **25**(4), 309–332 (2003). https://doi.org/10.1016/S0378-8733(03)00013-3
10. Doreian, P., Batagelj, V., Ferligoj, A., (eds.): Advances in Network Clustering and Blockmodeling. John Wiley & Sons, New York (2020)
11. Cutler, A., Breiman, L.: Archetypal analysis. Technometrics **36**(4), 338–347 (1994). https://doi.org/10.1080/00401706.1994.10485840
12. Ragozini, G., D'Esposito, M.R.: Archetypal networks. In: ASONAM '15 Proceedings of the 2015 IEEE/ACM International Conference on Advances in Social Networks Analysis and Mining 2015, pp. 807–814 (2015). https://doi.org/10.1145/2808797.2808837
13. Porzio, G.C., Ragozini, G., Vistocco, D.: On the use of archetypes as benchmarks. Appl. Stoch. Models Bus. Ind. **24**(5), 419–437 (2008). https://doi.org/10.1002/asmb.727
14. Harland, C.M., Lamming, R.C., Zheng, J., Johnsen, T.E.: A taxonomy of supply networks. J. Supply Chain Manag. **37**, 21–27 (2001). https://doi.org/10.1111/j.1745-493X.2001.tb00109.x
15. Onnela, J.-P., et al.: Taxonomies of networks from community structure. Phys. Rev. E **86**, 036104 (2012). https://doi.org/10.1103/PhysRevE.86.036104
16. D'Esposito, M. R., Ragozini, G., Vistocco, D.: Exploring data through archetypes. In: Locarek-Junge, H., Weihs, C., (eds.) Classification as a Tool for Research. Proceedings of the 11th IFCS Biennial Conference and 33rd Annual Conference of the Gesellschaft für Klassifikation eV, Dresden, March 13-18, 2009, pp. 287–298. Springer (2010). https://doi.org/10.1007/978-3-642-10745-0
17. D'Esposito, M.R., Palumbo, F., Ragozini, G.: Interval archetypes: a new tool for interval data analysis. Stat. Anal. Data Min. **5**(4), 322–335 (2012). https://doi.org/10.1002/sam.11140
18. Vinué, G., Epifanio, I., Alemany, S.: Archetypoids: a new approach to define representative archetypal data. Comput. Stat. Data Anal. **87**, 102–115 (2015). https://doi.org/10.1016/j.csda.2015.01.018
19. Ragozini, G., Palumbo, F., D'Esposito, M.R.: Archetypal analysis for data-driven prototype identification. Stat. Anal. Data Min. **10**(1), 6–20 (2017). https://doi.org/10.1002/sam.11325
20. Pacca, A., Ragozini, G., Vitale, M.P.: Measuring peer effects in a network perspective. Survey Design and Privacy Issues. In: Di Bella, E., Gioia, V., Lagazio, C., Zaccarin, S., (eds.) Statistics for Innovation IV: SIS 2025, Short Papers, Contributed Sessions 3, pp. 120–126. Springer, Cham (2025). https://doi.org/10.1007/978-3-031-96033-8
21. Sulis, I., Tocchioni, V., Vitale, M.P., (eds.): Dalla scuola superiore all'università. Valutare l'influenza dei pari sulle diseguaglianze educative e sulle performance. Franco Angeli, Milano (2026)
22. Knight, B., Harrison, N.: Teaching and learning as complex phenomena: implications for policy and teacher professional identity. In: Waller, R., Andrews, J., Clark, T., (eds.) Critical Perspectives on Educational Policies and Professional Identities: Lessons from Doctoral Studies, pp. 87–104. Emerald Publishing Limited, Bingley (2024). https://doi.org/10.1108/978-1-83753-332-920241006

Riemannian Principal Component Analysis

Oldemar Rodríguez(✉)

School of Mathematics and Research Center for Pure and Applied Mathematics (CIMPA), University of Costa Rica, San José, Costa Rica
oldemar.rodriguez@ucr.ac.cr
https://oldemarrodriguez.com/

Abstract. This paper proposes an extension of Principal Component Analysis (PCA) beyond the assumption of Euclidean data, enabling its application to Riemannian manifolds. A central challenge is the absence of vector space operations on such manifolds. Fletcher et al., in [1] and [4], introduced Principal Geodesic Analysis (PGA) as a geometric method for analyzing manifold-valued data, particularly effective for structured datasets such as medical images, where intrinsic geometry is explicit. However, PGA is limited for general datasets lacking an implicit notion of local distance. We introduce a generalized framework, termed *Riemannian Principal Component Analysis (R-PCA)*, extending PGA to any data endowed with a local distance structure. By equipping data tables with local metrics, we adapt PCA to incorporate manifold geometry. This approach provides a unified framework for dimensionality reduction and statistical analysis on manifolds, enabling the study of datasets with region-specific distance structures while preserving their intrinsic geometric and topological properties.

Keywords: Riemannian · manifold · topology · R-PCA · statistics · distances

1 Providing to a Classical Data Table with a Riemannian Manifold Structure

Riemannian statistics provides a rigorous framework for extending classical statistical analysis beyond the Euclidean setting, and a detailed and general treatment of this paradigm can be found in [7], where arbitrary data tables are endowed with a Riemannian manifold structure through locally defined metrics. In that work, the Euclidean space is transformed into a Riemannian manifold using the UMAP (Uniform Manifold Approximation and Projection) algorithm, following the construction of a k-nearest neighbor graph with normalized local distances as described by McInnes et al. [3]. By leveraging concepts from algebraic topology, such as simplicial complexes, Čech complexes, and the Nerve Theorem, UMAP defines meaningful local metric spaces at each data point,

P. Brito et al. (Eds.): IFCS 2026, STUDIES CLASS, pp. 243–251, 2026.
https://doi.org/10.1007/978-3-032-32009-4_30

assigning edge weights that capture intrinsic geometric separation. These locally induced distances allow the definition of Riemannian means, variances, covariances, and correlation measures, making it possible to study Riemannian statistics in detail even for data without an explicit manifold structure. See [7] for a detailed exposition.

Unlike traditional graph-weighted approaches, the proposed framework leverages the Nerve Theorem to induce a genuine Riemannian manifold structure from simplicial complexes. This allows us to define rigorous local metric spaces, moving beyond simple Euclidean differences to a truly intrinsic geometric representation.

UMAP algorithm provides the data table with local distance notions. To the graph construction, UMAP algorithm begins by constructing a weighted k-nearest neighbor graph from the dataset $X = \{x_1, \ldots, x_n\}$, where a distance metric d defines the distances between points. The process can be described as follows:

1. The dataset X consists of points in a high-dimensional space, and a distance metric d (e.g., Euclidean distance) is used to compute the similarity between points.
2. The algorithm identifies the k-nearest neighbors using the distance metric d. This step produces a directed graph where each point has outgoing edges to its k-nearest neighbors.
3. The local parameters, ρ_i and σ_i, are defined as follows:
   - ρ_i: Local Connectivity. For each x_i, ρ_i is the minimum distance to its nearest neighbor that is greater than zero, $\rho_i = \min\{d(x_i, x_{i_j}) \mid 1 \leq j \leq k,\ d(x_i, x_{i_j}) > 0\}$, this ensures that x_i connects to at least one other point with an edge of weight 1.
   - σ_i: Local Scaling. σ_i normalizes the distances of all neighbors, ensuring consistency in the local metric. It is computed by solving:

$$\sum_{j=1}^{k} \exp\left(-\frac{\max(0, d(x_i, x_{i_j}) - \rho_i)}{\sigma_i}\right) = \log_2(k).$$

4. To constructing the weighted directed graph we use ρ_i and σ_i, and the weight of each directed edge (x_i, x_{i_j}) is computed as:

$$w((x_i, x_{i_j})) = \exp\left(-\frac{\max(0, d(x_i, x_{i_j}) - \rho_i)}{\sigma_i}\right).$$

5. To symmetrized the graph we convert the directed graph into an undirected graph, the adjacency matrix A is symmetrized using $B = A + A^t - A \circ A^t$, where A is the adjacency matrix, $\circ$ denotes the Hadamard product. The resulting symmetric graph represents the unified global manifold structure of the dataset.
6. The symmetrized graph G forms a *fuzzy simplicial set* that captures both the local and global topological structure of the data. Also, this symmetrized graph G allows us to define a local similarity as $S_{\text{UMAP}}(x_i, x_j) = B_{ij}$. In UMAP,

the distance graph encodes normalized connection weights in the k-nearest neighbor graph, taking values in $[0, 1]$, where 0 indicates no connection and values closer to 1 represent stronger affinities between points.

To generalize Principal Component Analysis, it is important to understand how UMAP connects points from different neighborhoods, since this affects the projection of individuals lying on different Riemannian submanifolds. This is achieved through the graph symmetrization step, where directed neighborhood relations are combined to create undirected edges between points not mutually included in their k-nearest neighbor sets, using the weight $w_{ij} = w_{\overrightarrow{ij}} + w_{\overleftarrow{ij}} - w_{\overrightarrow{ij}} w_{\overleftarrow{ij}}$, thereby preserving both local and global structure while avoiding double-counting overlapping neighborhood information.

2 Riemannian Principal Component Analysis for Any Type of Data

In the paper [1], they aim to generalize Principal Component Analysis (PCA) to Principal Geodesic Analysis (PGA) a generalization of principal component analysis to manifolds using geodesic distances and geodesic submanifolds.

Let be $x_1, \ldots, x_n \in \mathbb{R}^p$ with zero mean. Principal component analysis find an orthonormal basis $\{v_1, \ldots, v_p\}$ of $\mathbb{R}^p$, which satisfies the recursive relationship:

$$v_1 = \arg \max_{\|v\|=1} \sum_{i=1}^{n} (v \cdot x_i)^2, \tag{1}$$

$$v_s = \arg \max_{\|v\|=1} \sum_{i=1}^{n} \sum_{j=1}^{s-1} (v_j \cdot x_i)^2 + (v \cdot x_i)^2, \quad s = 2, \ldots, p. \tag{2}$$

Then, the subspace $V_s = \text{span}(\{v_1, \ldots, v_s\})$ is the s-dimensional subspace that maximizes the variance of the data projected onto that subspace. As is well known, the basis $\{v_s\}$ is computed as the set of ordered eigenvectors of the sample covariance matrix of the data.

The lower-dimensional subspaces in PCA are linear subspaces, in [1] for general manifolds H, they extend the concept of a linear subspace to a geodesic submanifold. A geodesic is a curve that is locally the shortest path between points. In this way, a geodesic is the generalization of a straight line. Thus, it is natural to use a geodesic curve as the one-dimensional subspace, the analog of the first principal direction in PCA.

Let M be a manifold, the projection of a point $x \in M$ onto a geodesic submanifold H of M is defined as the point on H that is nearest to x in geodesic distance. So, the projection operator $\pi_H : M \to H$ is:

$$\pi_H(x) = \arg \min_{y \in H} d(x, y)^2. \tag{3}$$

Since projection is defined by a minimization, there is no guarantee that the projection of a point exists or that it is unique. However, by restricting to a small enough neighborhood about the mean, the projection is unique for any geodesic submanifold at the mean and the projection onto a geodesic submanifold can be approximated linearly in the tangent space of M, that is, in 3 to ensure that $\pi_H(x)$ is actually within H, an approximation of the projection is used, then they find a sequence of nested geodesic submanifolds that maximize the projected variance of the data. Finally, given a dataset $x_1, x_2, \ldots, x_n \in M$, they compute the intrinsic mean μ, which minimizes the sum of squared Riemannian distances, $\mu = \arg\min_{x \in M} \sum_{i=1}^{n} d(x, x_i)^2$. Each data point is mapped to the tangent space using $u_i = x_i - \mu$. The covariance matrix, $S = \frac{1}{n} \sum_{i=1}^{n} u_i u_i^T$, is constructed, and eigendecomposition yields principal directions v_k and variances are λ_k, see [1] for the details.

To generalize these ideas to any data table, we define a Local Manifold Approximation, let $X = \{x_1, x_2, \ldots, x_n\}$ be a dataset embedded in a high-dimensional space $\mathbb{R}^p$. For each data point x_i, UMAP defines a local neighborhood N_i consisting of its k-nearest neighbors under a metric d. The k-nearest neighbors of x_i are defined as the set $\mathcal{N}(x_i)$ such that:

$$\mathcal{N}(x_i) = \{x_j \in X \setminus \{x_i\} \mid d(x_i, x_j) \leq d(x_i, x_k),\ \forall x_k \notin \mathcal{N}(x_i)\},$$

where $d(x_i, x_j)$ is the distance between x_i and x_j and k is the number of neighbors (a hyperparameter).

Thus, $\mathcal{N}(x_i)$ is the set of the k closest points to x_i in $\mathbb{R}^p$. The *Patch* $\mathcal{P}_i$ is defined as a *fuzzy simplicial set* that approximates the local geometry of the manifold around x_i. Formally:

$$\mathcal{P}_i = \{(x_i, x_j, w_{ij}) \,|\, x_j \in N_i,\ w_{ij} > 0\},$$

where x_i is the anchor point, $x_j \in \mathcal{N}_i$ are the k-nearest neighbors of x_i, w_{ij} is the weight (or affinity) associated with the edge between x_i and x_j, defined in the previews section.

The individual patches $\mathcal{P}_i$ for all $x_i \in X$ can be combined to form the global fuzzy simplicial set o graph G, representing the entire manifold:

$$G = \bigcup_{i=1}^{n} \mathcal{P}_i.$$

Because the data resides on a Riemannian manifold, with local distances it is necessity to define something akin to a Riemannian correlation, requiring a Riemannian mean, and, more broadly, necessitating the development of *Riemannian Statistics*. By leveraging the one-to-one relationship given by the Nerve Theorem and its corollaries, see [7], we define the vector subtraction the Riemannian correlation on manifold as follows.

The core of R-PCA lies in the definition of a Riemannian subtraction operator ($\ominus$) and an associated distance (d_{UMAP}). While computations are performed in a local coordinate system, the method is inherently intrinsic; it approximates the manifold's geometry within the tangent space at each point. This ensures that the non-linear topological structure is preserved throughout the projection process, rather than being discarded as in classical linear methods.

Definition 1. *Let x_α and x_β rows of X, we define the subtraction induced by the UMAP algorithm as $x_\alpha \ominus x_\beta = \rho_{\alpha\beta}(x_\alpha - x_\beta)$, where $\rho_{\alpha\beta} = 1 - S_{\mathsf{UMAP}}(x_\alpha, x_\beta)$. Then, local distances generated by the UMAP algorithm can be defined as:*

$$d_{\mathsf{UMAP}}(x_\alpha, x_\beta) = \|x_\alpha \ominus x_\beta\| = \sqrt{\langle x_\alpha \ominus x_\beta, x_\alpha \ominus x_\beta \rangle},$$

where $\langle \cdot, \cdot \rangle$ is an inner product on the vector space $\mathbb{R}^p$. We also define the the addition induced by the UMAP algorithm as $x_\alpha = x_\beta \oplus x_\gamma$ if $x_\gamma = x_\alpha \ominus x_\beta$.

In the subsequent definition, we generalize Fréchet mean:

Definition 2. *Let $X \in M_{n \times p}$ the data table. We denote by $x_1, \ldots, x_n \in \mathbb{R}^p$ the rows of X and by $y_1, \ldots, y_p \in \mathbb{R}^n$ the columns of X. Each vector x_i can be also considered a point in the Riemannian manifold M induced by the simplicial complex. The Riemannian mean is the minimizer of the sum-of-squared distances to the data:*

$$g = \arg\min_{x \in M} \sum_{i=1}^{n} d_{\mathsf{UMAP}}(x, x_i)^2 .$$

Definition 3. *We defined the variance-covariance matrix $S \in M_{p \times p}$ of X as $S = \frac{1}{n}\sum_{i=1}^{n}(x_i \ominus g)(x_i \ominus g)^t$, where $(x_i \ominus g) = \rho_{i\lambda}\begin{bmatrix} x_{i1} - g_1 \\ \vdots \\ x_{ip} - g_p \end{bmatrix}$, note that g must be equal to x_λ for some λ. Then, we can define Riemannian correlation between y_i and y_j columns of X, that are in $\mathbb{R}^n$, as follows $R(y_i, y_j) = \frac{S_{ij}}{\sqrt{S_{ii}S_{jj}}}$.*

We can say that each patch $\mathcal{P}_i$ induces a *Riemannian Submanifold* H_i with the local metric d_{UMAP} generated by the UMAP algorithm, defined in previous paragraphs. We are now ready to define *Riemannian Principal Component Analysis* (R-PCA) for any data $x_1, \ldots, x_n \in \mathbb{R}^p$. Our goal, analogous to PCA, is to find a sequence of subspaces S_i de $\mathbb{R}^p$ that maximize the projected variance of the data, but, which also takes into account the local distances of each of the submanifods H_i that we have in the data. We find a sequence of subspaces, not a sequence submanifolds, because our total space is finite, then requiring the projections to be in there would greatly degrade the result, something that even happens in the PGA proposed in [1].

Taking into account that $G = \bigcup_{i=1}^{n} \mathcal{P}_i$ and the following properties of patches: 1. $\mathcal{P}_i$ is a weighted graph over the local neighborhood $\mathcal{N}_i$, where the weights $w_{ij} \in [0, 1]$ define the strength of the connection between x_i and x_j, 2. $\mathcal{P}_i$ represents the local structure of the manifold around x_i, capturing relationships with its neighbors, 3. the weight w_{ij} can be interpreted as a probability or affinity

Algorithm 1 Riemannian Principal Component Analysis (R-PCA)

Input: $x_1, \ldots, x_n \in \mathbb{R}^p$ and the number k of nearest neighbors in UMAP algorithm.
Output: Principal directions $v_s \in S_i$, variances $\lambda_s \in \mathbb{R}$.
Generate symmetric UMAP graph G.
Compute the matrix of UMAP similarities $S_{\text{UMAP}}(x_i, x_j)$ with $i, j = 1, \ldots, n$.
Compute the matrix $P = \rho_{ij}$ for $i, j = 1, \ldots, n$, as in Definition 1.
Compute the Riemannian distance matrix D, using Definition 1.
Compute the Riemannian mean g using D, as in Definition 2.
Compute the Riemannian variance-covariance matrix $S \in M_{p \times p}$ as

$$S = \frac{1}{n} \sum_{i=1}^{n} (x_i \ominus g)(x_i \ominus g)^t.$$

Compute the Riemannian correlation matrix $R \in M_{p \times p}$ where $R_{ij} = \frac{S_{ij}}{\sqrt{S_{ii} S_{jj}}}$.
Extract eigenvectors and eigenvalues of R: $\{v_s, \lambda_s\}$.
Compute the Riemannian Principal Components.
For the correlation circle, the Riemannian correlations between the original variables and the principal components are calculated.

that x_i and x_j are connected within the underlying manifold, the *Riemannian Principal Component Analysis* (R-PCA) that respects the local metrics of the submanifolds H_i, can be defined by the Algorithm 1.

In practice, in Algorithm 1 it is recommended to use at least $k = \lfloor \frac{n}{c} \rfloor$ (the whole part) where c is the number of clusters that the data table is suspected to have, or the number of clusters that you want to study, that is, k is at least the average number of individuals that each cluster has.

3 Applications with Simulated and Real Data

To demonstrate the empirical consistency and superiority of R-PCA, we compare it against classical PCA. Our results show that R-PCA captures significantly higher explained variancc (e.g., 43.38% vs. 25.58% in high-dimensional synthetic data) and achieves superior separation of non-linear nested clusters, proving that the induced Riemannian structure provides information that is inaccessible to Euclidean-based approaches.

3.1 Description of the Simulated Data in `Data10D.csv`

The file `Data10D.csv` contains 2900 observations with 10 variables, where the first two columns define five distinct clusters—explicitly labeled in the `cluster` column—including two pairs of nested (concentric) clusters and one separate parabolic cluster, while the remaining eight variables (`var1`–`var8`) were independently generated from a standard normal distribution $N(0,1)$ to enrich the dataset without affecting the original clustering structure. The data in the file `Data10D.csv` corresponds to the 2D plot, on the left in Fig. 1.

The Fig. 1 we have visualizations of the Principal Component Analysis (PCA) and the Riemannian Principal Component Analysis (R-PCA) applied to the dataset `Data10D.csv`. Here we are computing R-PCA with $k = \left\lfloor \frac{2900}{5} \right\rfloor = 580$. The principal plane of the R-PCA on the right of Fig. 1 preserves the structure of the original five clusters in `Data10D.csv` more effectively than the standard PCA, on the left. The nested circular clusters (Clusters 1, 2, 3, and 4) and the parabolic cluster (Cluster 5) are clearly separated and retain their geometric shapes. In contrast, the principal plane of the PCA on the center of Figure 1 shows more diffuse clusters, with less distinct shapes. Especially Cluster 5 (orange) which is a circle inside Cluster 1 (blue), is better captured in the R-PCA than the PCA showing the orange points mostly enclosed in the blue points. Similarly it happens between clusters 3 and 4 in green and red respectively.

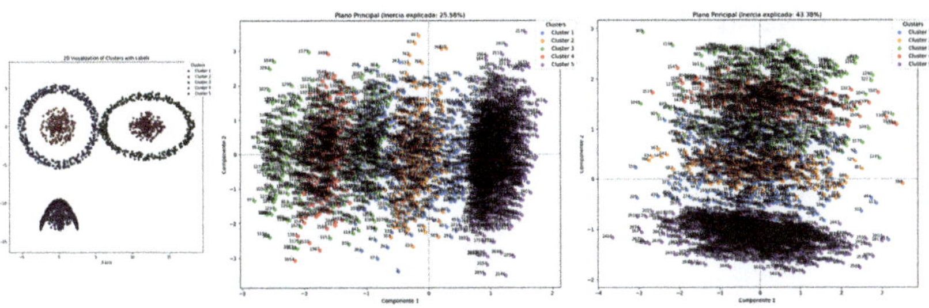

Fig. 1. PCA on the center and R-PCA on the right of `Data10D.csv`.

The R-PCA achieves an explained variance (inertia) of **43.38%**, which is significantly higher than the **25.58%** explained by the PCA. This indicates that the R-PCA captures a larger proportion of the variability in the dataset, making it a more effective tool for dimensionality reduction in this context.

3.2 Application to the Real Data Set Olivetti Faces

The *Olivetti Faces* dataset is a widely used benchmark in the fields of pattern recognition, principal component analysis (PCA), and dimensionality reduction. Collected by the AT&T Laboratories in Cambridge, the dataset contains 400 grayscale images of 40 different individuals, with 10 distinct images per person that vary in facial expression, detail, and lighting conditions. Each image has a resolution of 64×64 pixels, resulting in 4096 features per observation when flattened into a one-dimensional vector. The Olivetti Faces is shown in Fig. 2, we can see [8] for more details. When comparing the results of classical PCA, at the top in Fig. 2, with the Riemannian PCA, at the bottom in Fig. 2, differences arise in both variance explanation and cluster structure. First, the explained inertia in the first principal plane is higher in the R-PCA (39.59%) than in the classical PCA (39.22%), indicating that the Riemannian projection captures more information from the original data structure. Moreover, the 40 clusters appear more clearly and homogeneously separated in the R-PCA projection, suggesting that this method is more effective in preserving inter-individual differences in the lower-dimensional space. In contrast, the classical PCA shows a higher degree of overlap between groups, making them harder to distinguish visually.

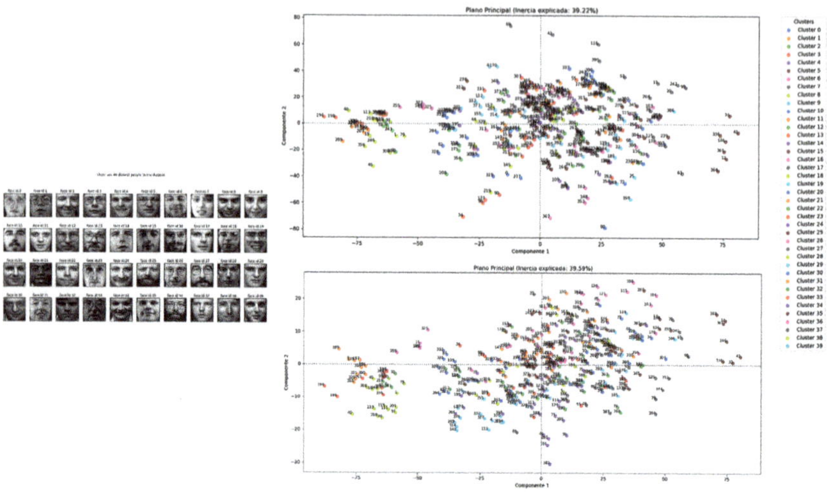

Fig. 2. PCA on top and R-PCA on bottom of Olivetti Faces data set.

4 Conclusions

In this paper, we successfully extend the ideas proposed by *Pennec et al.* in [4], broadening the scope to compute Riemannian statistical indices and Riemannian data analysis models to any data table. Unlike previous approaches, our methodology is not restricted to data with an intrinsic Riemannian manifold structure. This advancement opens up a new field of research, where diverse methods like regression, k-means, and more, can be generalized for broader applicability. Regarding the induced geometry's dependence on UMAP hyperparameters, such as the number of neighbors (k), we emphasize that k is a critical factor for structural stability. We recommend a heuristic of $k = \lfloor n/c \rfloor$ to ensure robust local connectivity. While this version focuses on the foundational framework, future work will provide extended analysis on the robustness and consistency of these algorithmic choices.

The Riemannian STATS Package, available on GitHub [5] or in PyPI [6] is a Python package developed to extend classical multivariate statistical methods to the Riemannian framework. It enables the transformation of a Euclidean data table into its Riemannian counterpart by applying UMAP algorithm. This tool is particularly valuable for analyzing datasets where the linear assumptions of traditional methods do not hold.

References

1. Fletcher, P.T., Lu, C., Pizer, S.M., Joshi, S.: Principal geodesic analysis for the study of nonlinear statistics of shape. IEEE Trans. Med. Imaging **23**(8), 995–1005 (2004). https://doi.org/10.1109/TMI.2004.831793
2. van der Maaten, L., Hinton, G.: Visualizing data using t-SNE. J. Mach. Learn. Res. **9**, 2579–2605 (2008). https://www.jmlr.org/papers/v9/vandermaaten08a.html
3. McInnes, L., Healy, J., Melville, J.: UMAP: uniform manifold approximation and projection for dimension reduction. arXiv:1802.03426 (2018). https://doi.org/10.48550/arXiv.1802.03426
4. Pennec, X., Sommer, S., Fletcher, P.T. (eds.): Riemannian Geometric Statistics in Medical Image Analysis. Academic Press, Elsevier (2020). https://doi.org/10.1016/C2017-0-01561-6
5. Rodríguez, O.: riemannian_stats: tools for statistical analysis of Riemannian data (Version 1.0). GitHub repository (2025). https://github.com/OldemarRodriguez/riemannian_stats
6. Rodríguez, O.: riemannian_stats: tools for statistical analysis of Riemannian data (Version 0.1.1). Python package, PyPI (2025). https://pypi.org/project/riemannian-stats/
7. Rodríguez, O.: Riemannian statistics for any type of data. In: Trejos, J., Chadjipadelis, T., Grané, A., Villalobos, M. (eds.) Data Science, Classification, and Artificial Intelligence for Modeling Decision Making. IFCS 2024. Studies in Classification, Data Analysis, and Knowledge Organization, pp. 151–160. Springer, Cham (2025). https://doi.org/10.1007/978-3-031-85870-3_17
8. Samaria, F.S., Harter, A.C.: Parameterisation of a stochastic model for human face identification. In: Proceedings of the 2nd IEEE Workshop on Applications of Computer Vision, pp. 138–142 (1994). https://doi.org/10.1109/ACV.1994.341300

Identifying Genes that Characterise Spatial Cellular Domains via Bayesian Mixtures with Shrinkage Priors

Andrea Sottosanti(✉)

Department of Statistical Sciences, University of Padova, via Cesare Battisti 241, 35121 Padova, Italy
andrea.sottosanti@unipd.it
https://andreasottosanti.github.io/

Abstract. Spatial transcriptomics enables the profiling of gene expression while preserving the spatial organisation of tissues, offering unprecedented insight into the architecture of complex organs. Identifying spatial domains—contiguous regions of the tissue with coherent transcriptional profiles—remains a central challenge. We present a methodological adaptation of the Bayesian model PERLA (Sottosanti et al., 2025), originally developed for spatial disease and mortality mapping, to high-resolution transcriptomic data. Unlike alternative approaches that rely on dimensionality reduction, PERLA operates directly on gene expression values and employs shrinkage priors to highlight the features driving cluster formation. We demonstrate its application to a mouse olfactory bulb dataset, showing that PERLA identifies spatially distinct clusters that are coherent with known anatomical layers. Moreover, the model reveals the gene signatures that characterise each domain, providing interpretable biological insight.

Keywords: areal data · mouse olfactory bulb · spatial clustering · spatial transcriptomics

1 Introduction

The advent of spatial transcriptomics (ST) has revolutionised our understanding of tissue architecture by providing gene expression profiles within their physical context. Unlike single-cell RNA sequencing, which requires tissue dissociation, ST preserves the spatial coordinates of each *spot*, offering a unique window into the organisational principles of complex organs such as the brain. In spot-based ST, a tissue slice is placed over a grid of spots, so that each spot contains a few neighbouring cells. The gene expression of each spot is then characterised, producing a dataset of tens of thousands of genes per spot along with their spatial locations.

The rise of this technology has prompted the integration of advanced statistical methods into the analysis of transcriptomic experiments. In particular,

P. Brito et al. (Eds.): IFCS 2026, STUDIES CLASS, pp. 252–258, 2026.
https://doi.org/10.1007/978-3-032-32009-4_31

retaining the spatial structure of the tissue requires models and methods from spatial statistics, as exploiting this additional information can provide new and biologically relevant insights [3,7].

A key question in single-cell studies—the identification of different cellular populations—translates in ST into the identification of spatial domains, *i.e.*, contiguous tissue regions where genes show similar levels of expression. To identify these clusters, it is natural to rely on methods that leverage the spatial information. An example is BayesSpace [11], which employs a Bayesian hidden Potts model to identify spatial domains. Despite its popularity, a key limitation of BayesSpace is its reliance on dimensionality reduction techniques, such as principal components analysis or GLMPCA [8], which reduces interpretability because each feature represents a combination of multiple genes. Moreover, the software is designed to process data only from two specific platforms, ST and Visium, which differ in technical characteristics such as pixel grid structure and pixel size. The software provides no options for handling data from other technologies, and its hyperparameter settings are tailored to the platform used. These technical constraints limit the flexibility and broader applicability of BayesSpace.

Recently, the PEnalised Regression with Localities Aggregation (PERLA, [6]) was introduced in the epidemiological literature to cluster spatial areas based on multivariate outcomes, optionally accounting for external covariates. In this work, we propose a methodological transfer of PERLA to high-resolution transcriptomics, showing how it can be adapted to identify distinct spatial domains. Unlike BayesSpace, PERLA operates directly on gene expression values without requiring dimensionality reduction, which improves interpretability. Additionally, the model introduces shrinkage priors that highlight which features drive the identification of specific clusters, allowing the identification of genes that characterise different spatial domains.

The remainder of the article is structured as follows. Section 2 introduces the Bayesian mixture model PERLA and its application to spatial transcriptomic data. Section 3 presents a case study on a mouse olfactory bulb, describing data preprocessing, model configuration, and the key spatial patterns discovered.

A dedicated online repository[1] contains the code to reproduce the results presented in this work.

2 Methods

2.1 Data Structure

A spatial transcriptomic experiment produces $\mathbf{Y}$, an $n \times d$ matrix collecting the expression of d genes in n spots. Each spot contains one or more cells, depending on the resolution of the technology. Thus, y_{ij} quantifies the (log-transformed, zero-centered) expression of gene j in spot i, with higher values indicating higher expression. Since the spatial coordinates of the spots are known,

[1] https://github.com/andreasottosanti/IFCS_2026.git.

the neighbourhood structure of the tissue can be determined. The number of neighbours for each spot depends on its position within the array (e.g., interior versus boundary spots) as well as on the underlying spatial technology. For example, the 10X Visium platform defines up to six neighbouring spots in a hexagonal layout [11]. Given this neighbourhood information, a proximity matrix can be constructed to encode the spatial relationships among spots.

2.2 Clustering Model

We now describe how the model under investigation can be adapted to spatial transcriptomic experiments. Briefly, we recall that PERLA is a multivariate spatial Bayesian model originally developed to cluster areal data based on d outcomes in epidemiological studies. There, the outcomes are zero-centred transformations of disease incidence or mortality, with positive values denoting an increase relative to the analysed region and negative values a decrease.

Let $\mathbf{y}_{i\cdot}$ denote the vector of d features for spot i. In spatial transcriptomics, $\mathbf{y}_{i\cdot}$ represents a zero-centred transformation of the d gene expressions at that spot. The model assumes that the spots can be grouped into K unknown clusters. Cluster membership is encoded by a label vector $\mathbf{Z}_{i\cdot}$, where $\mathrm{Z}_{ik} = 1$ if spot i belongs to cluster k, and 0 otherwise. Let also $\mathbf{x}_{i\cdot}$ denote a vector of p covariates associated with spot i. The PERLA model is then specified as:

$$\mathbf{y}_{i\cdot} \mid \mathrm{Z}_{ik} = 1, \boldsymbol{\mu}_k, \boldsymbol{\Sigma}, \boldsymbol{\beta} \sim \mathcal{N}_d(\boldsymbol{\mu}_k + \mathbf{x}_i^{\mathsf{T}}\boldsymbol{\beta}, \boldsymbol{\Sigma}), \qquad \mathbf{Z}_{i\cdot} \mid \boldsymbol{\pi}_{i\cdot} \sim \mathcal{M}ult(1, \boldsymbol{\pi}_{i\cdot}). \quad (1)$$

The spatial structure is incorporated into the clustering probabilities in three steps. First, a multinomial stick-breaking representation [4] defines $\tilde{\pi}_{ik} = \pi_{ik}/(1 - \sum_{k'<k} \pi_{ik'})$. Second, $\psi_{ik} = \mathrm{logit}(\tilde{\pi}_{ik})$ is defined. Finally, for $k = 1, \ldots, K-1$, $\boldsymbol{\psi}_{\cdot k}$ is assumed to follow an intrinsic autoregressive (IAR) model with scale parameter τ_k.

A key feature of PERLA is its use of shrinkage priors to detect which outcomes show meaningful deviations from zero in specific clusters. In the spatial transcriptomic context, this corresponds to identifying genes that are up- or down-regulated in particular spatial domains, while genes showing no evidence of differential expression are shrunk towards zero. For each cluster k and gene j, the model imposes a global-local shrinkage prior on the cluster-specific intercepts:

$$\mu_{kj} \mid \phi, \zeta_j, \gamma_{kj} \sim \mathcal{N}(0, \phi\, \zeta_j\, \gamma_{kj}), \quad (2)$$

where $\sqrt{\phi}$, $\sqrt{\zeta_j}$, and $\sqrt{\gamma_{kj}}$ follow a truncated Cauchy distribution $\mathcal{C}_{+}(0,1)$. Formula (2) extends the horseshoe prior [1] by including three shrinkage parameters: the global factor ϕ controls shrinkage across all $\{\mu_{kj}\}$, the gene-specific factor ζ_j regulates the overall contribution of gene j to cluster identification, and the local factor γ_{kj} adjusts the influence of gene j within cluster k. Large values of ζ_j or γ_{kj} indicate strong contribution to cluster separation, while small values shrink the parameter toward zero (*e.g.*, [2]). The idea of allowing each (kj)-intercept to have a variance parameter representing the relevance of the j-th outcome in the k-th cluster was also considered, for instance, by [10].

3 Case Study: Mouse Olfactory Bulb

In this section, we show how to apply PERLA to a spatial trascriptomic dataset. We applied our method to a mouse olfactory bulb tissue sample [7]. Gene expression levels were obtained from thin tissue sections placed on arrays of poly(dT) probes with spatially resolved DNA barcodes, forming a grid of circular spots with a diameter of 100 μm. At each spot, the mRNA abundance of approximately 10–100 cells was measured.

3.1 Data Pre-processing

The original dataset comprises expression measurements for 16,218 genes across 260 spatial spots, as shown in the left panel of Fig. 1. Each interior pixel has eight neighbouring pixels, whereas border pixels have fewer neighbours depending on their location. More than 50% of the spots have eight neighbours, while only one spot has three neighbouring spots.

First, we removed genes expressed in fewer than 5% of spots, and then we applied the multinomial null model [8]. The resulting binomial deviance allows us to quantify genes that substantially deviate from the null model and are therefore highly variable. We initially selected the top 500 genes according to this criterion. To reduce the computational burden and facilitate interpretation, we subsequently focused on the 20 most spatially variable genes, identified using the `nnSVG` model [9]. The initial choice of retaining the top 500 highly variable genes has little practical impact on the final set, since spatially variable genes are typically also highly variable in expression. This preliminary filtering step was therefore primarily introduced to substantially reduce the computational cost of the analysis. Finally, if c_{ij} denotes the counts in pixel i for gene j, we defined $y_{ij} = \log(c_{ij} + 1) - \overline{\log(c + 1)}$, thus rescaling the log-transformed counts by their mean. This transformation allows us to interpret $\mu_{kj} > 0$ as indicating an abundance of expression of gene j in cluster k, and $\mu_{kj} < 0$ as indicating the *vice versa*.

3.2 Data Analysis

PERLA was estimated via MCMC using five parallel chains of 10,000 iterations each, with the first 5,000 iterations of each chain discarded as burn-in. We fixed $K = 4$ based on prior knowledge of the number of spatial domains in the region under study, and set $\tau_k = 1$ for all k, as the model has been shown to be robust to this choice (see [6], Sect. 4). To address the label switching problem, we applied both the PRA and ECR relabelling algorithms described by [5]. The two procedures produced identical permutations of the posterior samples and yielded the same final partition, confirming the stability of the clustering solution. Convergence of the MCMC chains was assessed using the posterior draws of $\{\mu_{kj}\}_{k=1,\ldots,4;\, j=1,\ldots,20}$. The estimated median Potential Scale Reduction Factor (PSRF) was 1.011. All simulations and data processing were performed

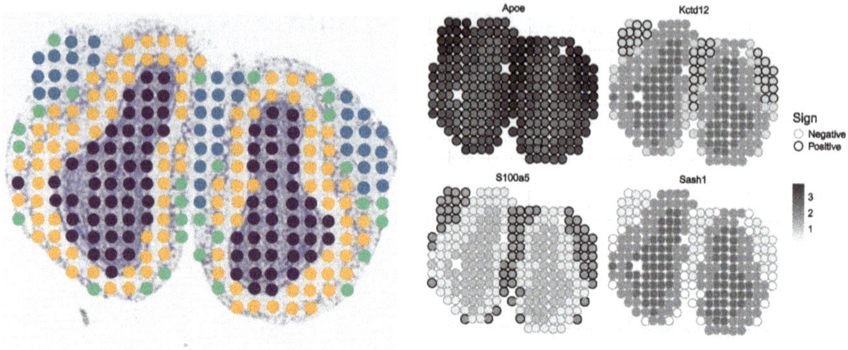

Fig. 1. Left: image of the analysed mouse olfactory bulb [7], with the 260 spots superimposed and grouped into four clusters. Right: estimated expression of four genes over the 260 spots.

on a Linux server equipped with an Intel Xeon Gold 6226R CPU (2.90 GHz) and 96 GB of RAM. Each MCMC chain required an average CPU time of 180 s on the specified hardware.

We estimated the model with $K = 4$, based on prior knowledge of the number of domains in this region. The estimated clusters are displayed in the left panel of Fig. 1. PERLA identifies four spatially distinct regions that align closely with known anatomical layers of the olfactory bulb. Cluster 1 (purple) corresponds to the granular cell layer (GCL), the innermost and most densely populated region, composed predominantly of inhibitory interneurons. Cluster 2 (blue) predominantly comprises cells from the olfactory nerve layer (ONL), the most superficial anatomical layer representing the entry point of sensory axons originating from the olfactory epithelium. Cluster 3 (green) captures the glomerular layer (GL), a key functional region where axons of olfactory sensory neurons form synapses within spherical neuropil structures. Finally, Cluster 4 (orange) maps to the mitral cell layer (MCL) and the outer plexiform layer (OPL), which together contain the principal output neurons responsible for transmitting processed sensory information to the olfactory cortex. The sharp spatial separation of these clusters highlights the ability of PERLA to recover the underlying anatomical organisation directly from gene expression data, without relying on prior histological segmentation.

Figure 2 shows the posterior distributions of the parameters μ_{kj} for $k = 1, \dots, 4$ clusters and $j = 1, \dots, 20$ genes. Several genes display substantial posterior mass both above and below zero, indicating that the selected genes contribute meaningfully to the identification of the spatial clusters. At the same time, a number of genes exhibit cluster-specific patterns, being either up-regulated or down-regulated in only a subset of clusters, while remaining close to the average expression level in others. To illustrate these heterogeneous expression patterns, we selected four genes displaying distinct spatial and cluster-specific behaviour, which are visualised in the right panel of Fig. 1. For each gene

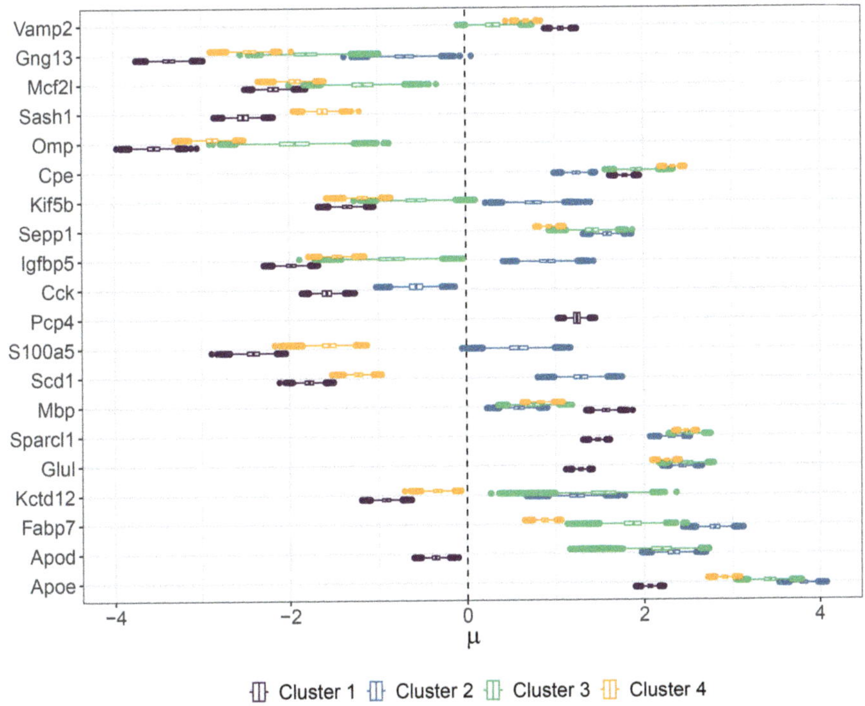

Fig. 2. Posterior distributions of the cluster intercepts μ_{kj} whose 99% posterior HPD intervals do not include zero. The colours in the plot resemble those used to denote the clusters in the left panel of Fig. 1.

j, spots assigned to cluster k by PERLA were coloured according to $\overline{\mu}_{kj}$, the posterior mean of the corresponding cluster-specific expression parameter.

The gene *Apoe* shows consistently high expression across the tissue, acting as a ubiquitous metabolic marker and reflecting the widespread distribution of glial cells involved in lipid transport. *Kctd12*, a key modulator of GABA_B receptor signalling, exhibits pronounced layer specificity, with down-regulation in Clusters 1–4 and up-regulation within the olfactory nerve layer (ONL) and the glomerular layer (GL). *S100a5*, a marker of mature sensory input, is up-regulated in the ONL and down-regulated in the central layers. Finally, *Sash1* displays a pattern complementary to that of *Kctd12*, with expression concentrated in the deeper neuronal strata, consistent with its proposed role as a scaffold protein in signalling pathways underlying the structural organisation of the inner layers of the olfactory bulb.

4 Conclusions

We have shown that PERLA can be effectively adapted to spatial transcriptomic data, allowing the identification of spatially coherent domains directly from gene expression values. By avoiding dimensionality reduction and leveraging shrinkage priors, the model highlights the genes that drive cluster formation, providing interpretable biological insight. Applied to the mouse cortical dataset, PERLA recovered known anatomical layers and revealed their characteristic gene signatures, demonstrating its potential as a flexible tool for the analysis of high-resolution spatial transcriptomics.

In our case study, no additional external information was available. Nevertheless, as shown in Formula (1), PERLA accommodates external covariates, extending its applicability to more complex scenarios, such as incorporating spot-level quality metrics or performing joint analyses across multiple samples.

References

1. Carvalho, C., Polson, N., Scott, J.: The horseshoe estimator for sparse signals. Biometrika, 465–480 (2010)
2. Denti, F., et al.: A horseshoe mixture model for Bayesian screening with an application to light sheet fluorescence microscopy in brain imaging. Ann. Appl. Stat. **17**, 2639–2658 (2023)
3. Emons, M., et al.: Harnessing the potential of spatial statistics for spatial omics data with pasta. Nucleic Acids Res. **53**, gkaf870 (2025)
4. Linderman, S., Johnson, M., Adams, R.: Dependent multinomial models made easy: stick-breaking with the Pólya-Gamma augmentation. Adv. Neural Inf. Process. Syst. **28** (2015)
5. Papastamoulis, P.: label. switching: an R package for dealing with the label switching problem in MCMC outputs. J. Stati. Softw. Code Snippets. **69**, 1–24 (2016)
6. Sottosanti, A., Bovo, E., Belloni, P., Boccuzzo, G.: Bayesian mapping of mortality clusters. Biostatistics. **26**, kxaf028 (2025)
7. Ståhl, P., Salmén, F., Vickovic, S., Lundmark, A., Navarro, J., Magnusson, J., et al.: Visualization and analysis of gene expression in tissue sections by spatial transcriptomics. Science **353**, 78–82 (2016)
8. Townes, F., Hicks, S., Aryee, M., Irizarry, R.: Feature selection and dimension reduction for single-cell RNA-Seq based on a multinomial model. Genome Biol. **20**, 295 (2019)
9. Weber, L., Saha, A., Datta, A., Hansen, K., Hicks, S.: nnSVG for the scalable identification of spatially variable genes using nearest-neighbor Gaussian processes. Nat. Commun. **14**, 4059 (2023)
10. Yau, C., Holmes, C.: Hierarchical Bayesian nonparametric mixture models for clustering with variable relevance determination. Bayesian Anal. **6**, 329–351 (2011)
11. Zhao, E., Stone, M., Ren, X., Guenthoer, J., Smythe, K., Pulliam, T., et al.: Spatial transcriptomics at subspot resolution with BayesSpace. Nat. Biotechnol. **39**, 1375–1384 (2021)

Tree-Of-Trees Inference: Visualization and Analysis of Phylogenetic Tree Sets

Nadia Tahiri[1(✉)] and Marc Le Pouliquen[2]

[1] Département d'informatique, Université de Sherbrooke, Sherbrooke, QC, Canada
Nadia.Tahiri@USherbrooke.ca
[2] LabSTICC, Université de Bretagne Occidentale, Brest, France
marc.lepouliquen@univ-brest.fr

Abstract. Comparing multiple phylogenetic trees is essential for studying variability in evolutionary histories, although consensus approaches can obscure structural diversity. This work proposes a more explicit representation of relationships among trees. Pairwise topological distances are computed using Robinson–Foulds and quartet metrics, and a weighted least squares criterion reconstructs a tree whose branch lengths approximate these distances, yielding a structure defined on the set of trees. The resulting tree separates major topology classes, including balanced, comb-like, and unresolved forms, and places the star tree centrally. Internal nodes capture consensus patterns within subsets of trees. Applied to biological data, the method organizes gene trees into coherent groups while revealing dispersion and overlap not captured by standard clustering. Overall, this tree-based representation provides an interpretable and informative view of relationships among phylogenetic trees.

Keywords: Consensus trees · Hierarchical clustering · Phylogenetic trees · Algorithms · Metrics

1 Introduction

To describe the evolutionary relationships among biological, textual, or other entities, phylogenetic trees can be used. In many cases, uncertainty about evolutionary history leads to the production of multiple competing trees. A central question then becomes: how can these trees be compared, summarized, and interpreted? In this context, measures of distance between phylogenetic trees play a fundamental role. They allow, among other things, the quantification of divergences between trees and the grouping of similar trees. These distances make it possible to construct a metric space of trees (or tree-space [1]), in which the relationships between trees can be represented and analyzed.

In this work, we propose formalizing and visualizing this space through a central object: the phylogenetic tree of phylogenetic trees. This meta-structure allows for the exploration of relationships among trees themselves and provides a better understanding of the chosen tree distance. An initial tree-of-trees is

P. Brito et al. (Eds.): IFCS 2026, STUDIES CLASS, pp. 259–267, 2026.
https://doi.org/10.1007/978-3-032-32009-4_32

constructed from the set of rooted trees with four leaves, whether fully resolved or not. This provides an overview of the diversity of possible trees and highlights natural clusters of trees sharing common properties under the chosen distance. The relevance of this approach is further illustrated through a real application, showing how the tree-of-trees representation enhances the interpretation of phylogenetic data in biological contexts. The proposed structure differs from classical approaches by explicitly constructing a phylogenetic structure on the entire set of trees themselves. This results in a hierarchical representation that preserves pairwise relationships between trees while remaining interpretable as a phylogenetic object. Unlike consensus trees or supertree methods, the tree of trees offers a comprehensive and structured view that allows for the identification of transition topologies that is, trees that share topological properties with several groups (partially resolved or balanced trees).

2 Definitions

2.1 Phylogenetic Tree Comparison Metrics

Several metrics exist to compare phylogenetic trees, some of which rely solely on tree topology, others incorporate branch lengths, and still others combine both topology and branch lengths. These metrics can also vary depending on different criteria, such as consideration of internal nodes, tree rooting, and whether they are defined on the same set of species. In this study, we restrict our analysis to rooted trees defined on the same set of species, focusing exclusively on topological distances.

Definition 1 (Rooted Phylogenetic Tree). A rooted phylogenetic tree is a tree $T = (V, E)$ with a distinguished node $r \in V$ called the root, where each node $v \in V\backslash\{r\}$ has exactly one parent. The leaves of T, denoted $L(T) \subseteq V$, correspond to a set of species. Each edge $(u, v) \in E$ may have an associated non-negative length $\ell(u, v) \geq 0$, representing the evolutionary distance between nodes u and v.

Definition 2 (Robinson-Foulds [11]). Let T_1 and T_2 be two phylogenetic trees. The Robinson-Foulds distance between T_1 and T_2, denoted $d_{\mathrm{RF}}(T_1, T_2)$, is defined as the number of differing elements in the partitions of the leaf set induced by the edges of the two trees:

$$d_{\mathrm{RF}}(T_1, T_2) = \frac{(|P(T_1)\backslash P(T_2)| + |P(T_2)\backslash P(T_1)|)}{2}, \tag{1}$$

where $P(T_1)$ denotes the set of partitions of the leaf set induced by the edges of tree T_1, and similarly for $P(T_2)$ with tree T_2.

Definition 3 (Quartet [3]). Let T_1 and T_2 be two phylogenetic trees. The quartet distance between T_1 and T_2, denoted $d_{\mathrm{Q}}(T_1, T_2)$, is defined as the number of distinct sets of four species (quartets) that are resolved differently in the two trees:

$$d_Q(T_1, T_2) = \sum_{Q \in \mathcal{Q}} \mathbf{1}_{\{P(Q,T_1) \neq P(Q,T_2)\}}, \tag{2}$$

where $\mathcal{Q}$ is the set of all quartets, equivalent to $\binom{n}{4}$ with $n = |L(T_1)| = |L(T_2)|$, and $\mathbf{1}_{\{\cdot\}}$ is the indicator function that equals 1 if the partitions of Q in T_1 and T_2 differ, and 0 otherwise.

The metrics defined in this section and applied in this study were implemented in Python (v. 3.11) in Jupyter Notebook format and are freely available on the GitHub repository of the project.

2.2 Consensus Tree Inference Methods

Definition 4 (Consensus Tree). Let $\mathcal{T}$ be a set of k phylogenetic trees, denoted $\mathcal{T} = \{T_1, \ldots, T_k\}$, defined on the same set of species $\mathcal{L} = L(T_1) = \cdots = L(T_k)$. A *consensus tree* of $\mathcal{T}$ is a phylogenetic tree T_c such that $L(T_c) = \mathcal{L}$ and that topologically summarizes the set of trees in $\mathcal{T}$.

Given a set of phylogenetic trees, several methods exist to define a consensus tree [2,6]. The most well-known types of consensus trees are the *strict consensus tree* [13,17], the *majority-rule consensus tree* [9,16], the *extended majority-rule consensus tree* (also called the greedy consensus tree in the literature) [2,6], and the *frequency difference consensus tree* [4,5,7] (called the plurality consensus tree in [15]). All consensus tree inference methods are based on the topology of the input phylogenetic trees. They primarily aim to represent each clade present in these trees. Let $|S_u|$ denote the number of trees supporting clade C_u; the consensus methods are defined as follows:

Definition 5 (Strict Consensus [13]). A strict consensus tree retains only the clades present in all trees of the collection, i.e., clades C_u such that $|S_u| = k$. This method often produces partially resolved trees, which may be minimally informative.

Definition 6 (Majority-Rule Consensus [9]). A majority-rule consensus tree contains all clades appearing in more than half of the trees, i.e., those for which $|S_u| > k/2$. This often results in a partially resolved tree.

Definition 7 (Extended Majority-Rule Consensus [2]). An extended majority-rule consensus tree includes all majority clades as well as additional compatible clades, added in decreasing order of frequency ($|S_u|$). This commonly used approach generally produces more resolved trees.

Definition 8 (Frequency Difference Consensus [5]). A frequency difference consensus tree contains each clade C_u supported by a greater number of trees than any clade incompatible with C_u.

3 Application to the Set of Rooted Four-Leaf Trees

Description of Phylogenetic Tree. We generated the complete set of rooted four-leaf trees, all rooted at node 0, including both binary and non-binary trees. This procedure yielded 20 distinct topologies, numbered 1 to 20 (see Fig. 1a, where the numbers appear on the right). To simplify processing, all branch lengths were initialized to 1.

Among these 20 phylogenetic trees, certain groupings warrant further analysis, which may enrich the interpretation of our results. Of the 20 topologies, five are unresolved (numbers 16–20), while the other 15 are fully resolved. One of the unresolved trees, tree 16, is completely unresolved: the four leaves emerge directly from the root, forming a star-like structure.

Regarding the 15 resolved trees, three of them (numbers 1, 2, and 9) are balanced, meaning that the heights of the subtrees originating from the root are similar. Note that only the star tree (number 16) is obtained using the different consensus methods.

Subsequently, the 20 trees were encoded in `Newick` format, and pairwise distances were computed. A tree-of-trees was then constructed from this distance matrix using the Weighted Least Squares (WLS) method of Makarenkov and Leclerc [8]. The weight associated with each pair of trees (T_i, T_j) is defined as:

$$W_{T_i,T_j} = \frac{1}{D^p_{T_i,T_j}}, \tag{3}$$

where D_{T_i,T_j} denotes the distance between trees T_i and T_j, and p is an adjustable parameter.

The method fits a tree T' whose path distances $d_{ij}(T')$ approximate the observed distances D_{T_i,T_j} by minimizing:

$$OF = \min_{T'} \left(\sum_{i<j} W_{T_i,T_j} \left(D_{T_i,T_j} - d_{ij}(T')\right)^2 \right). \tag{4}$$

Unlike clustering approaches such as hierarchical clustering, which group trees without explicitly reconstructing a metric structure, WLS produces a phylogenetic tree whose branch lengths directly reflect the input distances. In addition, internal nodes can be interpreted in relation to consensus structures of subsets of trees, which provides an interpretable representation consistent with the nature of the data. From a computational perspective, the WLS approach belongs to the class of distance-based phylogenetic reconstruction methods. Its complexity is polynomial in the number of input trees, typically dominated by the optimization procedure used to fit the tree metric. In practice, the implementation we used relies on established algorithms that efficiently approximate the optimal solution, making the approach tractable for moderate to large sets of trees. This computational behavior is comparable to that of other distance-based methods such as Neighbor-Joining, while providing additional flexibility through weighting.

In this study, the parameter p was set to 0, which corresponds to assigning a uniform weight to all tree pairs and leads to a global fit of the distance matrix. More generally, the parameter p controls the influence of distances: for $p > 0$, smaller distances are emphasized, favoring the preservation of local relationships between similar trees, whereas larger distances are downweighted. This parameter therefore governs a trade-off between local and global fidelity of the reconstructed structure. A systematic analysis of the sensitivity to p is left for future work. Optimization was performed globally using the Kimura two-parameter (K2P) substitution model. Positions with missing data were excluded, and a bootstrap of 100 replicates was conducted to assess the robustness of the results.

Inference of Phylogenetic Trees. From the 20 tree topologies considered (see Sect. 3), we applied the Robinson–Foulds distance and the quartet metric (see Sect. 2.1) to construct a tree representing the relationships among the trees in the set. Both metrics yielded identical results. The tree-of-trees obtained from the Robinson–Foulds distance is shown in Fig. 1a.

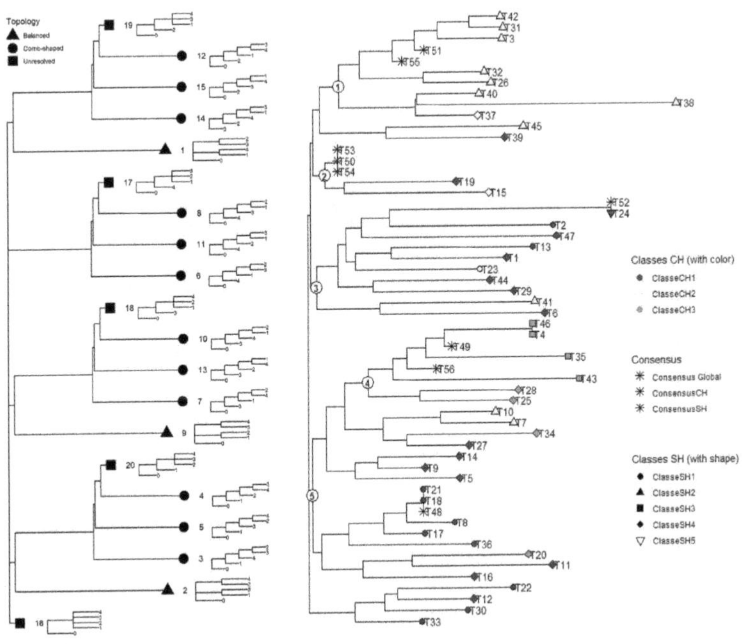

(a) (b)

Fig. 1. (a) Tree-of-trees constructed using the Robinson–Foulds distance from 20 rooted phylogenetic trees on four leaves, grouped into comb-like (black circle), balanced (black triangle), and unresolved (black square) classes. (b) Tree-of-trees for archaeal data constructed using the Robinson–Foulds distance from 56 phylogenetic trees on 17 leaves, grouped into 5 SH classes (colors) and 3 CH classes (shapes) [14].

This tree reflects the structure of the original topologies: it is divided into four main groups of similar trees. Three of these groups each contain five trees, while the fourth contains four. Within each group, the comb-like trees[1] can be interpreted as variants resulting from permutation of the most distal subtree relative to the root, preserving the partitions closest to it.

This observation was also made for three of the four groups for the balanced tree. Additionally, the star tree (which is the consensus tree according to all methods) logically appears at the root of our tree-of-trees.

In general, the internal nodes of the tree-of-trees (which is an unresolved tree) also correspond to consensus trees of the leaves they group.

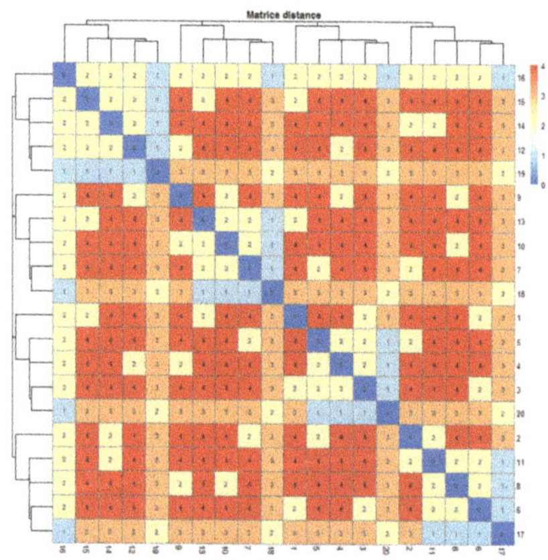

Fig. 2. Robinson–Foulds distance matrix computed between all pairs of trees from a set of 20 rooted phylogenetic trees defined on the same four taxa. Rows and columns of the matrix were reordered by simultaneous biclustering to highlight similarity structures between trees.

To facilitate interpretation of the reconstructed tree, the distance matrix was reordered by joint permutation of rows and columns to reflect the structure of the resulting tree (see Fig. 1a). The result is shown in Fig. 2. The comb-like tree groupings observed in the matrix correspond to those identified in the tree-of-trees. Similarly, the unresolved trees (except for the star tree) belong to the same groups in both representations. However, among the resolved trees, only tree 9 occupies a position consistent with the tree-of-trees; trees 1 and 2 differ in their placement. This divergence can be explained by the proximity of the balanced trees to multiple groups in the distance matrix.

4 Biological Data

[1] This binary (usually rooted) tree topology is highly unbalanced: at each node, one branch is a leaf, and the other recursively contains the remaining leaves.

Description. We studied the evolution of 47 ribosomal proteins from 14 archaeal organisms, including 11 species of Euryarchaeota and 3 species of Crenarchaeota. These data, initially analyzed by Matte-Tailliez *et al.* [10], were used to infer a single species phylogenetic tree based on concatenated protein sequences. However, the evolution of each protein can be represented by a distinct phylogenetic tree. Comparative analysis of these trees allowed identification of divergent evolutionary trajectories, suggesting horizontal gene transfer events or other evolutionary processes. The 47 phylogenetic trees were generated using `PHYML` from the complete sequence alignments. These trees, available in `Newick` format on the GitHub repository of the project, were compared using the two metrics described in Sect. 2.1. This analysis produced a square, symmetric distance matrix among the trees. The analysis in [14] identified two types of classes. The SH classes were obtained by applying the k-medoids algorithm to the 47 phylogenetic trees using the RF distance, with an optimal number of five classes determined by maximizing the Silhouette cluster validity index (SH) [12]. In parallel, the optimal number of CH classes was set to three based on the Calinski–Harabasz index (CH) [18], which measures the ratio of inter- to intra-cluster dispersion.

Results. For the experiment, the majority-rule consensus trees of each SH and CH class were added, along with the global consensus, resulting in a total of 56 trees (47 + 5 CH + 3 SH + 1 Global). Figure 1b visualizes the tree-of-trees, including classes and consensus trees. SH classes are color-coded, while CH classes are distinguished by shape (e.g., circle, square, triangle). Consensus trees are identified by both color and shape.

SH Classes. SH2 (blue) is grouped within the subtree obtained from node ④. Consensus T49 is closest to the root of this subtree. SH4 (purple) and SH1 (red) are mostly grouped in subtrees obtained from nodes ① with consensus T49, and ⑤ in the lower part with consensus T48, respectively. SH5 contains only tree T24 attached to its consensus T52. SH3, the largest class (20 trees), is dispersed throughout the tree, and its consensus T50 is identical to the global consensus T53.

CH Classes. CH2 (triangle) and CH3 (square) are mostly grouped in subtrees obtained from nodes ① with consensus T55, and ④ with consensus T56, respectively. CH1 (circle) is the largest class (25 trees) and is mainly dispersed in subtrees from ③ and the lower part of ⑤. Its consensus T55 is also identical to the global consensus.

5 Discussion

Set of Rooted Four-Leaf Trees. The study of all 20 possible topologies of trees shows that the resulting tree of trees corresponds to the structure that would naturally be constructed from the topological relationships between the trees. It reflects a consistent canonical classification, clearly distinguishing between resolved, unresolved, and intermediate forms, with the star tree logically occupying a central position. The observed groupings accurately reflect the proximity induced by the metrics, and the internal nodes correspond to consensus trees of the subsets considered. In particular, balanced and comb-shaped topologies form well-defined clusters, while unresolved trees act as transitional configurations connecting these groups. In contrast to consensus methods, the tree-of-trees representation allows for the identification of groups of similar trees, the detection of atypical topologies, and the visualization of gradual transitions between different tree structures.

Biological Data. Applied to biological data, this approach corroborates previous classifications based on ribosomal proteins and offers a comprehensive and structured view of all phylogenetic trees. The Fig. 1b shows a hierarchical tree of taxa T1T56 structured into five main clusters. The topology indicates heterogeneous branch lengths, with some groups compact and others more dispersed, suggesting variable internal similarity across clusters. Several taxa appear on long branches, indicating outlier behavior relative to neighboring groups. The SH classes (colors) and CH classes (shapes) show partial concordance with the tree structure. Some subtrees are dominated by a single SH color or CH shape, but neither classification is strictly monophyletic across the full topology. CH shapes appear more locally consistent in several subtrees, whereas SH colors are more frequently intermixed across adjacent branches. Consensus markers are located near internal nodes associated with relatively homogeneous subtrees. Their distribution indicates agreement within subsets of taxa but limited overlap between SH and CH consensus regions, which reflects methodological differences in class assignment. The tree of trees thus appears to be a relevant tool for summarizing and interpreting the diversity of evolutionary histories.

6 Limitations

This paper proposes an approach for visualizing and analyzing a set of phylogenetic trees through the construction of a tree-of-trees. By representing the topological distances between trees within a tree structure, one obtains a visualization of the proximities and natural groupings of trees, whether derived from simulations or real biological data. The inclusion of consensus trees confirms the special role of these trees in the visualization. Looking ahead, this approach could be extended to the complete set of five-leaf trees and to other sets of artificial trees where branch lengths are varied artificially. The construction of the distance between trees could also take branch lengths into account, allowing exploration of the impact of these variations on the resulting tree-of-trees. The sets of trees to be visualized could also be defined on partially overlapping sets of leaves.

References

1. Billera, L.J., Holmes, S.P., Vogtmann, K.: Geometry of the space of phylogenetic trees. Adv. Appl. Math. **27**(4), 733–767 (2001)

2. Bryant, D.: A classification of consensus methods for phylogenetics. DIMACS Ser. Discrete Math. Theoret. Comput. Sci. **61**, 163–184 (2003)
3. Estabrook, G.F., McMorris, F.R., Meacham, C.A.: Comparison of undirected phylogenetic trees based on subtrees of four evolutionary units. Syst. Zool. **34**(2), 193–200 (1985)
4. Goloboff, P.A.: Minority rule supertrees? MRP, compatibility, and minimum flip may display the least frequent groups. Cladistics **21**(3), 282–294 (2005)
5. Goloboff, P.A., Farris, J.S., Källersjö, M., Oxelman, B., Ramírez, M.J., Szumik, C.A.: Improvements to resampling measures of group support. Cladistics **19**(4), 324–332 (2003)
6. Jansson, J., Shen, C., Sung, W.-K.: Improved algorithms for constructing consensus trees. J. ACM (JACM) **63**(3), 1–24 (2016)
7. Jansson, J., Sung, W-K., Tabatabaee, S.A., Yang, Y.: A faster algorithm for constructing the frequency difference consensus tree. In: 41st International Symposium on Theoretical Aspects of Computer Science (STACS 2024), pp. 43–1. Schloss Dagstuhl–Leibniz-Zentrum für Informatik (2024)
8. Makarenkov, V., Leclerc, B.: An algorithm for the fitting of a tree metric according to a weighted least-squares criterion. J. Classif. **16**(1), 3–26 (1999)
9. Margush, T., McMorris, F.R.: Consensus n-trees. Bull. Math. Biol. **43**(2), 239–244 (1981)
10. Matte-Tailliez, O., Brochier, C., Forterre, P., Philippe, H.: Archaeal phylogeny based on ribosomal proteins. Mol. Biol. Evol. **19**(5), 631–639 (2002)
11. Robinson, D.F., Foulds, L.R.: Comparison of phylogenetic trees. Math. Biosci. **53**(1–2), 131–147 (1981)
12. Rousseeuw, P.J.: Silhouettes: a graphical aid to the interpretation and validation of cluster analysis. Comput. Appl. Math. **20**, 53–65 (1987)
13. Sokal, R.R., Rohlf, F.J.: Taxonomic congruence in the Leptopodomorpha reexamined. Syst. Zool. **30**(3), 309–325 (1981)
14. Tahiri, N., Willems, M., Makarenkov, V.: A new fast method for inferring multiple consensus trees using k-medoids. BMC Evol. Biol. **18**, 1–12 (2018)
15. Velasco, J.D.: The foundations of concordance views of phylogeny. Philos. Theory, Pract. Biol. **11**(20) (2019)
16. Wilkinson, M.: Majority-rule reduced consensus trees and their use in bootstrapping. Mol. Biol. Evol. **13**(3), 437–444 (1996)
17. Wilkinson, M., Thorley, J.L.: Efficiency of strict consensus trees. Syst. Biol. **50**(4), 610–613 (2001)
18. Caliński, T., Harabasz, J.: A dendrite method for cluster analysis. Commun. Stat. Theor. Methods **3**(1), 1–27 (1974)

A Forward Search Framework for Outlier Detection in Ordinal Data

Tamara Gaia Vasiljev(✉)

University of Milano-Bicocca, Milan, Italy
t.vasiljev@campus.unimib.it

Abstract. Outlying response patterns pose a serious challenge for the analysis of ordinal data in psychological and educational assessments, where even mild contamination can distort parameter estimation and model fit. This paper investigates the use of the Forward Search as a robust diagnostic tool for detecting atypical observations in ordinal Item Response Theory models. Forward plots of goodness-of-fit statistics are used to identify aberrant response patterns and departures from the assumed model. Methods for initializing, progressing, and monitoring the search are described, with individual likelihood contributions guiding the inclusion of observations. Simulation envelopes are constructed to assess whether changes in the monitored statistics can be attributed to random variation. A simulation study under partial misspecification illustrates the effectiveness of the proposed approach in identifying clusters of outlying response patterns, and an empirical application to a Big Five personality dataset confirms its practical utility.

Keywords: Forward Search · outlier detection · robust statistics · Item Response Theory · ordinal data · careless respondents · random answering · psychometrics

1 Introduction

Ordinal data are widely used in the psychological, educational, biomedical and social sciences, where responses are commonly collected through Likert-type or rating scales. In applied settings, however, the observed data may contain various forms of contamination, or the assumed statistical model may be misspecified [17]. In questionnaire data, typical examples include inattentive or careless responding, random answering, fatigue effects, and item misinterpretation [8,11]. Detecting outliers in ordinal data is particularly challenging. Unlike continuous variables, ordinal responses are discrete and ordered, making the notion of an outlier less straightforward and often obscured by the limited support of the response scale [13]. Nevertheless, aberrant ordinal response patterns can substantially bias parameter estimates, distort goodness-of-fit measures, and ultimately lead to incorrect substantive conclusions [18]. These issues are especially critical in applied contexts, where distorted measurement may have practical

P. Brito et al. (Eds.): IFCS 2026, STUDIES CLASS, pp. 268–275, 2026.
https://doi.org/10.1007/978-3-032-32009-4_33

consequences, such as misclassification of individuals or erroneous policy decisions [5].

As a consequence, there is a strong need for robust diagnostic tools capable of identifying atypical response patterns and mitigating their impact on statistical modeling of ordinal data. Several robust approaches have been proposed to address outlying or aberrant response patterns in this setting. Early contributions introduced weighted likelihood methods to down-weight atypical observations [12]. Subsequent work proposed Item Response Theory (IRT) based methods for detecting person-level outliers using likelihood-based statistics [6]. More recently, iterative purification approaches and methods designed to remain robust under partial model misspecification have been proposed to address aberrant or non-effortful responses, including procedures based on response time residuals and estimators that are stable when only an unknown fraction of observations deviates from the assumed model [9,16]. Most of these methods rely on estimating the model using the full dataset before identifying anomalous observations. When contamination is present, this strategy may lead to distorted parameter estimates, thereby reducing the reliability of diagnostic conclusions. This limitation motivates the use of sequential, forward procedures for robust inference in ordinal models.

Within this perspective, the Forward Search (FS) algorithm has emerged in the robust statistical literature as a powerful diagnostic tool for outlier detection. Originally introduced by Atkinson and Riani [1], the FS proceeds by fitting a model to a carefully selected initial subset of the data and then progressively enlarging this subset while monitoring diagnostic quantities. Unlike standard backward approaches—such as deletion diagnostics or influence measures computed on the full sample—the FS is specifically designed to mitigate the problems of masking and swamping, which can prevent the reliable identification of outliers when contamination is present [13]. The effectiveness of the FS has been demonstrated across a wide range of statistical models, including regression and multivariate analysis [2], as well as factor analysis for continuous and binary data [10,11].

Despite these developments, the application of the FS to ordinal response models remains largely unexplored. In this work, the FS is embedded in the IRT framework for ordinal data by adopting Samejima's Graded Response Model (GRM) [15]. The GRM extends the two-parameter logistic model (2PL) to ordered response categories through a cumulative logit formulation. Specifically, the probability that subject i responds with category y to item j is defined as

$$p_{jy}(\theta_i) = p^*_{jy}(\theta_i) - p^*_{j,y+1}(\theta_i),$$

where

$$p^*_{jy}(\theta_i) = \frac{1}{1 + \exp\{-a_j(\theta_i - d_{jy})\}},$$

θ_i denotes the latent trait, a_j the item discrimination parameter, and d_{jy} the threshold parameters. The estimation of the model is typically based on marginal maximum likelihood, relying on Gauss–Hermite quadrature to approximate the required integrals over the latent trait [14].

Within this framework, the FS is used as a diagnostic device to identify aberrant response patterns by monitoring model fit as observations are sequentially incorporated. Goodness-of-fit is assessed using limited-information statistics for ordinal IRT models, allowing departures from the assumed model to be detected in a robust and data-driven manner.

2 Forward Search

The FS starts from an initial subset of observations assumed to be free of contamination. It proceeds by progressively adding the closest observation to the currently estimated model and refitting it to the augmented subset, until all units are included. After each inclusion step, goodness of fit statistics are evaluated. While genuine observations aren't expected to produce changes in the chosen statistics, outliers do. The FS can be described by three main steps: (i) choosing the initial subset; (ii) progressing in the search; (iii) monitoring the statistics.

Step 1: Choosing the Initial Subset
Let $\mathbf{Y}$ denote the data matrix of order $n \times k$, collecting ordinal answers to k items in a survey from a sample of n subjects. An initial subset of size $g < n$ is defined as $S_g = (\mathbf{y}_{i_1}, \ldots, \mathbf{y}_{i_g})$, where $\mathbf{y}_i$ denotes the ith row of $\mathbf{Y}$ and $1 \leq i_1 < \cdots < i_g \leq n$. The size g of the initial subset should be large enough to ensure stable estimation, but sufficiently small to limit the inclusion of contaminated observations [1]. Since, in practice, an exhaustive search over all possible subsets of size g is infeasible, the initial subset is selected from a finite number H of randomly generated subsets of size g as the one providing the best model fit.

In the present work, model fit is assessed using the C_2 statistic proposed by [4], a limited-information goodness-of-fit measure based on first- and second-order marginal information. Let $\widehat{\boldsymbol{\sigma}}$ and $\boldsymbol{s}$ denote, respectively, the vector of model-implied first- and second-order marginal quantities and the corresponding observed proportions. The C_2 statistic is defined as

$$C_2 = N(\widehat{\boldsymbol{\sigma}} - \boldsymbol{s})' \widehat{\boldsymbol{\Omega}} (\widehat{\boldsymbol{\sigma}} - \boldsymbol{s}),$$

where N is the sample size and $\widehat{\boldsymbol{\Omega}}$ is a weight matrix accounting for the covariance structure of the marginal residuals. Under the null hypothesis of correct model specification, C_2 is asymptotically chi-squared with $q - d$ degrees of freedom, where q is the number of first- and second-order marginal quantities and d is the number of estimated model parameters [4].

At the beginning of the search, the selected subset S_g initializes the *basic set*, while the remaining $n - g$ observations form the *non-basic set*. The basic and non-basic sets are mutually exclusive throughout the FS.

Step 2: Progressing in the Forward Search
Starting from the initial basic set, the FS proceeds by sequentially updating the basic and non-basic sets and re-estimating the model at each step, until all observations are included. At each iteration, the model is fitted on the current

basic set, and observations in the non-basic set are ranked by their plausibility under the fitted model. The most likely observation is then added to the basic set.

Here, plausibility is assessed through the individual likelihood contribution which, for subject i, is defined as

$$LC_i = \sum_{j=1}^{k} \sum_{c=0}^{C_j} I(Y_{ij} = c) \log[P(Y_{ij} = c \mid \theta_i, a_j, d_j)],$$

where Y_{ij} is the response of subject i on item j, c indexes the ordered categories, and $I(Y_{ij} = c)$ is an indicator function. Observations with higher LC_i are more consistent with the current model and enter the basic set early, whereas poorly fitted patterns with low LC_i typically enter in the final steps.

Step 3: Monitoring the Forward Search

Throughout the progression of the FS, diagnostic quantities are monitored as functions of the search step. These include parameter estimates, individual likelihood contributions, and global goodness-of-fit measures. Here, the C_2 statistic is computed at each step of the search and used to assess the stability of the model fit as the basic set expands. Since the distribution of the test statistics is unknown for data-dependent subsets, direct reference to asymptotic distributions is not appropriate within the FS framework [3]. To address this issue, simulation envelopes are constructed under the null hypothesis of clean data. Specifically, envelopes are obtained via parametric bootstrapping. First, B datasets of the same size as the original sample are generated under the model assumed for clean data. Then, the FS is applied to each dataset under the same conditions (initial subset size and search progression). This yields, at each step h, a simulated distribution of B values of the C_2 statistic. Selected quantiles (e.g., 0.025 and 0.975, as in Fig. 1) are then used to define the envelopes at each h. These allow assessment of whether changes in the statistic are compatible with random variation or indicate departures from the model. Observations entering at steps where the statistic exceeds the envelope are flagged as potentially atypical, providing a principled tool for outlier detection within the sequential FS.

3 Simulation Study

The present simulation study investigates the ability of the FS to detect individuals who adopt atypical response strategies in questionnaire data. Specifically, attention is focused on a *partial misspecification* scenario [16], in which only a subset of items deviates from the assumed latent structure. This form of misspecification reflects realistic situations in which respondents progressively lose attention toward the end of a survey. As a consequence, responses to the final items may no longer be driven by the intended latent traits, but instead follow erratic or quasi-random patterns, while earlier items remain well explained by the model.

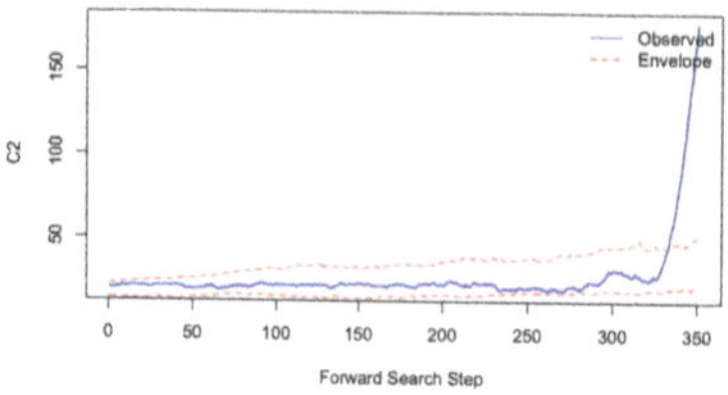

Fig. 1. Forward plot of the C_2 statistics (blue line) together with its 2.5 % and 97.5% simulation envelopes (red dashed lines). (Color figure online)

Table 1. Observed response patterns for the last steps of the FS.

step	Y_1	Y_2	Y_3	Y_4	Y_5	Y_6	Y_7	Y_8	Y_9	Y_{10}	label
341	3	1	2	3	2	2	4	0	4	0	478
342	2	1	2	2	0	1	4	0	4	0	486
343	2	1	1	3	2	1	4	0	4	0	481
344	1	2	1	2	2	0	4	0	4	0	496
345	1	3	3	2	1	2	4	0	4	0	500
346	3	1	1	1	1	3	4	0	3	0	498
347	4	3	2	4	4	3	4	0	4	0	490
348	1	3	1	1	0	2	4	0	4	0	492
349	2	4	2	1	3	1	4	0	4	0	483
350	2	0	3	0	1	1	4	0	4	0	482

Clean data are generated for $n = 500$ subjects and $k = 10$ ordinal items, each measured on a 5-point Likert scale with categories $\{0, 1, 2, 3, 4\}$. Latent continuous responses follow a multivariate normal distribution with zero mean vector and equicorrelated covariance matrix, with unit variances and pairwise correlations $\rho = 0.5$. Observed ordinal responses are obtained by discretizing the latent variables using fixed thresholds (-1.5, -0.5, 0.5, 1.5), resulting in approximately balanced category frequencies. Partial misspecification is introduced by contaminating the last four items for 5% of the observations. Specifically, the last 25 units ($i = 476, \ldots, 500$) have their responses to items 7–10 replaced by values generated independently from a multivariate normal distribution with mean $(2, -2, 2, -2)$ and variance 0.2, while all other items follow the clean generating mechanism. This creates localized deviations with an overrepresentation of extreme category combinations in the contaminated items, while the dependence structure of the first six items remains intact.

The FS is initialized by randomly drawing 1000 candidate subsets from the full dataset and selecting, among them, the subset that minimizes the C_2 goodness-of-fit statistic to obtain the initial subset. With items having five ordered response categories, a relatively large initial subset ($g = 150$) was chosen to ensure a sufficiently rich set of response patterns, yielding $C_2 = 19.92$ with a corresponding p-value of 0.98, which indicates a good model fit at the initial step. Importantly, no contaminated observations were included in this subset. At each step of the algorithm, one observation is added to the basic set according to its plausibility with the fitted model, as measured by its individual likelihood contribution. Monitoring is performed using the C_2 goodness-of-fit statistic. As shown in Fig. 1, the model fit remains largely stable throughout the search as clean observations are progressively added to the basic set. A noticeable deterioration occurs only in the final steps: around step 333, C_2 rises sharply with the inclusion of contaminated response patterns, marking the first departure of the observed trajectory beyond the 97.5% simulation envelope. According to the FS rationale [3], such departures signal behavior unlikely under the assumed model, and the observations entering at these stages are classified as outliers. The final units form a distinct cluster of contaminated patterns; the last ten, reported in Table 1, correspond to the injected contamination.

4 Empirical Application

The proposed method is illustrated using an empirical psychometric dataset based on the Big Five personality traits.[1] The data consist of self-reported Likert-type items rated on a five-point ordinal scale. Each dimension includes ten items formulated in alternating directions, so that consistent responses tend to exhibit a regular pattern (e.g., lowhighlow), while deviations may indicate inattentive behavior.

In this study, we focus on the ten items associated with the extraversion dimension, assuming a unidimensional latent structure consistent with the GRM. A subset of 500 respondents is selected to obtain a manageable sample for analysis. To assess the proposed diagnostic procedure under realistic conditions, the sample includes a proportion of 5% of response patterns deviating from the dominant structure, reflecting potential inattentive responding.

The FS is performed under the same setting adopted in the simulation study, with the exception that observations are allowed to enter and exit the basic set at each step. This modification accounts for the possibility that the initial subset may include atypical observations, allowing the procedure to progressively refine its composition. Figure 2 displays the polychoric correlation matrices computed on the initial subset (left panel) and on the full dataset (right panel). In the initial subset, the correlation structure clearly reflects the expected pattern induced by items formulated in alternating directions, resulting in a pronounced checkerboard structure. In contrast, this pattern appears substantially attenuated when computed on the full dataset, suggesting the presence of response irregularities that obscure the underlying latent structure.

Further insight is provided by monitoring the C_2 statistic along the search (Fig. 3): an initial decrease is observed at the early stages of the search, indicating an improvement in model fit due to the exclusion of observations with low likelihood contributions. For most of the search, the curve of the C_2 remains relatively stable, suggesting that the observations entering the basic set are compatible with the fitted model. In the final steps, a sharp increase in the statistic is observed, corresponding to the inclusion of response patterns that deviate from the dominant structure. Table 2 reports the observations entering the basic set in the last ten steps of the search. These units are characterized by inconsistent response patterns and are identified as potentially atypical.

[1] The dataset contains 1,015,342 questionnaire responses collected online by Open Psychometrics and is freely available at https://www.kaggle.com/datasets/tunguz/big-five-personality-test. It is based on the IPIP Big-Five Factor Markers [7].

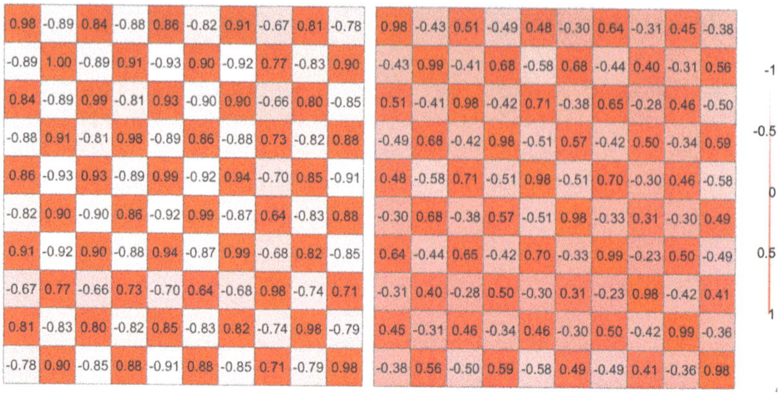

Fig. 2. Polychoric correlations computed on the initial subset (left), and the entire dataset (right).

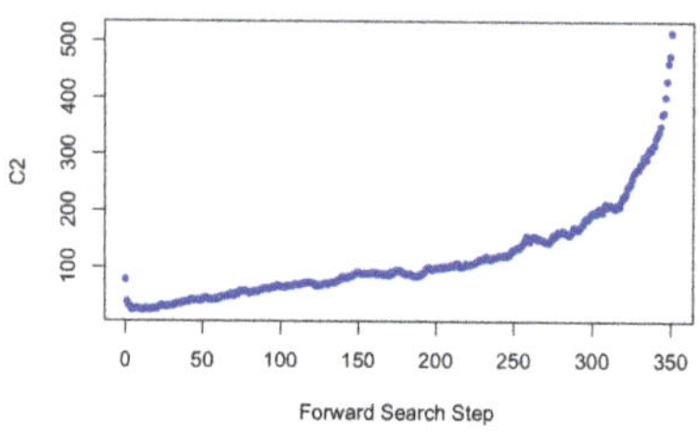

Fig. 3. Forward plot of the C_2 statistics in the Big Five dataset.

Table 2. Observed response patterns for the last steps of the FS.

step	Y_1	Y_2	Y_3	Y_4	Y_5	Y_6	Y_7	Y_8	Y_9	Y_{10}
341	4	2	4	3	4	4	4	2	1	4
342	3	0	1	2	0	0	0	2	4	2
343	0	4	4	2	4	4	4	4	4	2
344	4	4	4	4	4	4	4	2	4	2
345	4	0	4	4	4	4	4	4	0	4
346	4	4	4	4	4	4	4	4	4	4
347	4	4	4	4	4	4	4	4	4	4
348	4	4	4	4	4	4	4	4	4	4
349	4	4	4	4	4	4	4	1	0	4
350	0	0	0	0	0	0	0	0	0	0

5 Conclusions

This paper has explored the use of the FS as a robust diagnostic tool for detecting atypical response patterns in ordinal data within the IRT framework. By monitoring limited-information goodness-of-fit measures during the search, the proposed approach provides a principled way to identify clusters of aberrant observations while preserving stable estimation for the majority of the data.

Both the simulation and the empirical study illustrate that the FS effectively isolates atypical response patterns, with aberrant units entering the search only in its final stages. Simulation envelopes further aid interpretation by distinguishing departures from the assumed model from random variation, confirming that the procedure can detect irregular responses without compromising estimation for the majority of observations. Overall, the results support the FS as a valuable addition to the toolbox of robust methods for ordinal data analysis. Beyond improving model diagnostics, the approach provides insight into heterogeneous response behavior and yields evidence of the presence of inconsistent respondents in the sample.

This work represents a first step toward adapting the FS to ordinal data. Future research will investigate the joint monitoring of additional diagnostic measures, such as adjusted residuals and Cook's distance, to better identify individual departures and assess the influence of atypical observations on parameter estimates. Further developments may also consider extensions to more complex ordinal IRT models and applications to real-world assessment data.

References

1. Atkinson, A.C., Riani, M.: Robust Diagnostic Regression Analysis, vol. 2. Springer, Heidelberg (2000)
2. Atkinson, A.C., Riani, M., Cerioli, A.: Multivariate data and the forward search. In: Exploring Multivariate Data with the Forward Search, pp. 31–88. Springer, Cham (2004)
3. Atkinson, A.C., Riani, M.: Distribution theory and simulations for tests of outliers in regression. J. Comput. Graph. Stat. **15**(2), 460–476 (2006)
4. Cai, L., Monroe, S.: A new statistic for evaluating Item Response Theory models for ordinal data. No. 839 CRESST Report (2014)
5. Curran, P.G.: Methods for the detection of carelessly invalid responses in survey data. J. Exp. Soc. Psychol. **66**, 4–19 (2016)
6. Felt, J.M., Castaneda, R., Tiemensma, J., Depaoli, S.: Using person fit statistics to detect outliers in survey research. Front. Psychol. **8**, 863 (2017)
7. Goldberg, L.R.: The development of markers for the Big-Five factor structure. Psychol. Assess. **4**(1), 26 (1992)
8. Johnson, J.A.: Ascertaining the validity of individual protocols from web-based personality inventories. J. Res. Pers. **39**(1), 103–129 (2005)
9. Liu, Y., Liu, H.: Detecting noneffortful responses based on a residual method using an iterative purification process. J. Educ. Behav. Stat. **46**(6), 717–752 (2021)
10. Mavridis, D., Moustaki, I.: Detecting outliers in factor analysis using the Forward Search algorithm. Multivar. Behav. Res. **43**(3), 453–475 (2008)
11. Mavridis, D., Moustaki, I.: The forward search algorithm for detecting aberrant response patterns in factor analysis for binary data. J. Comput. Graph. Stat. **18**(4), 1016–1034 (2009)
12. Mislevy, R.J., Bock, R.D.: Biweight estimates of latent ability. Educ. Psychol. Meas. **42**(3), 725–737 (1982)
13. Riani, M., Torti, F., Zani, S.: Outliers and robustness for ordinal data. In: Modern Analysis of Customer Surveys: With Applications Using R, pp. 155–169 (2011)
14. Rizopoulos, D.: ltm: an R package for latent variable modeling and item response analysis. J. Stat. Softw. **17**, 1–25 (2007)
15. Samejima, F.: Estimation of latent ability using a response pattern of graded scores. Psychometrika **34**(Suppl. 1), 1–97 (1969)
16. Welz, M., Mair, P., Alfons, A.: Robust estimation of polychoric correlation. arXiv preprint arXiv:2407.18835 (2024)
17. Welz, M.: Robust estimation and inference for categorical data. arXiv preprint arXiv:2403.11954 (2024)
18. Woods, C.M.: Careless responding to reverse-worded items: implications for confirmatory factor analysis. J. Psychopathol. Behav. Assess. **28**(3), 186–191 (2006)

Not All Harms Hurt Equally: Understanding AI Safety Through Ordinal Severity

Sofia Vei[1(✉)], Paolo Giudici[2], Pavlos Sermpezis[1], Athena Vakali[1], and Adelaide Emma Bernardelli[3]

[1] School of Informatics, Aristotle University of Thessaloniki, 54124 Thessaloniki, Greece
{sofiavei,sermpezis,avakali}@csd.auth.gr

[2] Department of Economics and Management, University of Pavia, 27100 Pavia, Italy
paolo.giudici@unipv.it

[3] University School for Advanced Studies (IUSS) Pavia, 27100 Pavia, Italy
adelaide.bernardelli@iusspavia.it

Abstract. AI systems increasingly shape high-stakes societal domains, yet many risk assessment approaches remain compliance-driven and provider-centric, offering limited insight into how harms are experienced by affected stakeholders. We introduce *AI HarmSync*, a stakeholder-centric framework for prioritizing AI harms using real-world incident data and ordinal severity information. The framework embeds external human annotations into risk assessment and introduces *AIHS*, a harm concentration metric that extends the Gini framework to ranked severity data, avoiding unreliable or absent numerical severity scores. Experiments on AIAAIC indicate that *Political & Economic* and *Physical* harms exhibit the highest severity concentration, highlighting priorities for mitigation and governance.

Keywords: AI harms prioritization · AI risks · Ethical AI · Gini coefficient · Criticality Index

1 Introduction and Related Work

Artificial intelligence (AI) is increasingly embedded in critical decision-making across domains such as healthcare, finance, employment, and governance. While AI enables substantial societal and economic benefits, real-world deployments have already produced harmful outcomes that vary widely in type and intensity, and often affect stakeholders unevenly [3,7]. In this work, we distinguish *AI harms*, realized negative impacts (e.g., physical, psychological, social, or economic loss), from *AI risks*, which describe uncertainty about whether and how such harms may occur.

Despite rapid regulatory and governance developments, most **AI risk assessment frameworks** still adopt an internal, compliance-oriented perspective, emphasizing predefined checklists and provider-centric controls [5]. This

P. Brito et al. (Eds.): IFCS 2026, STUDIES CLASS, pp. 276–284, 2026.
https://doi.org/10.1007/978-3-032-32009-4_34

can under-represent affected stakeholders and miss emergent harms [8]. Coarse severity schemes (e.g., binary) also blur the distinction between minor and high-stakes failures, a key issue for risk-based regulation such as the European AI Act [6,9].

A key obstacle is that severity information in incident repositories is often noisy or unavailable. When numerical scores exist, they may be inconsistently assigned across sources, and annotators may disagree due to incomplete information and epistemic uncertainty. Many repositories therefore report ordinal severity (e.g., minor, moderate, severe), which supports ranking without imposing unjustified distance assumptions [10]. Similar ordinal approaches are used in cyber and operational risk where quantitative evidence is limited [4].

Motivated by these gaps, we pose two research questions: **RQ1** How can AI risk assessment embed human perspectives on harm severity? **RQ2** How can harms be prioritized when precise numerical severity data are unavailable or unreliable? We answer these questions with *AI HarmSync*, a harm-centric and stakeholder-adaptive framework grounded in incident repositories such as the ones identified by the popular AIAAIC repository [2]. Initially, we introduce a stakeholder-centered methodology that leverages external human annotations to characterize harm distributions across affected groups, complementing internal audit practices. Then, we propose *AI HarmSync (AIHS)*, a novel ordinal harm concentration that extends the Gini framework to purely ordinal severity data, enabling inequality-style concentration analysis without numerical scoring assumptions, and benchmark it against the Criticality Index [4]. Finally, we empirically demonstrate the framework on the AIAAIC dataset, showing that political and physical harms exhibit the highest severity concentration, motivating targeted mitigation strategies and policy attention.

2 AIAAIC Incident Data

We evaluate our approach on the AIAAIC repository, an openly shared dataset of AI-, algorithmic-, and automation-related incidents annotated with structured harm information [1,2]. We use AIAAIC as a benchmark because it (i) documents real-world incidents across domains, grounding prioritization in observed outcomes, (ii) provides a stable harm taxonomy and stakeholder groups for consistent aggregation [1], and (iii) relies on human-reviewed annotations, aligning with our goal of embedding external perspectives [1,2].

The raw data consist of incident-level annotations. In our snapshot, each annotation includes an incident identifier, one or more affected stakeholder groups, a harm category, and additional metadata (e.g., notes). Stakeholders follow a predefined taxonomy (e.g., *Users, Vulnerable Groups, Workers, General Public, Government/Public Sector, Business, Artists/Content Creators*), and harm categories include high-level classes such as *Autonomy, Financial & Business, Human Rights & Civil Liberties, Physical, Political & Economic, Reputational, Societal & Cultural* [1]. Since harms are not mutually exclusive, a single incident may contribute to multiple stakeholder–harm pairs.

Our method requires only aggregated counts (or frequencies) of stakeholder exposure within each harm category. Let h_i denote a harm category and g_j a stakeholder group. From the annotation table we compute the frequency n_{ij} of pairs (h_i, g_j) and normalize within each category to obtain $f_{ij} = n_{ij} / \sum_j n_{ij}$. Thus, the input to our analysis is a compact table of the form *(harm category, stakeholder group, frequency)*, which is sufficient to construct the ordinal concentration measures introduced in Sect. 3. Table 1 illustrates this aggregation.

Table 1. Example of the aggregated input used by our method (illustrative).

Harm Category	Stakeholder Group	Count
Autonomy	Users	29
Autonomy	Artists/Content Creators	15
Physical	Vulnerable Groups	22
Physical	Users	14

Figure 1 reports the resulting contingency table for our snapshot, visualizing how frequently each stakeholder group appears within each harm category. The distribution is uneven across stakeholders, motivating the concentration-based analysis that follows.

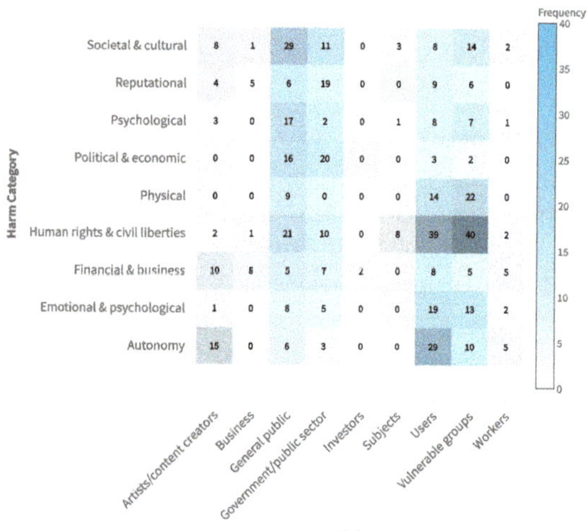

Fig. 1. Incidents by harm category and stakeholder group (AIAAIC). Each cell shows the number of annotations for a category–stakeholder pair.

3 Methodology

In this section, we present the core component of the *AI HarmSync* framework: the *AIHS* metric, which quantifies how concentrated severe harms are across stakeholder groups.

Let h_i denote a harm category and let g_j, $j \in \{1, \dots, N\}$, denote stakeholder groups. For each category h_i, we consider the distribution of stakeholders through frequencies f_{ij}, which represent how harm in category h_i is allocated across stakeholder groups g_j.

Ordinal Severity Ordering. Because severity information is often unavailable as reliable numerical values, we treat severity as *ordinal*. Stakeholder groups are sorted by increasing severity,

$$s_j < s_{j+1}, \quad \forall j,$$

where s_j represents a rank (order) rather than a cardinal magnitude.

AIHS: Harm Concentration. Harms are not experienced uniformly: some categories may distribute mild harms broadly, while others concentrate severe impacts on a few high-severity stakeholder groups. Inspired by the Gini framework but adapted to ordinal settings, we define the *AIHS* metric as an area-based summary of a rank-based ("derivative") Lorenz curve, while Fig. 2 illustrates the conceptual difference between the classic Lorenz curve and the derivative Lorenz curve used for ordinal severity in HarmSync framework.

Theorem 1. *The AIHS metric for a category h_i with stakeholders g_j, $j \in \{1, \dots, N\}$ sorted by severity and having frequency f_{ij}, is defined as*

$$AIHS_i = \int_0^1 \ell_i(x)\, dx, \tag{1}$$

where, for each category h_i, ℓ_i is the curve defined by the points

$$\ell_i \rightarrow \left\{ \sum_{j=0}^{k} f_{ij}, \frac{k}{N} \right\}, \quad \forall k \in \{0, \dots, N\},$$

with $f_{i0} = 0$.

The curve depends only on the ordering of severities, not on their numeric spacing, enabling concentration analysis under purely ordinal information.

Relation to the Criticality Index (CI). To place *AIHS* in context with prior ordinal risk metrics, we also report the Criticality Index (CI) from the cyber risk literature [4]. Under purely ordinal inputs, CI and *AIHS* are closely aligned (up to a linear transformation), so CI serves as a benchmark for interpreting *AIHS* values in an established ordinal criticality framework.

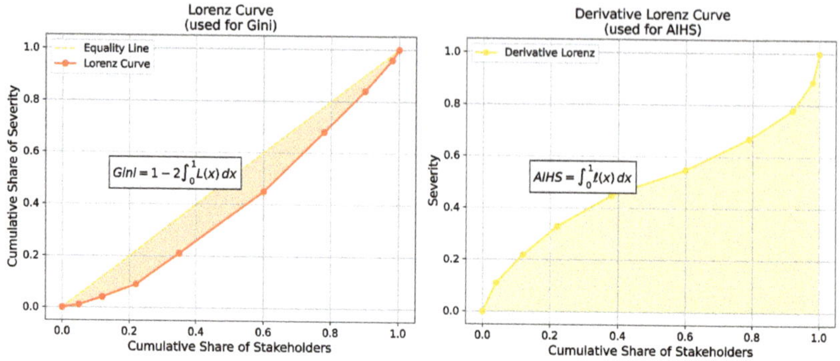

Fig. 2. Left: Classic Lorenz curve $L(x)$ (numeric severity). **Right:** Derivative Lorenz curve $\ell(x)$ (ordinal severity); the shaded area defines *AIHS*.

4 Empirical Results

We apply the proposed methodology to the AIAAIC dataset to examine how AI-related harms are distributed across stakeholder groups and to identify which harm categories exhibit the strongest concentration of severe impacts. As described in Sects. 2 and 3, our analysis is based on aggregated category–stakeholder frequencies and an ordinal severity ordering over stakeholder groups.

Stakeholder Severity Ordering. Because AIHS operates on ordinal rather than numerical severity information, the method requires an explicit ordering of stakeholders according to their relative exposure to harm. In our empirical analysis we adopt the following ordering from least to most severely affected: *Artists/Content Creators* < *Subjects* < *Business* < *Investors* < *Workers* < *Users* < *Vulnerable Groups* < *Government/Public Sector* < *General Public*. The framework is flexible to alternative stakeholder orderings, and prior sensitivity analyses [11] indicate that the resulting rankings remain stable under reasonable perturbations. Table 2 encodes this order for reproducibility; only the ranking matters.

Harm Concentration Across Categories. For each harm category, we compute the AIHS score defined in Sect. 3, which summarizes how strongly harms concentrate on higher-severity stakeholder groups. Figure 3 illustrates the rank-based Lorenz-type curves for selected categories.

Financial & Business exhibits a relatively flat curve, indicating a more balanced distribution of harms across stakeholders. By contrast, the curve for *Autonomy* lies visually closer to the diagonal but shows its largest deviations around the middle of the ranking. This indicates that harms are not evenly distributed but rather concentrated around specific stakeholder groups, even though the curve does not deviate strongly at the highest severity ranks. *Political & Economic* shows a steep curve concentrated toward the highest severity ranks, meaning that harms disproportionately affect the most severely impacted stakeholders.

Table 2. Stakeholder groups and ordinal severity ranking (numeric values encode order only).

Stakeholder Group	Severity Rank
Artists/Content Creators	1
Subjects	2
Business	3
Investors	4
Workers	5
Users	6
Vulnerable Groups	7
Government/Public Sector	8
General Public	9

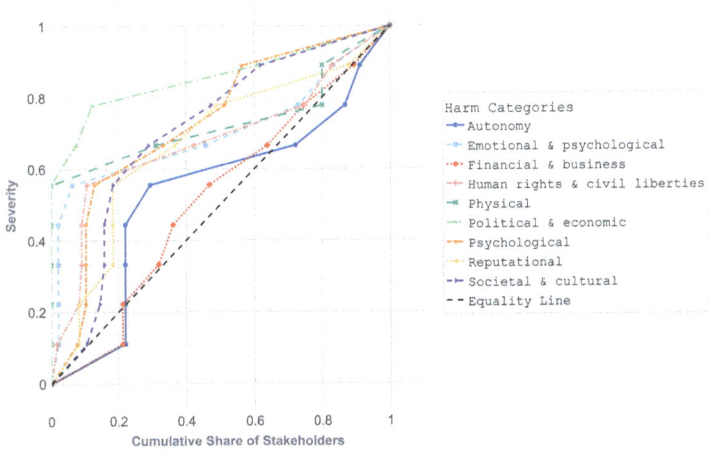

Fig. 3. Rank-based Lorenz-type curves for harm categories, illustrating differences in harm concentration across stakeholder severity levels.

AIHS and CI Rankings. Table 3 reports the AIHS and Criticality Index (CI) values for all harm categories, while Fig. 4 additionally illustrates their strong agreement across categories. Both metrics yield consistent rankings: *Political & Economic* harms exhibit the highest concentration (AIHS = 0.85), followed by *Physical* and *Psychological* harms, indicating that severe impacts disproportionately affect the most vulnerable stakeholders. In contrast, *Financial & Business* (AIHS = 0.51) and *Autonomy* (AIHS = 0.53) show the lowest concentration, reflecting more balanced harm distributions. From a governance perspective, these results are consistent with the broader risk-based logic of the EU AI Act, which gives particular attention to AI uses that may adversely affect safety, fundamental rights, and democratic processes [6,9]. In this sense, the high concentration observed for *Political & Economic* harms highlights categories that may require closer oversight because they can be associated with manipulation, public-interest distortions, or impacts on democratic life, while the concentration of *Physical* harms is aligned with regulatory concern for safety-critical applications that may affect health or bodily integrity. The alignment between AIHS and CI confirms that the proposed ordinal metric captures harm concentration patterns in line with established criticality measures.

Table 3. AIHS and CI values across harm categories. Higher values indicate stronger concentration of severe harms.

Harm Category	AIHS	CI
Autonomy	0.53	0.54
Emotional & Psychological	0.70	0.72
Financial & Business	**0.51**	0.51
Human Rights & Civil Liberties	0.67	0.70
Physical	0.73	0.76
Political & Economic	**0.85**	0.89
Psychological	0.73	0.75
Reputational	0.67	0.69
Societal & Cultural	0.70	0.73

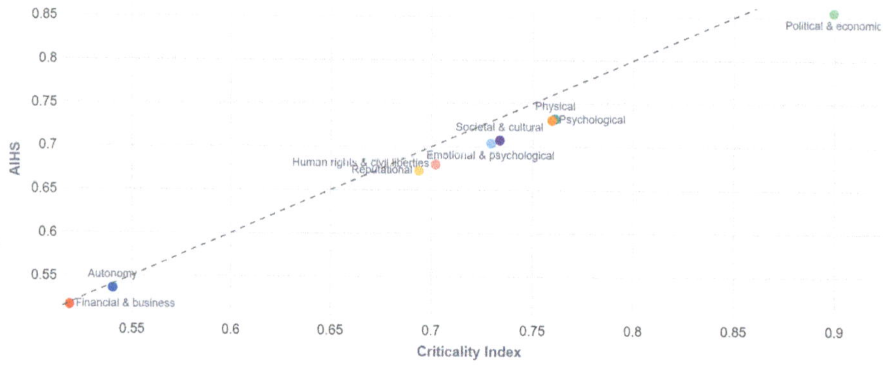

Fig. 4. Comparison of AIHS and CI values across harm categories. The dashed line indicates perfect agreement.

5 Conclusions

We introduced *AI HarmSync*, a stakeholder-centric approach to prioritizing AI harms in incident repositories under ordinal severity information. We proposed *AIHS*, an ordinal concentration metric that captures how strongly harms accumulate on higher-severity stakeholders without requiring numerical severity scores.

On AIAAIC, AIHS indicates that *Political & Economic* and *Physical* harms are most concentrated, while *Financial & Business* and *Autonomy* are more balanced, supporting actionable prioritization for mitigation and governance. A more comprehensive version of this work, including additional analyses and methodological details, is available at: https://arxiv.org/abs/2509.10104.

Acknowledgments. The work has been supported by the EXCALIBUR project (EXplainable and demoCrAtized pipeLInes for unBiased, trUstworthy, and Responsible human-centric AI), funded by the General Secretariat for Research and Technology (GSRT) and the Hellenic Foundation for Research and Innovation (HFRI), under the National Recovery and Resilience Plan "Greece 2.0", funded by the European Union – NextGenerationEU.

References

1. Abercrombie, G., et al.: A collaborative, human-centred taxonomy of AI, algorithmic, and automation harms. arXiv preprint arXiv:2407.01294 (2024)
2. AI, Algorithmic and Automation Incident and Controversy (AIAAIC): AIAAIC Repository (2022). https://www.aiaaic.org/aiaaic-repository, Licensed under CC BY-SA 4.0. https://creativecommons.org/licenses/by-sa/4.0/
3. Avila-Rieger, J.F., et al.: Early life exposure to structural sexism and late-life memory trajectories among black and white women and men in the United States. Alzheimer's Dementia (2024)
4. Giudici, P., Raffinetti, E.: Explainable AI methods in cyber risk measurement. Qual. Reliab. Eng. Int. (2023)
5. Hoffman, F., Kesharwani, R., Maynard, J.: How an outside perspective can help an organization enhance their supply chain risk mitigation strategy. Issues Inf. Syst. **24**(1) (2023)
6. Jariwala, M.: A comparative analysis of the EU AI act and the Colorado AI act: regulatory approaches to artificial intelligence governance. Int. J. Comput. Appl. **975**, 8887 (2024)
7. Jones, J.: Don't fear artificial intelligence, question the business model: how surveillance capitalists use media to invade privacy, disrupt moral autonomy, and harm democracy. J. Commun. Inq. **49**(1), 6–26 (2025)
8. Lancaster, C.M., Schulenberg, K., Flathmann, C., McNeese, N.J., Freeman, G.: "It's everybody's role to speak up... but not everyone will": understanding AI professionals' perceptions of accountability for AI bias mitigation. ACM J. Responsible Comput. **1**(1), 1–30 (2024)
9. Novelli, C., Casolari, F., Rotolo, A., Taddeo, M., Floridi, L.: Taking AI risks seriously: a new assessment model for the AI act. AI Soc. **39**(5), 2493–2497 (2024)

10. Robinson, J.: Likert scale. In: Encyclopedia of Quality of Life and Well-Being Research, pp. 3917–3918. Springer (2024)
11. Vei, S., Giudici, P., Sermpezis, P., Vakali, A., Bernardelli, A.E.: AI harmonics: a human-centric and harms severity-adaptive AI risk assessment framework. arXiv preprint arXiv:2509.10104 (2025)

Estimation of High-Dimensional Copula-Based CUB Models Using Composite Likelihood

Matteo Ventura[1(✉)], Dimitris Karlis[2], Julien Jacques[3], and Paola Zuccolotto[4]

[1] Department of Economics, Statistics and Finance, University of Calabria, Rende, Italy
matteo.ventura@unical.it

[2] Department of Statistics, Athens University of Economics and Business, Athens, Greece
karlis@aueb.gr

[3] ERIC Laboratory, Université Lumière Lyon 2, Lyon, France
julien.jacques@univ-lyon2.fr

[4] Department of Economics and Management, University of Brescia, Brescia, Italy
paola.zuccolotto@unibs.it

Abstract. Multivariate extensions of CUB (Combination of discrete Uniform and shifted Binomial) models via copulas face severe computational challenges due to the high number of terms required to evaluate the joint probability mass function. In this study, we propose a pairwise likelihood approach to overcome these limitations by focusing on bivariate marginal distributions. The methodology involves a two-step estimation procedure: individual bivariate models are fitted for each pair, and the resulting estimates are then summarized to obtain the parameters of the full joint model. Simulation results using Archimedean copulas demonstrate that the method is computationally efficient and provides accurate parameter estimates.

Keywords: CUB models · Copulas · Pairwise likelihood · Ordinal data

1 Introduction

In recent decades, research across various fields, including psychology, marketing, sociology, and behavioral sciences, has focused extensively on measuring and analyzing latent constructs such as perceptions, opinions, attitudes, and tastes. Information regarding these hidden traits is commonly gathered by administering questionnaires where respondents express their evaluations by selecting a score from a predefined set of ordered alternatives. The resulting data are typically classified as ordinal categorical variables, which pose significant statistical challenges. Their discrete and bounded nature requires specialized methods and

P. Brito et al. (Eds.): IFCS 2026, STUDIES CLASS, pp. 285–292, 2026.
https://doi.org/10.1007/978-3-032-32009-4_35

models that can accommodate their unique properties, moving beyond the common but often inappropriate practice of treating them as numerical or continuous variables.

Among the models proposed for rating data, the CUB (Combination of a discrete Uniform and a shifted Binomial random variable) class is particularly notable [13]. Originally proposed by Piccolo and D'elia [6,12], this framework posits that the final rating is the outcome of a joint latent decision process involving two distinct components: feeling and uncertainty. Statistically, these are modelled through a mixture distribution, where a shifted Binomial random variable models the individual's subjective sensation or agreement, the feeling; while a discrete Uniform distribution accounts for the inherent indecision and potential randomness in the choice process, the uncertainty.

Formally, the CUB model is defined as follows:

$$P(R = r \mid \pi, \xi) = \pi \binom{m-1}{r-1} \xi^{m-r}(1-\xi)^{r-1} + (1-\pi)\frac{1}{m}, \quad r = 1, \dots, m \tag{1}$$

where m is the number of levels, $1-\xi$ is the feeling parameter, and $1-\pi$ is the uncertainty parameter.

While the univariate CUB model has been widely extended to include covariates and different response styles, its extension to a multivariate setting remains a complex frontier. In practice, it is common to collect multiple ratings from the same individual across several items, leading to multivariate ordinal data where the dependence between responses must be appropriately modeled. Since this is a complex task, some authors proposed multivariate CUB models assuming independence among the variables, to easily fit the model in a model-based clustering context [15]. However, the idea of coupling CUB models with copulas is not new. In the bivariate context, Andreis and Ferrari [1] explored the use of the Archimedean family to capture the dependence between two CUB random variables, while Corduas [3,4] proposed an approach based on the Plackett distribution to model the association between ordinal responses, also allowing for the inclusion of subject-specific covariates. Colombi and Giordano [2] proposed a bivariate construction using the Sarmanov approach. For a multivariate extension, Corduas [5] utilized a Plackett distribution as a copula together with a discrete vine pair-copula construction to achieve computational efficiency. In contrast, the model suggested by Ip and Wu [9] is very different, extending the mixture of a binomial and a uniform distribution to two dimensions through a mixture of bivariate discrete uniforms and bivariate binomials. However, as the number of dimensions d increases, the full likelihood estimation for copula-based CUB models becomes computationally intensive due to the 2^d terms required to evaluate the joint probability mass functions.

In this paper, we address these severe computational limitations by proposing a pairwise likelihood approach. This methodology simplifies the estimation process by focusing exclusively on bivariate marginal distributions, which provides a robust and efficient alternative for modeling high-dimensional dependency structures in rating data.

2 Modelling Complex Interdependencies with Copulas

The complexity of defining a multivariate CUB model through copulas stems primarily from the discrete and ordinal nature of the data. While the use of a d-dimensional copula for continuous variables is computationally straightforward, in the case of discrete variables, the joint probability mass function must be derived through finite differences of the joint distribution function. Following Panagiotelis et al. [11], the probability of observing a vector of responses $\mathbf{r} = (r_1, \ldots, r_d)'$ with $r_1 \in \{1, \cdots, m_1\}, r_d \in \{1, \cdots, m_d\}$, is given by:

$$P(\mathbf{r}) = \sum_{\mathbf{v} \in \mathcal{V}} \operatorname{sgn}(\mathbf{v}) C\left(G_1(v_1), \ldots, G_d(v_d)\right), \tag{2}$$

where each $G_j(\cdot)$ represents the marginal cumulative distribution function of a CUB model with parameters (π_j, ξ_j) and $C(\cdot, \ldots, \cdot)$ is any d-dimensional copula function. The vector $\mathbf{v} = (v_1, \ldots, v_d)'$ represents the vertices of the unit hypercube defined by $v_j \in \{r_j, r_j - 1\}$, and the sign function is defined as:

$$\operatorname{sgn}(\mathbf{v}) = \begin{cases} 1, & \text{if } v_j = r_j - 1 \text{ for an even number of indices } j \\ -1, & \text{if } v_j = r_j - 1 \text{ for an odd number of indices } j \end{cases}. \tag{3}$$

Following Panagiotelis et al. [11], the joint probability of observing a vector of responses $\mathbf{r}$ is obtained by evaluating the copula at the vertices of a d-dimensional unit hypercube. This involves the summation of 2^d terms, as shown in Eq. 2. Consequently, the Full Likelihood (FL) is defined as the product of these joint probabilities over the sample of n independent observations and its logarithm is given by:

$$\ell_{FL}(\boldsymbol{\theta}) = \sum_{i=1}^{n} \ln P(r_{i1}, \ldots, r_{id}; \pi_1, \ldots, \pi_d, \xi_1, \ldots, \xi_d, \boldsymbol{\alpha}) \tag{4}$$

where θ denotes the vector of all CUB (π, ξ) and copula parameters (α).

For a model in dimension d, the evaluation of ℓ_{FL} becomes extremely burdensome. Consider, for instance, a psychometric questionnaire with $d = 10$ items: the estimation would require calculating $2^{10} = 1024$ terms for every single observation at each iteration of the maximization process. In high-dimensional settings, this makes the maximization of the usual likelihood often prohibitive in terms of time and computational resources.

2.1 A Pairwise Likelihood Approach for Multivariate CUB Models with Copulas

The fundamental idea that allow us to overcome these difficulties is that, rather than defining and estimating the full d-dimensional model, it may be sufficient to consider only the bivariate marginal distributions. To do that, we propose to adopt a pairwise likelihood (PL) approach [10,14], which for a d-dimensional model is defined as follows:

$$\ell_{PL}(\boldsymbol{\theta}_{jk}) = \sum_{i=1}^{n} \sum_{j<k} \ln P(r_{ij}, r_{ik}; \pi_j, \pi_k, \xi_j, \xi_k, \alpha_{jk}) \tag{5}$$

where each term $P(r_{ij}, r_{ik}; \pi_j, \pi_k, \xi_j, \xi_k, \alpha_{jk})$ is the bivariate distribution defined by CUB marginals for the j-th and k-th item, and a chosen copula with dependence parameter α_{jk}.

This formulation clearly shows that each component of the sum depends only on the parameters of two items at a time, significantly reducing the computational complexity compared to the full likelihood.

This approach offers a two-fold advantage: it drastically reduces computational complexity, limiting the calculation to only $2^2 = 4$ terms per pair, and it allows for the use of a wide range of copula families to model different types of interdependencies between pairs of variables, an operation that would be far more complex in a purely multivariate setting. However, the simultaneous optimization of $\ell_{PL}(\theta)$ can still be challenging as the number of parameters grows. Following Fieuws and Verbeke [7], we adopt a two-step estimation procedure. In the first step, we independently estimate each of the $d(d-1)/2$ bivariate models:

$$\hat{\theta}_{jk} = \arg\max_{\theta_{jk}} \sum_{i=1}^{n} \ln P(r_{ij}, r_{ik}; \pi_j, \pi_k, \xi_j, \xi_k, \alpha_{jk}). \tag{6}$$

It is important to note that the parameter vector θ_{jk} contains both the marginal parameters of the CUB models $(\pi_j, \xi_j, \pi_k, \xi_k)$ and the copula dependence parameter α_{jk}. Consequently, each marginal parameter appears in $d-1$ different bivariate models. In the above setting, α_{jk} denotes the association parameter specific to the pair of responses (j, k). By default, we allow for a saturated association structure where each pair is characterized by a unique α_{jk}. However, for the sake of parsimony, equality constraints can be imposed (e.g., $\alpha_{jk} = \alpha$ for all $j \neq k$) to reduce the number of parameters to be estimated, particularly in high-dimensional settings. In the simulation study, the equality constraint $\alpha_{jk} = \alpha$ for all $j \neq k$ is adopted for parsimony. At the second stage, to obtain a unique estimate for each parameter of the full joint model, we summarize the obtained estimates. The most effective solution consists of calculating the average (simple or weighted by the Hessian matrix) of the individual estimates for each parameter across all its occurrences in the bivariate models. Note that standard errors for the full pairwise likelihood method can be estimated via the Godambe information matrix [14], while for our approach with the method described in [8].

3 Simulation Study

To evaluate the performance of the proposed pairwise likelihood approach and compare it with the standard full likelihood estimation, we carried out two distinct simulation experiments. The first experiment aims at evaluating the accuracy of estimates and the efficiency of the PL relative to the FL approach, while the second assesses computational scalability as the number of items increases.

In all scenarios, data were generated from a multivariate CUB distribution coupled with three Archimedean copula families: Clayton, Frank, and Gumbel. These families were selected to represent a broad range of dependence structures: the Clayton copula exhibits lower-tail dependence, the Gumbel copula upper-tail dependence and the Frank copula symmetric dependence with no tail concentration.

3.1 Parameter Estimates and Efficiency

We considered a set of 5 items, since for $d = 5$ the full likelihood approach begins to reach its practical limit for intensive simulations due to the $2^5 = 32$ terms required for each joint probability evaluation. To focus the comparison on the effect of copula family and dependence strength, the marginal parameters were set equal across all items: $\pi_j = 0.6$ and $\xi_j = 0.7$ for $j = 1, \ldots, 5$. Dependence strength was calibrated in terms of Kendall's τ, set to 0.3 (weak) and 0.6 (strong), and converted into the copula-specific parameter for each family. We tested sample sizes $n \in \{100, 500, 1000\}$ with $B = 100$ replicates for each scenario. In the second stage of the estimation procedure, the marginal parameter estimates are synthesised via averaging across all bivariate models in which each parameter appears. The performance was evaluated through the estimates' mean, standard deviation (SD), and mean squared error (MSE), which is computed for each parameter θ as the average of the squared differences between the estimates $\hat{\theta}_b$ obtained in each of the B replications and the true parameter value, namely $\text{MSE}(\hat{\theta}) = \frac{1}{B}\sum_{b=1}^{B}(\hat{\theta}_b - \theta)^2$. By including this metric, we account for both the precision and the accuracy of the estimates. Both PL and FL are consistent across all copula families considered, with MSE values decreasing as n grows. A slightly higher variability in the estimation of π parameters compared to ξ is observed, a known characteristic of the CUB family arising from the mixture nature of the model. To assess the loss of efficiency when moving from FL to PL, we calculated the relative efficiency (RE), defined as RE $=$ MSE_{FL}/MSE_{PL}. The results, reported in Fig. 1, show that RE values remain consistently close to 1 across all copula families and dependence levels, confirming that the pairwise approach incurs negligible loss of estimation accuracy regardless of the dependence structure.

Standard errors for the PL estimator were obtained via the Godambe information matrix following [8], while for the FL estimator they were derived from the observed Hessian. To assess the reliability of the uncertainty quantification, we compared, for each scenario, the average estimated Standard Errors with the empirical Standard Deviations across replications. Table 1 reports this comparison at $n = 500$ for π_1, ξ_1, and α across all copula families and dependence levels. The two quantities are in close agreement throughout, indicating that the Standard Errors estimator proposed in [8] provides well-calibrated standard errors for the PL approach.

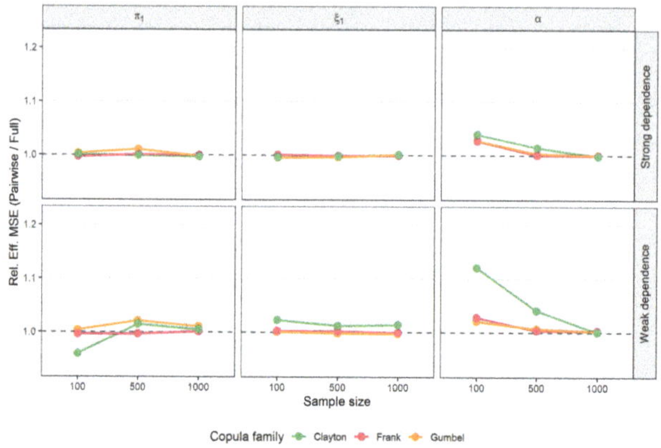

Fig. 1. Relative efficiency for the $d = 5$ multivariate CUB model across copula families (Clayton, Frank, Gumbel), dependence levels (weak: $\tau = 0.3$, strong: $\tau = 0.6$), and sample sizes $n \in \{100, 500, 1000\}$. Values close to 1 indicate negligible loss of efficiency of PL relative to FL.

3.2 Computational Scalability

The second experiment aimed at quantifying the computational gain offered by the pairwise likelihood approach.

We measured the execution time as a function of the model's dimensionality, ranging from $d = 2$ to $d = 15$, using a single replication per scenario with $m = 5$ categories for each ordinal variable and dependence calibrated at $\tau = 0.6$. A timeout of 10 min was imposed for each replication: when FL estimation exceeded this threshold the curve was interrupted, which explains why the FL lines terminate before $d = 15$ for all families.

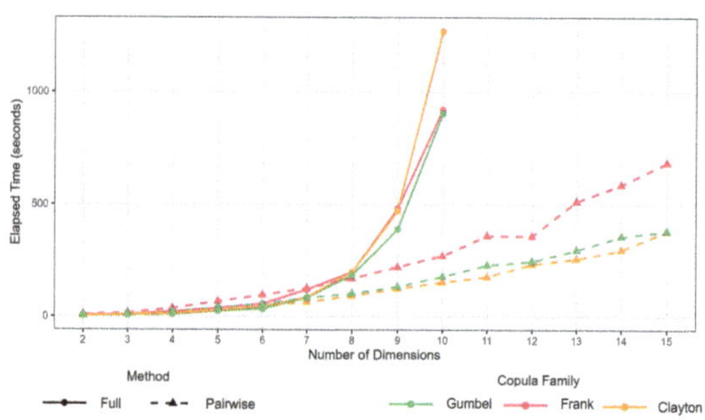

Fig. 2. Execution time (seconds) for FL vs PL as a function of the dimension d. An upper limit of 10 min was used.

The results reported in Fig. 2 consistently highlight the curse of dimensionality affecting the joint full likelihood across all Archimedean copula families

Table 1. Empirical standard deviations (SD) and average estimated standard errors (SE) for PL and FL at $n = 500$, across copula families and dependence levels, for parameters π_1, ξ_1, and α.

Param		Gumbel		Frank		Clayton	
		Weak	Strong	Weak	Strong	Weak	Strong
Pairwise Likelihood (PL)							
π_1	SD	0.0541	0.0559	0.0618	0.0530	0.0547	0.0602
	SE	0.0542	0.0534	0.0551	0.0535	0.0552	0.0538
ξ_1	SD	0.0211	0.0188	0.0183	0.0174	0.0204	0.0211
	SE	0.0189	0.0185	0.0188	0.0172	0.0187	0.0180
α	SD	0.0401	0.0974	0.1730	0.3192	0.0746	0.2152
	SE	0.0401	0.1000	0.1793	0.3232	0.0683	0.1907
Full Likelihood (FL)							
π_1	SD	0.0511	0.0532	0.0615	0.0517	0.0541	0.0598
	SE	0.0521	0.0514	0.0547	0.0518	0.0550	0.0530
ξ_1	SD	0.0206	0.0164	0.0174	0.0152	0.0196	0.0206
	SE	0.0184	0.0175	0.0181	0.0159	0.0183	0.0170
α	SD	0.0349	0.0940	0.1670	0.3151	0.0691	0.2078
	SE	0.0372	0.0960	0.1711	0.3122	0.0655	0.1794

considered. The FL execution time grows steeply on the log scale, reaching the timeout threshold at $d = 10$ for all the families, whereas the PL curves continue smoothly up to $d = 15$ in all cases. The two methods perform comparably up to $d = 8$, after which the FL execution time increases sharply, reaching over 900 s at $d = 10$ for all families while the PL method remains below 250 s. The observed computational times are consistent with expectations, as they reflect the exponential growth in the number of terms for the first model and the quadratic growth for the second.

4 Discussion

In this study, we proposed a multivariate CUB model framework that accounts for interdependencies through copula functions. To address the computational complexity inherent in high-dimensional settings, we introduced an innovative approach based on pairwise likelihood estimation.

Preliminary simulation studies demonstrate that this method offers significant computational advantages as the number of dimensions increases, while maintaining high estimation accuracy. Specifically, it reveals only a negligible loss of efficiency in parameter estimation when compared to the full likelihood approach. The flexibility of the proposed method is a key strength, as it accommodates different numbers of categories for each dimension and supports various copula families to capture diverse dependency structures.

Future research will involve more extensive simulation studies to further validate the model's robustness. We also note that, while the proposed framework supports a saturated association structure with pair-specific parameters, the simulation study adopts a common dependence parameter across pairs for parsimony; this parameter is estimated separately for each bivariate model and then synthesised into a single final estimate. The proposed framework can be extended to accommodate richer dependence structures by adopting multivariate copulas with more flexible parameterisations, such as the Gaussian copula. Additionally, we plan to apply this procedure to data from the European Social Survey, where multiple thematic modules collect batteries of Likert-scale items across a large number of respondents, representing a natural high-dimensional setting for the proposed methodology.

Acknowledgements.. Matteo Ventura was partially supported by the Master in Artificial Intelligence & Data Science (funded by "Patti Territoriali dell'Alta Formazione per le imprese"—CUP: H52C23000090001) and by the BODaI Lab.

References

1. Andreis, F., Ferrari, P.A.: On a copula model with CUB margins. Quaderni Stat. **15**, 33–51 (2013)
2. Colombi, R., Giordano, S.: A class of mixture models for multidimensional ordinal data. Stat. Model. **16**(4), 322–340 (2016)
3. Corduas, M.: Analyzing bivariate ordinal data with CUB margins. Stat. Model. **15**(5), 411–432 (2015)
4. Corduas, M.: Modelling correlated consumer preferences. In: Advances in Latent Variables: Methods, Models and Applications, pp. 27–36 (2015)
5. Corduas, M.: Joint modelling of ordinal data: a copula-based method. Adv. Stat. Modell. Ordinal Data 85 (2018)
6. D'Elia, A., Piccolo, D.: A mixture model for preferences data analysis. Comput. Stat. Data Anal. **49**(3), 917–934 (2005)
7. Fieuws, S., Verbeke, G.: Pairwise fitting of mixed models for the joint modeling of multivariate longitudinal profiles. Biometrics **62**(2), 424–431 (2006)
8. Hui, F.K., Müller, S., Welsh, A.: Sparse pairwise likelihood estimation for multivariate longitudinal mixed models. J. Am. Stat. Assoc. **113**(524), 1759–1769 (2018)
9. Ip, R.H., Wu, K.: A mixture distribution for modelling bivariate ordinal data. Stat. Pap. 1–36 (2024)
10. Lindsay, B.G.: Composite likelihood methods. Contemp. Math. **80**(1), 221–239 (1988)
11. Panagiotelis, A., Czado, C., Joe, H.: Pair copula constructions for multivariate discrete data. J. Am. Stat. Assoc. **107**(499), 1063–1072 (2012)
12. Piccolo, D.: On the moments of a mixture of uniform and shifted binomial random variables. Quaderni Stat. **5**(1), 85–104 (2003)
13. Piccolo, D., Simone, R.: The class of CUB models: statistical foundations, inferential issues and empirical evidence. Stat. Methods Appl. **28**, 389–435 (2019)
14. Varin, C., Reid, N., Firth, D.: An overview of composite likelihood methods. Stat. Sin. 5–42 (2011)
15. Ventura, M., Jacques, J., Zuccolotto, P.: Model-based clustering of multivariate rating data accounting for feeling and uncertainty. J. Classif. 1–22 (2025)

Multiclass Functional Classification Through Aggregated Higher–Order Moments

Marc Vidal[1,2(✉)]

[1] Department of Data Analysis, Ghent University, 9000 Ghent, Belgium
[2] Department of Neurology, Max Planck Institute for Human Cognitive and Brain Sciences, 04103 Leipzig, Germany
marc.vidalbadia@ugent.be

Abstract. Classical second–order moment methods provide limited insight into multiclass functional classification problems. We propose an approach based on aggregated higher–order moment information that combines multiple pairwise contrasts through a tensor–based criterion while yielding a single functional discriminant direction. Under this framework, near–perfect classification is attainable through the joint alignment of multiple contrasts. The approach is motivated by measure singularity and captures departures from mutual equivalence across class pairs.

Keywords: FDA · measure singularity · classification · tensor aggregation

1 Introduction

Empirical classifiers of functional data can asymptotically achieve perfect separation due to the underlying singularity of the class measures. The phenomenon itself dates back to Rao and Varadarajan (1963) [5], who introduced the notion of *perfect discrimination* as a manifestation of Gaussian measure singularity. Operator–valued analogues of higher–order moments—such as kurtosis–type operators acting on L^2—play an important role in this respect. It has been recently shown that their spectral properties provide a geometric characterization of dependence and singularity between Gaussian measures. Weak shifts in measure are reflected as perturbations in the spectral tail of their finite-dimensional empirical counterparts, yielding an operator–theoretic interpretation of near–perfect classification [6,7]. While in the bimodal functional setting the role of measure singularity in classification is well characterized at the level of Gaussian measures through the Feldman–Hájek dichotomy [1,2], its extension

The author acknowledges the support from the FWO project G013024N and thanks the anonymous reviewers for their helpful comments and suggestions.

P. Brito et al. (Eds.): IFCS 2026, STUDIES CLASS, pp. 293–301, 2026.
https://doi.org/10.1007/978-3-032-32009-4_36

beyond two classes remains less well understood. In the multiclass case, discrimination naturally involves several pairwise contrasts. Classical functional discriminant approaches encode this information through a single between–class covariance operator, which aggregates all contrasts into a unified covariance structure [4]. Although this representation is sufficient for linear discrimination, it does not make explicit how individual pairwise contrasts contribute to the discriminant geometry, nor how potentially competing contrasts interact in near–singular regimes.

Motivated by this limitation, we seek multiclass procedures that aggregate several pairwise contrasts without collapsing them a priori into a single second–order moment operator, while still returning a single functional discriminant. At the population level, one natural way to formalize the simultaneous balancing of multiple contrasts is through tensorial criteria that combine several pairwise moment operators. We restrict our attention to solutions in which this aggregation yields a single functional direction, so that tensor structure serves only as a mechanism for contrast aggregation. Under this restriction, stationarity is characterized by a single first–order condition, providing a variational description of how pairwise operators may be jointly aligned by one direction. On the empirical side, this perspective leads to two complementary implementations. The first is a non–tensorial formulation based on aggregated univariate kurtosis contrasts evaluated over a data–driven family of candidate directions. The second is a tensorial formulation in which pairwise class kurtosis operators are assembled into a low–rank tensor representation, used to estimate a common discriminant direction via standard tensor decompositions. Together, these approaches bridge higher–order moment information and functional discriminant analysis in a manner that is both statistically coherent and computationally tractable.

2 Formulation of Multiclass Fisher-Type Discriminants

Let $T \subset \mathbb{R}$ be a compact interval and set $H := L^2(T)$, endowed with the inner product $\langle f, g\rangle = \int_T f(t)g(t)\,\mathrm{d}t$. Throughout the paper we use the standard identification between elementary tensors and rank–one operators: the tensor $f \otimes g$ corresponds to the operator $h \mapsto \langle h, f\rangle g$ on H, and finite sums of such tensors are finite–rank operators. We use the same symbol $\otimes$ for tensor products of operators acting on H.

Consider a random pair (G, X) taking values in $\{1, \dots, G\} \times H$, where G denotes a class label and X is a square–integrable random element. Let $\mathbb{P}_\ell$ denote the conditional law of X given $\mathsf{G} = \ell$, and write $X_\ell \sim \mathbb{P}_\ell$ for a generic draw from class ℓ. Assume that $\mathsf{Pr}(\mathsf{G} = \ell) = \alpha_\ell > 0$ for all ℓ and $\sum_{\ell=1}^{G} \alpha_\ell = 1$. Denote by $m_\ell(t) = \mathbb{E}\{X(t) \mid \mathsf{G} = \ell\}$ the class–specific mean functions, and by $m(t) = \mathbb{E}X(t) = \sum_{\ell=1}^{G} \alpha_\ell m_\ell(t)$ the overall mean. For the classical Fisher criterion, we adopt the standard homoscedastic assumption that all classes share a common within–class covariance structure, given by the kernel

$$K_W(s,t) = \mathbb{E}\{(X_\ell(s) - m_\ell(s))(X_\ell(t) - m_\ell(t))\}, \qquad \ell = 1, \dots, G; \quad t, s \in T,$$

with associated covariance operator $(\mathsf{K}_W f)(t) = \int_T K_W(s,t)f(s)\,\mathrm{d}s$. Between–class variation is represented through the kernel

$$K_B(s,t) = \sum_{\ell=1}^{G} \alpha_\ell \{m_\ell(s) - m(s)\}\{m_\ell(t) - m(t)\},$$

with associated operator denoted by K_B, which measures how the class means are dispersed around the global mean in H.

For a direction $\beta \in H$ such that $\langle \beta, \mathsf{K}_W \beta \rangle > 0$, consider the functional Fisher ratio

$$\mathsf{F}_1(\beta) = \langle \beta, \mathsf{K}_B \beta \rangle / \langle \beta, \mathsf{K}_W \beta \rangle.$$

This quantity represents the ratio of between–class variance to within–class variance after projecting the functional data onto the one–dimensional subspace spanned by β. Maximizers of F_1 yield the classical functional linear discriminant directions, and are given by eigenfunctions of $\mathsf{K}_W^{-1}\mathsf{K}_B$, whenever this operator is well defined. Importantly, although the criterion involves all G classes, it pools their effect through the single second–order operator K_B. As a consequence, the resulting discriminant directions depend on the overall dispersion of class means and do not make explicit the contribution of individual pairwise contrasts $m_\ell - m_{\ell'}$. This observation motivates the extensions developed in the following lines.

Restriction to Elementary Tensors with Equal Factors. In the multiclass setting, discrimination typically involves $r = \binom{G}{2} = G(G-1)/2$ pairwise class contrasts. An aggregation of these contrasts leads naturally to tensor–valued criteria acting on $H^{\otimes r}$, the tensor product space of r copies of H endowed with the inner product $\langle f_1 \otimes \cdots \otimes f_r,\, g_1 \otimes \cdots \otimes g_r \rangle_{H^{\otimes r}} = \prod_{k=1}^{r} \langle f_k, g_k \rangle$. However, tensor–valued discriminants are neither interpretable as functional directions in H, which complicates their use within standard functional classification frameworks.

To retain a single functional discriminant while allowing for the joint influence of multiple contrasts, we restrict attention to directions of the form $\beta^{\otimes r} := \beta \otimes \cdots \otimes \beta$, with r identical tensor factors and $\beta \in H$. This restriction ensures that aggregation occurs at the level of the criterion, while the resulting solution itself remains a single function in H. Formally, the criterion is considered over the subset $\mathcal{M}_r = \{\beta^{\otimes r} : \beta \in H,\ \|\beta\| = 1\} \subset H^{\otimes r}$.

Tensorial Aggregation of Pairwise Fisher Criteria. Let $\mathsf{K}_W^{[k]}$ and $\mathsf{K}_B^{[k]}$ denote the within– and between–class covariance operators associated with the kth selected pairwise contrast (ℓ_k, ℓ'_k), with $\ell_k \neq \ell'_k$. At the population level, homoscedasticity implies $\mathsf{K}_W^{[k]} \equiv \mathsf{K}_W$ for all k, but the tensorial formulation does not require this restriction; in practice, heterogeneity across classes or pairwise standardization may lead to distinct operators. To combine these contrasts, we introduce the tensor product operators

$$\mathsf{K}_W^{\otimes r} = \mathsf{K}_W^{[1]} \otimes \cdots \otimes \mathsf{K}_W^{[r]}, \qquad \mathsf{K}_B^{\otimes r} = \mathsf{K}_B^{[1]s} \otimes \cdots \otimes \mathsf{K}_B^{[r]},$$

which act on $H^{\otimes r}$. For $\beta^{\otimes r} \in \mathcal{M}_r$, we define the aggregated Fisher ratio

$$\mathsf{F}_r(\beta^{\otimes r}) = \frac{\langle \beta^{\otimes r}, \mathsf{K}_B^{\otimes r} \beta^{\otimes r} \rangle_{H^{\otimes r}}}{\langle \beta^{\otimes r}, \mathsf{K}_W^{\otimes r} \beta^{\otimes r} \rangle_{H^{\otimes r}}}. \tag{1}$$

For elementary tensors, the inner product factorizes as

$$\langle \beta^{\otimes r}, (\mathsf{A}_1 \otimes \cdots \otimes \mathsf{A}_r) \beta^{\otimes r} \rangle_{H^{\otimes r}} = \prod_{k=1}^{r} \langle \beta, \mathsf{A}_k \beta \rangle,$$

so that $\mathsf{F}_r(\beta^{\otimes r})$ reduces to the product of the individual pairwise Fisher ratios. Consequently, optimization of F_r amounts to balancing several contrast–specific criteria through a single direction β.

Stationary Points of the Aggregated Criterion. The following result characterizes stationary points of the aggregated criterion under the restriction $\beta^{\otimes r} \in \mathcal{M}_r$.

Theorem 1. *Let $\mathsf{K}_B^{[k]}$ and $\mathsf{K}_W^{[k]}$ be bounded self–adjoint positive operators on H. For $\beta \in H$ define $b_k(\beta) = \langle \beta, \mathsf{K}_B^{[k]} \beta \rangle, w_k(\beta) = \langle \beta, \mathsf{K}_W^{[k]} \beta \rangle$, and assume $b_k(\beta) > 0$ and $w_k(\beta) > 0$ for $k = 1, \ldots, r$. Consider the functional on the unit sphere $\mathbb{S} = \{\beta \in H : \|\beta\| = 1\}$,*

$$\mathsf{F}_r(\beta^{\otimes r}) = \prod_{k=1}^{r} \frac{b_k(\beta)}{w_k(\beta)}.$$

If $\beta \in \mathbb{S}$ is a stationary point of F_r under the constraint $\|\beta\| = 1$, then there exists $\lambda \in \mathbb{R}$ such that

$$\sum_{k=1}^{r} \left(\frac{\mathsf{K}_B^{[k]} \beta}{b_k(\beta)} - \frac{\mathsf{K}_W^{[k]} \beta}{w_k(\beta)} \right) = \lambda \beta.$$

Proof. Define the log-functional on the domain where all $b_k(\beta), w_k(\beta)$ are positive,

$$\Phi(\beta) := \log \mathsf{F}_r(\beta^{\otimes r}) = \sum_{k=1}^{r} \big(\log b_k(\beta) - \log w_k(\beta) \big).$$

Since $\log(\cdot)$ is strictly increasing, β is stationary for F_r on $\mathbb{S}$ if and only if it is stationary for Φ on $\mathbb{S}$.

Fix $\beta \in \mathbb{S}$ and a direction $h \in H$. Because $\mathsf{K}_B^{[k]}$ is bounded and self-adjoint,

$$b_k(\beta + \varepsilon h) = \langle \beta + \varepsilon h,\, \mathsf{K}_B^{[k]} (\beta + \varepsilon h) \rangle = b_k(\beta) + 2\varepsilon \langle h, \mathsf{K}_B^{[k]} \beta \rangle + O(\varepsilon^2),$$

hence the derivative is $Db_k(\beta)[h] = 2\langle h, \mathsf{K}_B^{[k]} \beta \rangle$. Similarly, $Dw_k(\beta)[h] = 2\langle h, \mathsf{K}_W^{[k]} \beta \rangle$. By the chain rule for log,

$$D(\log b_k)(\beta)[h] = \frac{Db_k(\beta)[h]}{b_k(\beta)} = \frac{2\langle h, \mathsf{K}_B^{[k]} \beta \rangle}{b_k(\beta)}, \qquad D(\log w_k)(\beta)[h] = \frac{2\langle h, \mathsf{K}_W^{[k]} \beta \rangle}{w_k(\beta)}.$$

Therefore

$$D\Phi(\beta)[h] = 2\left\langle h, \sum_{k=1}^{r}\left(\frac{\mathsf{K}_B^{[k]}\beta}{b_k(\beta)} - \frac{\mathsf{K}_W^{[k]}\beta}{w_k(\beta)}\right)\right\rangle.$$

Now impose the constraint $\|\beta\| = 1$. The tangent space to $\mathbb{S}$ at β is $T_\beta\mathbb{S} = \{h \in H : \langle h, \beta\rangle = 0\}$. Stationarity of Φ on $\mathbb{S}$ means $D\Phi(\beta)[h] = 0$ for all $h \in T_\beta\mathbb{S}$, i.e.

$$\left\langle h, \sum_{k=1}^{r}\left(\frac{\mathsf{K}_B^{[k]}\beta}{b_k(\beta)} - \frac{\mathsf{K}_W^{[k]}\beta}{w_k(\beta)}\right)\right\rangle = 0 \quad \text{for all } h \perp \beta,$$

which yields the stated orthogonality condition. Equivalently, the vector

$$v(\beta) := \sum_{k=1}^{r}\left(\frac{\mathsf{K}_B^{[k]}\beta}{b_k(\beta)} - \frac{\mathsf{K}_W^{[k]}\beta}{w_k(\beta)}\right),$$

must be orthogonal to every $h \perp \beta$, hence $v(\beta) \in \operatorname{span}\{\beta\}$. Thus there exists $\lambda \in \mathbb{R}$ such that $v(\beta) = \lambda\beta$, which is the first condition.

Unlike the classical Fisher criterion, the aggregated functional does not lead to a single eigenvalue problem but to a nonlinear balance condition across contrasts. The characterization holds irrespective of homoscedasticity and therefore accommodates heterogeneous within–class structures.

3 Multiclass Extensions of Kurtosis–Based Contrast Rules

The role played by fourth–order moment operators in binary functional discrimination admits a natural analogue in multiclass settings, but its interpretation requires additional probabilistic considerations. Let $\mathbb{P}_\ell$ denote the law of the random element $X \mid \mathsf{G} = \ell$, with mean function $m_\ell = \mathbb{E}(X \mid \mathsf{G} = \ell)$ and common covariance operator K_X. For each pair of classes (ℓ, ℓ'), the Feldman–Hájek dichotomy yields a characterization of their mutual relation.

Theorem 2. *For every $\ell \neq \ell'$, exactly one of the following holds:*

$$\mathbb{P}_\ell \sim \mathbb{P}_{\ell'} \iff m_\ell - m_{\ell'} \in \operatorname{Im}(\mathsf{K}_X^{1/2}), \qquad \mathbb{P}_\ell \perp \mathbb{P}_{\ell'} \quad \textit{otherwise},$$

where $\sim$ denotes equivalence and $\perp$ mutual singularity of measures.

Proof. The result follows from applying the Feldman–Hájek dichotomy to each pair of Gaussian measures with common covariance.

Theorem 2 implies that near–perfect classification in multiclass problems must be understood in a *pairwise* sense. If all mean differences belong to $\operatorname{Im}(\mathsf{K}_X^{1/2})$, the measures are mutually absolutely continuous and no asymptotically perfect separation is possible. If all pairs are singular, full near–perfect classification can be achieved. Intermediate regimes, where singularity holds only for some pairs, lead to partial separability: certain groups can be asymptotically separated, while others remain indistinguishable.

As shown in [6], in the binary case deviations from equivalence under standardization are detected through fourth–order moment operators acting on H, whose spectral behaviour can characterize near–singular regimes in the tail of the spectrum. In particular, discriminant directions correspond to projections along which the univariate kurtosis approaches the reference value 1, a limit associated with maximal bimodality. Let $\mathcal{P} = \{(\ell, \ell') : 1 \leq \ell < \ell' \leq G\}$. For each $(\ell, \ell') \in \mathcal{P}$, denote by $X_{(\ell,\ell')}$ the (possibly standardized) sample restricted to classes ℓ and ℓ'. A natural multiclass criterion is therefore to seek a direction ψ that jointly drives the pairwise contrasts, as measured through the kurtosis coefficient kurt$(\cdot)$ of the projected samples $\langle X_{(\ell,\ell')}, \psi\rangle$, towards the reference value 1. This leads to the aggregated functional

$$\kappa_0(\psi) = \sum_{(\ell,\ell')\in\mathcal{P}} \left(\mathrm{kurt}(\langle X_{(\ell,\ell')}, \psi\rangle) - 1\right)^2, \tag{2}$$

which vanishes in the ideal singular limit and is minimized in near–singular regimes, providing a fourth–order analogue of the tensorial aggregation principle.

4 Implementation Details and Data Example

We consider two complementary empirical implementations of the proposed multiclass kurtosis–based methodology: a non–tensorial formulation based on aggregated univariate contrasts, and a tensorial formulation based on the aggregation of pairwise kurtosis operators. Both approaches yield a single discriminant direction, but differ in how pairwise contrast information is combined.

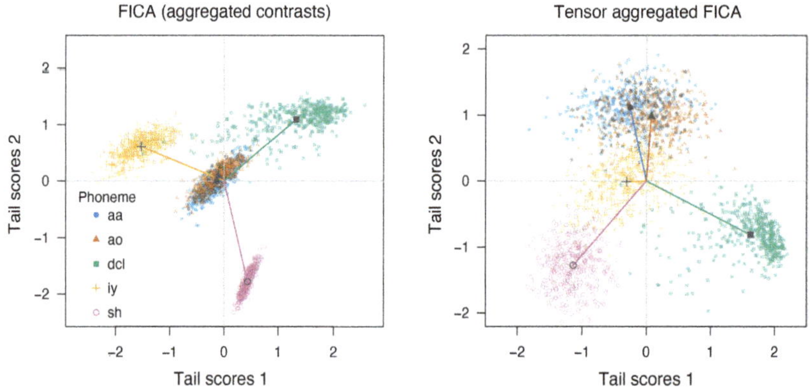

Fig. 1. Scatter plots of the last two components. Class barycentres are shown.

Non–tensorial Formulation. Discretely observed curves are represented in a q–dimensional subspace $H^{(q)} \subset H$, defined by a prescribed basis (see, for example, [4]). Candidate directions are generated from functional principal or independent component analysis (FPCA/FICA). In practice, these constructions depend on representation parameters, selected by numerical optimization or grid search. The contrast functional $\kappa_0(\psi)$ is then evaluated over the resulting finite family $\mathcal{D} \subset H^{(q)}$, and the selected direction is

$$\psi^* \in \arg \min_{\psi \in \mathcal{D},\ \|\psi\|=1} \kappa_0(\psi).$$

Tensorial Formulation. This formulation aggregates pairwise kurtosis operators rather than scalar contrasts. For each group pair (ℓ, ℓ'), the corresponding samples are pooled, whitened, and used to compute a kurtosis operator in coefficient space; see [6] for details. The resulting collection of pairwise kurtosis matrices is stacked into a third–order tensor $\mathfrak{K} \in \mathbb{R}^{q \times q \times \binom{G}{2}}$ whose frontal slices correspond to unordered group pairs. To summarize common structure across these pairwise contrasts, we perform a Tucker decomposition, $\mathfrak{K} \approx [\![\mathcal{C}; U_1, U_2, U_3]\!]$, where $\mathcal{C} \in \mathbb{R}^{q \times q \times s}, s \leq \binom{G}{2}$, is the core tensor, $U_1, U_2 \in \mathbb{R}^{q \times q}$ contain the column directions associated with the first and second tensor modes in coefficient space, and $U_3 \in \mathbb{R}^{\binom{G}{2} \times s}$ captures variation across the $\binom{G}{2}$ group pairs. Since the tensor consists of symmetric matrices, we typically find $U_1 \approx U_2$. Therefore, the discriminant direction is constructed using the last column of U_1 as a coefficient vector in the chosen basis, with a correction applied if the basis is not orthonormal. Projecting onto a specific pairwise–whitened space would bias the analysis toward particular contrasts, and a universal whitening operator is incompatible with the pairwise construction of the tensor. For interpretability and simplicity, we therefore project the resulting direction back onto the original, unwhitened functional data.

4.1 Data Example

We illustrate the proposed methodology using the phoneme data set, a well-known reference in the functional data analysis literature. The data consist of log–periodograms representing short speech frames, with the task of discriminating among five phonemes: [sh], [dcl], [iy], [aa], and [ao]. Following standard practice, we analyze 4,000 curves (800 per phoneme) observed on the first 150 frequency points, represented using B-splines of dimension 16 (no penalization is applied).

We compare several methods: FICA with a kurtosis criterion [6]; the proposed non–tensorial formulation based on FICA–derived aggregated kurtosis contrasts, where class representations are constructed via FPCA as in [6, §5.2], and the numbers of retained principal components are treated as class–specific tuning parameters selected by quasi–Newton optimization; the tensor–aggregated FICA (TA-FICA) via Tucker decomposition; Partial Least Squares (PLS) [3]; and a pipeline combining PLS, followed by either Linear Discriminant Analysis (LDA) or a kurtosis–based ICA applied to the resulting PLS scores. The misclassification rate (Err_0) is computed as the proportion of incorrectly predicted class labels, using a nearest-centroid rule applied to the resulting score vector. Sensitivity for each class is computed as the proportion of correctly classified observations within that class, i.e., $\text{Sensitivity}_\ell = \text{TP}_\ell/(\text{TP}_\ell+\text{FN}_\ell)$, where TP_ℓ denotes the number of true positives and FN_ℓ the false negatives for class ℓ.

Results (see Fig. 1, Table 1) show that the aggregated FICA formulation substantially improves performance over the baseline FICA, reducing the misclassification rate from 37.9% to 26.1%. TA-FICA yields a comparable overall error rate of 27.8%, with a more balanced sensitivity across classes. Among the PLS-based approaches, PLS–ICA achieves the best trade-off, outperforming both plain PLS and PLS–LDA, particularly for the vowel phonemes (aa, ao, iy). The per-class sensitivities further indicate that methods such as FICA with optimization and TA-FICA are better suited for capturing phoneme-specific structure, especially for the highly variable phoneme iy.

Table 1. Reported are the overall misclassification rate (Err_0), its standard deviation σ_{Err_0}, and per-class sensitivity for the five phonemes. All values are proportions.

Method	Err_0	σ_{Err_0}	Sensitivity				
			aa	ao	dcl	iy	sh
FICA	0.3790	0.0108	0.5175	0.5400	0.8975	0.1725	0.9775
FICA (aggregated)	0.2610	0.0098	0.5800	0.5100	0.8625	0.7425	1.0000
TA–FICA	0.2775	0.0100	0.6975	0.5050	0.9075	0.7350	0.7675
PLS	0.3940	0.0109	0.5450	0.5525	0.5600	0.4525	0.9200
PLS–LDA	0.3575	0.0107	0.6450	0.5025	0.6875	0.3925	0.9850
PLS–ICA	0.3185	0.0104	0.7375	0.6400	0.5225	0.9275	0.5800

References

1. Feldman, J.: Equivalence and perpendicularity of Gaussian processes. Pac. J. Math. **8**(4), 699–708 (1958)
2. Hájek, J.: A property of J-divergences of marginal probability distributions. Czech. Math. J. **8**, 460–462 (1958)
3. Preda, C., Saporta, G., Lévéder, C.: PLS classification of functional data. Comput. Stat. **22**, 223–235 (2007). https://doi.org/10.1007/s00180-007-0041-4
4. Ramsay, J. O., Silverman, B. W.: Functional Data Analysis. Springer Series in Statistics, 2nd edn. Springer, New York (2005). https://doi.org/10.1007/b98888
5. Rao, C.R., Varadarajan, V.S.: Discrimination of Gaussian processes. Sankhyā Ser. A **25**, 303–330 (1963)
6. Vidal, M., Leman, M., Aguilera, A.M.: Functional independent component analysis by choice of norm: a framework for near-perfect classification. Adv. Data Anal. Classif. **20**, 521–549 (2025). https://doi.org/10.1007/s11634-024-00622-5

7. Vidal, M.: A family of moment operators for functional data and its discriminative properties. In: Aneiros, G., Bongiorno, E.G., Goia, A., Hušková, M. (eds.) New Trends in Functional Statistics and Related Fields. Contributions to Statistics, pp. 557–564. Springer, Cham (2025). https://doi.org/10.1007/978-3-031-92383-8_66

Author Index

P. Brito et al. (Eds.): IFCS 2026, STUDIES CLASS, pp. 303–304, 2026.
https://doi.org/10.1007/978-3-032-32009-4

GPSR Compliance
The European Union's (EU) General Product Safety Regulation (GPSR) is a set of rules that requires consumer products to be safe and our obligations to ensure this.

If you have any concerns about our products, you can contact us on

ProductSafety@springernature.com

In case Publisher is established outside the EU, the EU authorized representative is:

Springer Nature Customer Service Center GmbH
Europaplatz 3
69115 Heidelberg, Germany

www.ingramcontent.com/pod-product-compliance
Ingram Content Group UK Ltd.
Pitfield, Milton Keynes, MK11 3LW, UK
UKHW022107180926
475695UK00001B/40
* 9 7 8 3 0 3 2 3 2 0 0 8 7 *